Bent-Shaped Liquid Crystals

Structures and Physical Properties

Bent-Shaped Liquid Crystals

Structures and Physical Properties

Hideo Takezoe and Alexey Eremin

CRC Press
Taylor & Francis Group
Boca Raton London New York

CRC Press is an imprint of the
Taylor & Francis Group, an **informa** business

CRC Press
Taylor & Francis Group
6000 Broken Sound Parkway NW, Suite 300
Boca Raton, FL 33487-2742

First issued in paperback 2019

© 2017 by Taylor & Francis Group, LLC
CRC Press is an imprint of Taylor & Francis Group, an Informa business

No claim to original U.S. Government works

ISBN-13: 978-1-4822-4759-6 (hbk)
ISBN-13: 978-0-367-24691-4 (pbk)

Library of Congress Cataloging-in-Publication Data

Names: Takezoe, Hideo. | Eremin, Alexey, 1976-
Title: Bent-shaped liquid crystals : structures and physical properties /
Hideo Takezoe, Alexey Eremin.
Description: Boca Raton : CRC Press, [2017] | Series: Liquid crystals book
series | Includes bibliographical references and index.
Identifiers: LCCN 2016057407 | ISBN 9781482247596 (hardback : alk. paper) |
ISBN 9781315372723 (ebook)
Subjects: LCSH: Liquid crystals. | Liquid crystals--Surfaces. | Nonlinear
optics. | Surfaces (Physics)
Classification: LCC QC173.4.L55 T35 2017 | DDC 530.4/29--dc23
LC record available at https://lccn.loc.gov/2016057407

Visit the Taylor & Francis Web site at
http://www.taylorandfrancis.com

and the CRC Press Web site at
http://www.crcpress.com

Contents

Preface

Bent-shaped liquid crystals opened a new era in liquid crystal science, particularly in the aspects of polarity and chirality. They provide us with the first polar liquid crystal composed of nonchiral molecules and with examples of the spontaneous symmetry breaking to form macroscopic chiral structures. New phenomena caused mainly by polarity and chirality have been discovered. A variety of novel phase structures have also been disclosed. There are many excellent review articles on bent-shaped mesogens, but no books have been published so far. After about two decades since the discovery of polar switching in a bent-shaped liquid crystal, which triggered extensive research on this topic, we decided to write a book on bent-shaped liquid crystals. The research activity in this field is expanding rapidly. Hence, new progress will be made by the time this book comes out. Yet we tried to include the most recent topics and discuss the most recent advances made in the research on bent-shaped liquid crystals. Our planning and writing this book started when HT stayed in Magdeburg as a Humboldt awardee. Since then, HT stayed as a visiting professor in different places such as Warsaw and Stuttgart. HT deeply acknowledges Professors Ralf Stannarius (Otto von Guericke University Magdeburg), Ewa Gorecka (University of Warsaw), and Frank Giesselmann (University of Stuttgart) for letting him concentrate on writing during his stay. HT is also indebted to Riken (Wako, Japan), which allows him to access electric journals as a research consultant. We thank Professor Ralf Stannarius, Dr. Fumito Araoka (Riken), Professor Frank Giesselmann, and Professor Carsten Tschierske (Martin Luther University Halle-Wittenberg) for fruitful discussions. We are grateful to Professors Ewa Gorecka and Mikhail Osipov (University of Strathclyde) for critically reading the draft of the book and giving invaluable comments.

Hideo Takezoe
Visiting fellow at the Toyota Physical and Chemical Research Institute (Japan)
Professor Emeritus of Tokyo Institute of Technology

Alexey Eremin
Associate Professor (apl.-Prof.) at the Otto von Guericke University Magdeburg
(Germany)

1 Introduction

Liquid crystals formed by bent-shaped molecules opened up a vast variety of morphologies and structures that became an exciting new research area in the last decade of the twentieth century. In this book, we first introduce liquid crystals particularly focusing on the shape of molecules and introduce a brief history of mesogens consisting of bent-shaped molecules. It is important to know how bent-shaped liquid crystals were discovered and what the scientific background was at that time.

1.1 BRIEF INTRODUCTION TO THE LIQUID CRYSTAL STATE

Liquid crystal is an intermediate state of matter between liquid and crystalline states. Because of this, the liquid crystalline state is characterized by fluidity and anisotropy, which are the main features of fluids and crystals, respectively. Hence, to realize liquid crystal states in a condensed molecular system, intermolecular interactions should not be too strong, and molecules have to be at least anisometric. Otherwise, they form crystals or isotropic liquids, respectively. In soft matter, entropic contribution to the free energy of the system, that is, repulsive interaction based on excluded volume effect, is particularly important. In the late 1940s, Onsager showed that the isotropic (Iso)-nematic (N) transition can be driven by entropy without considering intermolecular attractive interaction (Onsager 1949). Actually, Onsager explained the Iso-N transition in an aqueous solution of Tabaco mosaic virus, which is of a cylinder shape, by considering only repulsive force (entropy) due to the excluded volume effect without considering attractive force. In this sense, molecular shapes are very important for mesogenic phase formation. As for the molecular shape, rods and discs are typical mesogenic designs in conventional liquid crystals. Actually, rod- and disc-shaped molecules form a variety of liquid crystalline phases such as nematic, smectic, discotic nematic, columnar nematic, and various columnar phases, which are summarized in Figure 1.1. Here, the long axes of rod-shaped molecules and the disc normals of disc-shaped molecules align along a certain direction, which is called the director. We also define an orientational order parameter, which gives a measure of how well molecules align; the order parameter is unity when molecules perfectly align and zero when the orientation is random. Smectic and columnar phases are subdivided on the basis of molecular tilt with respect to layer's and column's axes, intralayer and intercolumnar orders, respectively. To optimize intermolecular interactions, molecules usually bear flexible chains and intermolecular interactions through flat and large π orbitals. Sometimes the introduction of substituted groups is also effective to realize liquid crystallinity.

In this way, molecular structure and shape are very important to form liquid crystalline phases. Actually, various mesophase structures can be created by the particular chemical design of the mesogenic molecules. Not only is the molecular shape important in the formation of the phase structure, but it also determines its physical properties. One of the examples is the flexoelectric effect that may be enhanced by

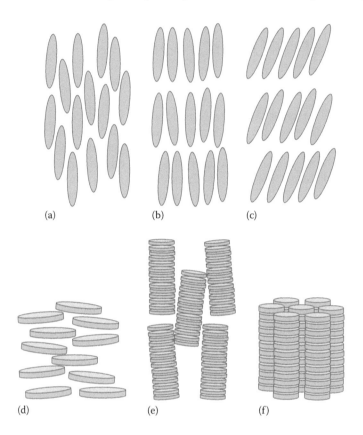

FIGURE 1.1   Typical liquid crystalline phases formed by rod- and disc-shaped molecules. (a) Nematic, (b) smectic A, (c) smectic C, (d) discotic nematic, (e) columnar nematic, and (f) hexogonal columnar.

slightly asymmetric molecular shapes, such as banana shape and pear shape with transverse and longitudinal electric dipoles, respectively. In an unperturbed nematic phase, molecules uniformly align along a particular direction called the director. The molecules freely rotate about their long axes. Fast flips about the short axes warrant head-to-tail symmetry. Thus, no macroscopic polarization may occur in the bulk of the uniform nematic phase. However, if bend (to the banana-shaped molecular system) or splay (to the pear-shaped molecular systems) deformation is applied, macroscopic polarization emerges. This is the flexoelectric effect and induced polarization is called flexoelectric polarization. See Section 4.1 and Figure 4.1 for details.

Chirality of molecules and geometric objects in general is defined in entities that have non-superimposable mirror images by translations or rotations. For instance, right and left hands and right- and left-handed screws are both mirror images and are chiral shapes. Chirality plays an important role in the phase structures of some liquid crystal phases (Kitzerow and Bahr 2001). For instance, uniform molecular orientations in the nematic and tilted smectic C (SmC) phases are perturbed to form helical structures, if these molecular systems are made of chiral molecules or contain chiral molecules. They are sometimes designated using different phase names, cholesteric (or chiral nematic N*) and chiral SmC (SmC*), respectively. It is noted that macroscopic symmetries are different between N and N* and between SmC and SmC*, although the local molecular orientation structures are the same. There are three types of molecular chirality: (a) central chirality, (b) axial chirality, and (c) plane chirality (see Figure 5.1) (Takezoe 2012). Central chirality is mostly due to the existence of chiral carbon(s). Any atom that has four different groups bonded to it, such as N, Si, and P, could be also a chiral center. In contrast, axial chirality originates from twisted structures of molecules such as binaphthyl and heptahelicene, which are not mutually interconverted in ambient conditions. Plane chirality arises in molecules with a chiral plane (O–C–C–Br in Figure 5.1c), where the substituent group destroys a perpendicular plane of symmetry. Chirality also arises from a chiral arrangement of nonchiral units. Even if constituent assemblies or molecules are not chiral, their two-dimensional (2D) packing structures could be chiral, as shown in Figure 1.2 (Takezoe 2012). In such cases, the vector products of directional

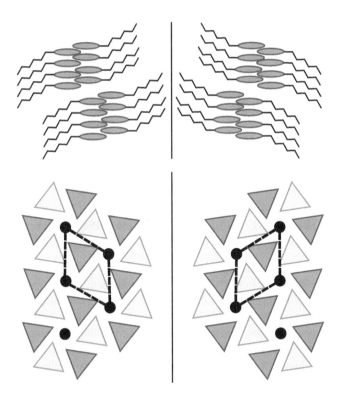

FIGURE 1.2 Chirality emerged by geometrical origin observed in two-dimensional systems. Molecules themselves are not chiral, but the structures on the right- and left-hand sides are mirror images and cannot be superimposable by translation and rotation about the surface normal. (With kind permission from Springer Science+Business Media: *Top. Curr. Chem.*, Spontaneous achiral symmetry breaking in liquid crystalline phases, 318, 2012, 303–330, Takezoe, H.)

unit vectors result in vectors normal to the surface with opposite directional senses. This kind of chirality has been observed also in 2D liquid crystal (LC) arrangements consisting of nonchiral molecules of disc shape (Li et al. 2002), rod shape (Berg and Patrick 2005), and even bent shape (Eremin et al. 2008).

In this book, we describe bent-shaped liquid crystals. These liquid crystal systems are sometimes called bent-core liquid crystals, since the central part of a mesogen providing a bent shape often consists of a phenyl or a naphthalene core. However, even bent-shaped molecules consisting of two mesogens linked by flexible chains such as an alkylene spacer show similar phases and phenomena as those shown in bent-core molecules, if a bent shape is realized by a spacer with an odd number of carbons. Hence, "bent-shaped" includes both kinds of mesogens: bent-core and dimers-linked by a flexible spacer. People sometimes call banana-shaped molecules and banana molecules as a nickname. The name "banana phases" is sometimes used for characteristic phases exhibited by bent-shaped molecules as well.

In view of a shape, bent-shaped molecules have been believed to be unfavorable molecules for liquid crystals, as far as the steric interaction is concerned, because of an enhanced excluded volume due to the free rotations of molecules about their long axes. However, we now realize that bent-shaped liquid crystals provide a variety of new phases including new phase structures and new phenomena, which have never been observed in rod- and disc-shaped liquid crystals. Just by bending molecular shape, we found a "treasure island" for liquid crystal science. Among many novel phenomena, the most important discovery is the polar (ferroelectric and antiferro-electric) LC phases and chiral phases in molecular systems consisting of nonchiral molecules. In this way, polarity and chirality are the central features of bent-shaped LCs, which should be highlighted. Bent-shaped liquid crystals also supply an inter-esting systems, for theoreticians too: free rotation about the molecular long axes in bent-shaped LCs is strongly hindered. This makes the system more complex but brings about new phases and new phenomena.

1.2   HISTORICAL OUTLINE TOWARD THE DISCOVERY OF POLAR BENT-SHAPED LIQUID CRYSTALS

Although intensive research on bent-core LCs started in 1996, the first mesogens were synthesized by Vorländer in the beginning of the twentieth century (Vorländer 1929, Vorländer and Apel 1932). He synthesized more than 2000 mesogenic compounds including mesogenic dimers with flexible spacers, metallomesogens, main-chain polymers, ferroelectric chiral smectic C (SmC*), and columnar mesogens. However, he could not clarify the phase structures and left them unassigned or just designated as mesophases. Many of these LC compounds are still kept in glass tubes stored in colorful cigar boxes (Figure 1.3). Among them bent-core mesogens, where a central aromatic core links two two-ring mesogenic units in the o- and m-positions, were included (Bruce et al. 1997, Pelzl et al. 2001). Thus, Vorländer is believed to be the first who synthesized bent-core liquid crystals. However, he also mentioned that the thermal stability of these mesogens is low compared with that of the rod-shaped analogs, and the liquid crystalline state is obtainable by an utmost linear shape of the molecules but not by a nonlinear shape.

FIGURE 1.3 (**See color insert.**) A cigar box, in which many compounds synthesized by the Vorländer's group are stored.

The first synthesis of bent-core mesogens by Vorländer remained unnoticed, and bent-core liquid crystals received no attention for more than 60 years until Matsunaga et al. (Kuboshita et al. 1991, Matsuzaki and Matsunaga 1993, Akutagawa et al. 1994) started to systematically synthesize bent-core molecules. Matsunaga's group was interested in examining the incompatibility of nonlinear molecules with liquid crystallinity but not in studying their physical properties. They first synthesized 1,2-phenylene bis[4-(4-alkoxybenzylideneamino)benzoates] particularly with long alkoxy end chains, which were not synthesized by Vorländer. They found the nematic (N) phase in the homologues with the shorter chain lengths, the smectic A (SmA) phase in those with longer chain lengths, and both in those with middle chain lengths (Kuboshita et al. 1991, Matsuzaki and Matsunaga 1993). They also examined the phase behavior by the substitution of the central aromatic ring. Then, they extended their systematic works of bent-core molecules with an acute-angled configuration to those with an obtuse-angle configuration. Akutagawa et al. (1994) identified two mesogenic phases, SmC, and more ordered smectic phases in the compound. Figure 1.4 shows a chemical structure they synthesized. One of them (compound I with alkyl chains instead of alkoxy chains) was a compound, in which a polar switching was shown for the first time by Niori et al. (1996). At the same period, Janietz et al. (1993) examined the liquid crystallinity of bent-core molecules containing a 1,3,5-triazine central core with a 6-alkoxy unit. They reported the existence of the N phase in most of the homologues. A smectic phase also emerged in one of the homologues with a 6-methyloxy in the central core and long terminal chains.

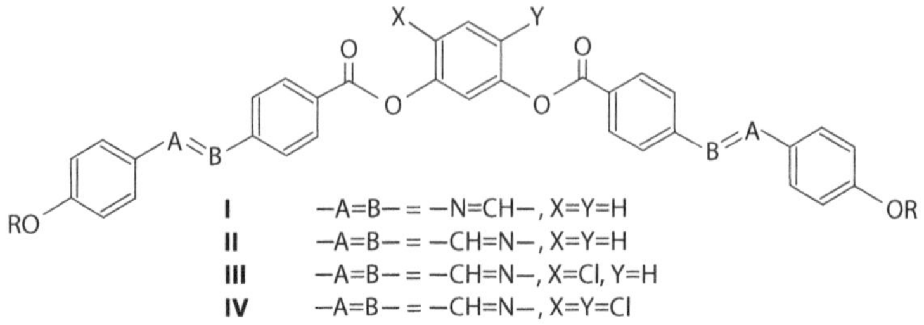

RO — [structure] — OR

I	−A=B− = −N=CH−, X=Y=H	
II	−A=B− = −CH=N−, X=Y=H	
III	−A=B− = −CH=N−, X=Cl, Y=H	
IV	−A=B− = −CH=N−, X=Y=Cl	

FIGURE 1.4 Chemical structures of the compounds, which Matsunaga's group synthesized.

However, they were interested only in the liquid crystallinity of molecules whose structure is far from a rod shape.

During this period when the syntheses of bent-core molecules were revived, very important theoretical works were published in 1992 by Brand et al. (1992) and Cladis and Brand (1993). They discussed symmetry and defects in the biaxial and fluid orthogonal SmC_M phase (Figure 1.5a), which emerges in liquid crystalline side-chain (Leube and Finkelmann 1990, 1991) and main-chain (Watanabe and Hayashi 1988, 1989) polymers. At this stage, there were no nonpolymeric liquid crystals, which show the SmC_M phase. In addition, they also discussed the biaxial and fluid orthogonal SmC_P phases (Figure 1.5b), which is exactly the structure of the bent-shaped $SmAP_F$ phase (see Section 3.2) later discovered. Actually, they pointed out that the SmC_P phase would be a possible first nonchiral ferroelectric phase. The SmC_M and SmC_P phases have D_{2h} and C_{2v} symmetries, respectively, whereas the conventional SmC phase has C_{2h} symmetry. Here, Schönflies notations for point groups are used: C_n shows a symmetry with an n-fold rotation, D_n with additional twofold (C_2) axis perpendicular to C_n axis. The subscripts h and v are added if mirror symmetry exists in planes horizontal and vertical to the main rotation axes, respectively. Note that ferroelectricity arises only in systems with C_n and C_{nv} symmetries.

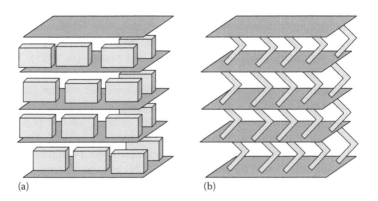

(a)   (b)

FIGURE 1.5 Structures of orthogonal biaxial smectic phases proposed by Brand et al. (a) SmC_M phase and (b) polar SmC_P phase.

The defect (disclination) line of the in-plane director is important to identify the phase structures. Around the defect line, the director rotates by $2\pi S$, where S defines the strength of the line. The SmC_M is special in the sense that this phase allows $S = \pm 1/2$ defects, although the SmC and SmC_P do not because of the head-tail asymmetry of the director field. It is noted that the anticlinic SmC_A and the antiferroelectric SmC_A^* can exhibit $S = \pm 1/2$ defects as a dispiration (Takanishi et al. 1992a,b), which is a combined defect of a screw dislocation and an $S = \pm 1/2$ edge disclination.

In this way, in the early 1990s, synthetic and theoretical backgrounds for the discovery of polar order in bent-shaped mesogens had been provided. Meantime, Watanabe and Hayashi (1988) had been engaged in main-chain liquid crystalline polymers based on biphenyl mesogens, BBn polyesters (Figure 1.6a). Depending on the number of methylene groups in diol n, they found using x-ray diffraction (XRD) measurements that the mesogenic groups orient parallel to each other for even n and in a zigzag fashion for odd n, exhibiting SmA and SmC_A, respectively (Watanabe and Hayashi 1989). This SmC_A phase was first designated as SmC_2 (Watanabe et al. 1994). Due to the discovery of the antiferroelectric SmC_A^* phase (Chandani et al. 1989), this phase is now designated as SmC_A (Watanabe et al. 1997).

In the course of studies on BBn main-chain polymers, Watanabe et al. (1993) also synthesized dimeric compounds, di-BBn (Figure 1.7a), as model compounds for main-chain liquid crystalline polyesters, BBn. They found that these dimers exhibit SmA for even n and SmC_A for odd n. As shown in Figure 1.7b, the structure of the SmC_A phase was found to be an intercalated one, where the thickness of a single smectic layer corresponds to one mesogenic unit, because the layer spacing obtained

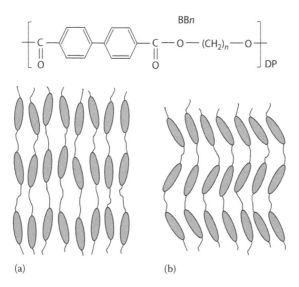

(a)   (b)

FIGURE 1.6   Chemical structure of main-chain liquid crystalline polyesters BBn and molecular arrangement of polymers with even n and odd n. (a) SmA (even n) and (b) SmC_A (odd n).

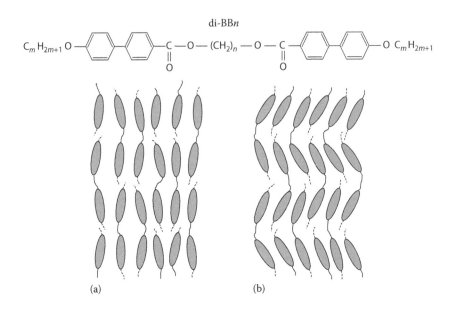

FIGURE 1.7 Chemical structure of dimeric mesogens di-BBn and molecular arrangement of dimers with even n and odd n. Flexible spacers and end chains were drawn by solid and dotted curves, respectively. (a) SmA (even n) and (b) SmC$_A$ (odd n).

by XRD is about half of the molecular length. In this structure, alkyl end groups and methylene spacer groups are randomly mixed, since their lengths are not significantly different. Later, a systematic study on the mesogenic properties of dimers with a variety of combinations of end chain length and spacer length was conducted (Choi et al. 1999, Watanabe et al. 2000, Izumi et al. 2006a,b, Takanishi et al. 2006). Details are described in Section 3.6. From a historical point of view, it is important to note that already in 1993 di-BBn were synthesized and their molecular orientation structures were analyzed by XRD. Actually, soon after the discovery of polar switching in bent-core mesogens (Niori et al. 1996), some of di-BBn (12AM5AM12 (Watanabe et al. 1998, Choi et al. 1999) and 8OAM5AMO8 (Choi et al. 1999)) were shown to also exhibit the so-called banana phases. Here, in m(O)AMnAM(O)m, m and n (odd number) stand for the number of carbons in both end chains and the methylene spacer, respectively, and O means alkoxy tails instead of alkyl tails. Di-BBn mentioned earlier is 4OAMnAMO4 ($n = 4–9$). More importantly, 12AM5AM12 is the first orthogonal polar biaxial smectic phase (SmAP$_A$; see Section 3.2), which was proved by the following observations in planar cells: (1) the extinction direction between crossed polarizers remained the same by applying an electric field, but only birefringence color changed, and (2) two switching current peaks were observed and the spontaneous polarization obtained was about 600 nC/cm^2.

Watanabe was motivated by Matsunaga's work because of the similarity in shape of his bent-shaped dimers and those of the homologous series of 1,3-phenylene bis[4-(4-n-alkylphenyliminomethyl)benzoates] (P-n-PIMB) and 1,3-phenylene bis[4-(4-n-alkyloxyphenyliminomethyl)benzoates] (P-n-O-PIMB) (see Figure 1.8),

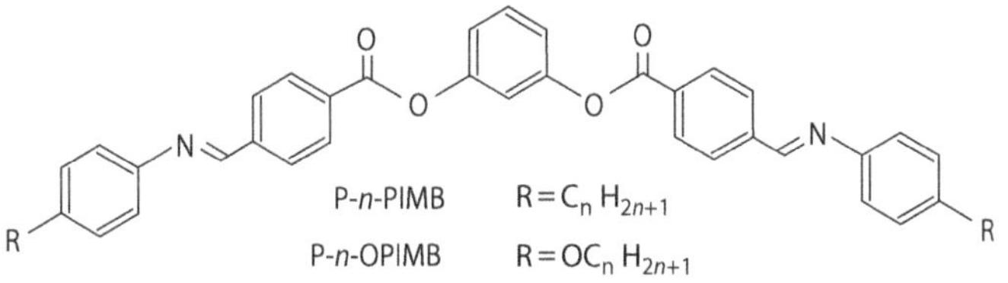

P-n-PIMB   R = $C_n H_{2n+1}$

P-n-OPIMB   R = $OC_n H_{2n+1}$

FIGURE 1.8 The first series of bent-core molecules synthesized by Watanabe's group, P-n-PIMB and P-n-OPIMB. Some of them are the same as compounds synthesized by Vorländer and Matsunaga.

where P stands for a phenyl in the central core to distinguish it from other core structures such as a naphthalene. Sometimes, different abbreviations are used, for example, NOBOW for P-9-O-PIMB.

Watanabe asked Takezoe for electric measurements such as switching current and dielectric measurements. The switching current measurements of P-8-PIMB using a triangular method (Miyasato et al. 1983) clearly showed a single switching current peak, suggesting the ferroelectric phase. Later, it was proven that the phase was not a ferroelectric but an antiferroelectric (Diele et al. 1998, Pelzl et al. 1999). Takezoe and Watanabe group also reexamined the switching behavior and recognized that polar switching in the original material was antiferroelectric (Zennyoji et al. 1999). The reason of the observation of a single peak suggesting a ferroelectric switching is not known. A similar mistake was also made by another group. Shen et al. (1998) reported a single switching current peak in an asymmetric bent-core compound, but it was actually the antiferroelectric phase (Shen et al. 2000). The real ferroelectric SmC_SP_F structure (see Section 3.3) was discovered later in 2000 using two chiral bent-core mesogens (Gorecka et al. 2000, Walba et al. 2000).

In the first paper by Niori et al. (1996), the obtained spontaneous polarization was reported to be about 50 nC/cm². This value is also about one order of magnitude smaller than that in later measurements (Heppke et al. 1999, Zennyoji et al. 2000). They pointed out that a large dielectric strength of about 10 is also an evidence of ferroelectricity (Niori et al. 1996). However, the dielectric strength in the real ferroelectric phase discovered later (Gorecka et al. 2000) was 10 times more than that reported by Niori et al. (1996). In addition, the polar phase was erroneously assigned to a nontilted (orthogonal) SmA-type phase (Niori et al. 1996) partly because of the use of shear-aligned cells. A small layer spacing obtained by XRD was interpreted by a molecular conformation (Niori et al. 1997). In this way, the first paper on polar switching in a bent-core material by Niori et al. has many mistakes. But the concept of polar layer formation, that is, hindered free rotation due to the packing constrains of bent-shaped molecules, is clearly written and is important.

The first oral presentation on the ferroelectric behavior of bent-core mesogens was made in the workshop of Kyoto Prize, November 1995. Takezoe presented "Ferroelectricity and antiferroelectricity in liquid crystals." He introduced a ferroelectric switching in a bent-core mesogen without showing the chemical structure according to Watanabe's suggestion, that is, many people may immediately get into the research of bent-core mesogens because of easiness of the synthesis, once we show the chemical structure. This became actually a reality. After publishing the paper (Niori et al. 1996), Takezoe gave a talk at the international liquid crystal conference in Kent (USA) in June 1996. Immediately after the presentation, many synthetic chemists started to synthesize this compound and an enormous number of papers soon appeared.

The most important work at the first stage is the emergence of chirality in a nonchiral tilted polar smectic phase, which was later designated as the SmCP phase with C and P for tilted and polar, respectively (Link et al. 1997). In this phase, there are three axes: layer normal direction, polar (bend) direction, and molecular long axis, which is tilted from the layer normal. The relationship among three axes gives right-handed and left-handed. Hence, the uniform tilt of a molecule with a transverse dipole with respect to a smectic layer makes the layer chiral, even if constituent molecules themselves are nonchiral. The details will be described in Section 3.3. Since then, polarity and chirality in nonchiral molecular systems have been two major topics, as reviewed in several articles (Pelzl et al. 1999, Reddy and Tschierske 2006, Takezoe and Takanishi 2006, Etxebarria and Ros 2008, Eremin and Jákli 2013).

REFERENCES

Akutagawa, T., Y. Matsunaga, and K. Yasuhara. Mesomorphic behavior of 1,3-phenylene bis[4-(4-alkoxyphenyliminomethyl)benzoates] and related compounds. *Liq. Cryst.* 17 (1994): 659–666.

Berg, A. M. and D. L. Patrick. Preparation of chiral surfaces from achiral molecules by controlled symmetry breaking. *Angew. Chem. Int. Ed.* 44 (2005): 1821–1823.

Brand, H. R., P. E. Cladis, and H. Pleiner. Symmetry and defects in the C_M phase of polymeric liquid crystals. *Macromolecules* 25 (1992): 7223–7226.

Bruce, D. W., K. Heyns, and V. Vill. Vorländer's wheel. *Liq. Cryst.* 23 (1997): 813–819.

Chandani, A. D. L., E. Gorecka, Y. Ouchi, H. Takezoe, and A. Fukuda. Antiferroelectric chiral smectic phases responsible for the tristable switching in MHPOBC. *Jpn. J. Appl. Phys.* 28 (1989): L1265–L1268.

Choi, S.-W., M. Zennyoji, Y. Takanishi, H. Takezoe, T. Niori, and J. Watanabe. Structure and switching in bent-shaped molecular liquid crystal systems with two mesogenic groups linked by alkylene spacer. *Mol. Cryst. Liq. Cryst. Sci. Technol.* 328 (1999): 185–192.

Cladis, P. E., Brand H. R., Electrooptic response of smectic O and smectic O*. *Liq. Cryst.* 14 (1993): 1327–1349.

Diele, S., S. Grande, H. Kruth, Ch. Lischka, G. Pelzl, W. Weissflog, and I. Wirth. Structure and properties of liquid crystalline phases formed by achiral banana-shaped molecules. *Ferroelectrics* 212 (1998): 169–177.

Eremin, A. and A. Jákli. Polar bent-shape liquid crystals—From molecular bend to layer splay and chirality. *Soft Matter* 9 (2013): 615–637.

Eremin, A., A. Nemeş, R. Stannarius, G. Pelzl, and W. Weissflog. Spontaneous bend patterns in homochiral ferroelectric SmCP films: Evidence for a negative effective bend constant. *Soft Matter* 4 (2008): 2186–2191.

Etxebarria, J. and M. B. Ros. Bent-core liquid crystals in the route to functional materials. *J. Mater. Chem.* 18 (2008): 2919–2926.

Gorecka, E., D. Pociecha, F. Araoka, D. R. Link, M. Nakata, J. Thisayukta, Y. Takanishi, K. Ishikawa, J. Watanabe, and H. Takezoe. Ferroelectric phases in a chiral bent-core smectic liquid crystal: Dielectric and optical second-harmonic generation measurements. *Phys. Rev. E* 62 (2000): R4524–R4527.

Heppke, G., A. Jákli, S. Rauch, and H. Sawade. Electric-field-induced chiral separation in liquid crystals. *Phys. Rev. E* 60 (1999): 5575–5579.

Izumi, T., S. Kang, T. Niori, Y. Takanishi, H. Takezoe, and J. Watanabe. Smectic mesophase behavior of dimeric compounds showing antiferroelectricity, frustration and chirality. *Jpn. J. Appl. Phys.* 45 (2006a): 1506–1514.

Izumi, T., Y. Naitou, Y. Shimbo, Y. Takanishi, H. Takezoe, and J. Watanabe. Several types of bilayer smectic liquid crystals with ferroelectric and antiferroelectric properties in binary mixture of dimeric compounds. *J. Phys. Chem. B* 110 (2006b): 23911–23919.

Janietz, D., F. Sundholm, and J. Leppanen. Liquid crystalline aromatic esters containing a 1,3,5-triazine moiety. *Liq. Cryst.* 13 (1993): 499–505.

Kitzerow, H.-S. and C. Bahr. *Chirality in Liquid Crystals.* New York: Springer-Verlag, 2001.

Kuboshita, M., Y. Matsunaga, and H. Matsuzaki. Mesomorphic behavior of 1,2-phenylene bis[4-(4-alkoxybenzylideneamino)benzoates]. *Mol. Cryst. Liq. Cryst. Sci. Technol.* 199 (1991): 319–326.

Leube, H. F. and H. Finkelmann. New liquid-crystalline side-chain polymers with large transversal polarizability. *Macromol. Chem. Phys.* 191 (1990): 2707–2715.

Leube, H. F. and H. Finkelmann. Optical investigations on a liquid-crystalline side-chain polymers with biaxial nematics and biaxial smectic A phase. *Macromol. Chem. Phys.* 192 (1991): 1317–1328.

Li, C., Q. Zeng, P. Wu, S. Xu, C. Wang, Y. Qiao, L. Wan, and C. Bai. Molecular symmetry breaking and chiral expression of discotic liquid crystals in two-dimensional systems. *J. Phys. Chem. B* 106 (2002): 13262–13267.

Link, D. R., G. Natale, R. Shao, J. E. Maclennan, N. A. Clark, E. Korblova, and D. M. Walba. Spontaneous formation of macroscopic chiral domains in a fluid smectic phase of achiral molecules. *Science* 278 (1997): 1924–1927.

Matsuzaki, H. and Y. Matsunaga. New mesogenic compounds with unconventional molecular structures 1,2-phenylene and 2,3-naphthylene bis[4-(4-alkoxyphenyliminomethyl)benzoates] and related compounds. *Liq. Cryst.* 14 (1993): 105–120.

Miyasato, K., S. Abe, H. Takezoe, A. Fukuda, and E. Kuze. Direct method with triangular waves for measuring spontaneous polarization in ferroelectric liquid crystals. *Jpn. J. Appl. Phys.* 22 (1983): L661–L663.

Niori, T., T. Sekine, J. Watanabe, T. Furukawa, and H. Takezoe. Distinct ferroelectric smectic liquid crystals consisting of banana shaped achiral molecules. *J. Mater. Chem.* 6 (1996): 1231–1233.

Niori, T., T. Sekine., J. Watanabe, T. Furukawa, and H. Takezoe. Distinct ferroelectric smectic liquid crystals consisting of achiral molecules with banana shape. *Mol. Cryst. Liq. Cryst. Sci. Technol.* 301 (1997): 337–342.

Onsager, L. The effects of shape on the interaction of colloidal particles. *Ann. N. Y. Acad. Sci.* 51 (1949): 627–659.

Pelzl, G., S. Diele, and W. Weissflog. Banana-shaped compounds—A new field to liquid crystals. *Adv. Mater.* 11 (1999): 707–724.

Pelzl, G., I. Wirth, and W. Weissflog. The first 'banana phase' found in an original Vorländer substance. *Liq. Cryst.* 28 (2001): 969–972.

Reddy, R. A. and C. Tschierske. Bent-core liquid crystals: Polar order, superstructural chirality and spontaneous desymmetrisation in soft matter systems. *J. Mater. Chem.* 16 (2006): 907–961.

Shen, D., S. Diele, I. Wirt, and C. Tschierske. A novel class of non-chiral banana-shaped liquid crystals with ferroelectric properties. *Chem. Commun.* 1998 (1998): 2573–2574.

Shen, D., A. Pegenau, S. Diele, I. Wirth, and C. Tschierske. Molecular design of nonchiral bent-core liquid crystals with antiferroelectric properties. *J. Am. Chem. Soc.* 122 (2000): 1593–1601.

Takanishi, Y., H. Takezoe, and A. Fukuda. Visual observation of dispirations in liquid crystals. *Phys. Rev. B* 45 (1992a): 7684–7689.

Takanishi, Y., H. Takezoe, A. Fukuda, H. Komura, and J. Watanabe. Simple method for confirming the antiferroelectric structure of smectic liquid crystals. *J. Mater. Chem.* 2 (1992b): 71–73.

Takanishi, Y., M. Toshimitsu, M. Nakata, N. Takada, T. Izumi, K. Ishikawa, H. Takezoe, J. Watanabe, Y. Takahashi, and A. Iida. Frustrated smectic layer structures in bent-shaped dimer liquid crystals studied by x-ray microbeam diffraction. *Phys. Rev. E* 74 (2006): 051703-1–051703-10.

Takezoe, H. Spontaneous achiral symmetry breaking in liquid crystalline phases. *Top. Curr. Chem.* 318 (2012): 303–330.

Takezoe, H. and Y. Takanishi. Bent-core liquid crystals: Their mysterious and attractive world. *Jpn. J. Appl. Phys.* 45 (2006): 597–625.

Vorländer, D. Die richtung der kohlenstoff-valenzen in benzol-abkommlingen. *Ber. Dtsch. Chem. Ges.* 62 (1929): 2831–2835.

Vorländer, D. and A. Apel. Die Richtung der Kohlen-stoff-Valenzen in Benzolabkommlingen (II.). *Ber. Dtsch. Chem. Ges.* 65 (1932): 1101–1109.

Walba, D. M., E. Korblova, R. Shao, J. E. Maclennan, D. R. Link, M. A. Glaser, and N. A. Clark. A ferroelectric liquid crystal conglomerate composed of racemic molecules. *Science* 288 (2000): 2181–2184.

Watanabe, J. and M. Hayashi. Thermotropic liquid crystals of polyesters having a mesogenic p,p′-bibenzoate unit. 1. Smectic mesophase properties of polyesters composed of p,p′-bibenzoic acid and alkylene glycols. *Macromolecules* 21 (1988): 278–280.

Watanabe, J. and M. Hayashi. Thermotropic liquid crystals of polyesters having a mesogenic p,p′-bibenzoate unit. 2. X-ray study on smectic mesophase structures of BB5 and BB-6. *Macromolecules* 22 (1989): 4083–4088.

Watanabe, J., M. Hayashi, A. Morita, and T. Niori. Structural characteristics of smectic liquid crystals in main-chain polyesters. *Mol. Cryst. Liq. Cryst. Sci. Technol.* 254 (1994): 221–240.

Watanabe, J., M. Hayashi, Y. Nakata, T. Niori, and M. Tokita. Smectic liquid crystals in main-chain polymers. *Prog. Polym. Sci.* 22 (1997): 1053–1087.

Watanabe, J., T. Izumi, T. Niori, M. Zennyoji, Y. Takanishi, and H. Takezoe. Smectic mesophase properties of dimeric compounds. 2. Distinct formation of smectic structures with antiferroelectric ordering and frustration. *Mol. Cryst. Liq. Cryst. Sci. Technol.* 346 (2000): 77–86.

Watanabe, J., H. Komura, and T. Niori. Thermotropic liquid crystals of polyesters having a mesogenic 4,4-bibenzoate unit: Smectic mesophase properties and structures in dimeric model compounds. *Liq. Cryst.* 13 (1993): 455–465.

Watanabe, J., T. Niori, S.-W. Choi, Y. Takanishi, and H. Takezoe. Antiferroelectric smectic liquid crystal formed by achiral twin dimer with two mesogenic groups linked by alkylene spacer. *Jpn. J. Appl. Phys.* 37 (1998): L401–L403.

Zennyoji, M., Y. Takanishi, K. Ishikawa, J. Thisayukta, J. Watanabe, and H. Takezoe. Partial mixing of opposite chirality in a bent-shaped liquid crystal molecular system. *J. Mater. Chem.* 9 (1999): 2775–2778.

Zennyoji, M., Y. Takanishi, K. Ishikawa, J. Thisayukta, J. Watanabe, and H. Takezoe. Electrooptic and dielectric properties in bent-shaped liquid crystals. *Jpn. J. Appl. Phys.* 39 (2000): 3536–3541.

2 Theories of Bent-Shaped Liquid Crystals

Polarity and structural chirality are among the most remarkable features of bent-shaped mesogens. The chirality of "banana" phases is not intrinsic but occurs as a result of a spontaneous symmetry breaking. In this chapter, we shall briefly review several theoretical models that highlight the main conditions for the formation of polar order and the resulting rich polymorphism of "banana" phases.

Bent-shaped mesogens are strongly anisometric molecules. Their rotation around their mean long axes may be strongly hindered. This means that a uniaxial cylinder-shape-model adopted for rod-shaped mesogens cannot be applied for bent-shaped molecular systems. The orientation of a bent-shaped molecule is given by three vectors: long molecular axis $\mathbf{n}$, the direction of the bow $\mathbf{p}$, and the vector $\mathbf{m}$, which is orthogonal to the molecular plane. In addition, the direction of the smectic layer normal is designated by $\mathbf{k}$. It is assumed that on average the mesogens have C_{2v} symmetry and the invariance $\mathbf{n} \rightarrow -\mathbf{n}$ is valid. An exhaustive symmetry classification of bent-shaped smectic phases has been given by Brand et al. (Brand et al. 1998, 2000, Cladis et al. 2000). At least four types of biaxial smectic phases have been suggested by the authors and designated as smectic-C subphases following de Gennes' notation (Table 2.1 and Figure 2.1) (De Gennes and Prost 1995).

These phases include orthogonal and tilted phases, in which the long molecular axes are parallel and tilted to the layer normal, respectively. In spite of the biaxiality of the structure, the former became customable to also designate the biaxial orthogonal phases as smectic A but not C. As it appears from the symmetry analysis, the biaxiality of the molecular shape and polar order can give rise to a manifold of smectic phases with fluid order including chiral phases, ferro-, antiferro-, and helielectric phases.

2.1 PHENOMENOLOGICAL THEORY OF THE NEMATIC ORDER

Exhaustive theory of nematic-type phases (phases with translational symmetry) formed by bent-shaped mesogens was developed by Lubensky and Radzihovsky (2002). In this section, we briefly outline the main principles of the construction of the mean-field phase diagram developed in Lubensky and Radzihovsky (2002). As in the case of standard nematogens, it is the entropic interactions rather than electric ones that drive the phase transitions (Chaikin and Lubensky 1995). A dominating ordering factor in the stabilization of the nematic phase is the shape of mesogens rather than electric dipoles. Later on, we show that electrostatic interactions are yet

TABLE 2.1

Possible Smectic Phases of Bent-Shaped Mesogens and Their Symmetries

Phase	Symmetry
C_M	D_{2h}
C_P	C_{2v}
C_Q	C_{1h}
C_R	C_1
C	C_{2h}
C_T	C_i
C_{B2}	C_2
C_{B1}	C_{1h}
C_G	C_1

Source: After Cladis, P. et al., *Ferroelectrics*, 243, 221, 2000.

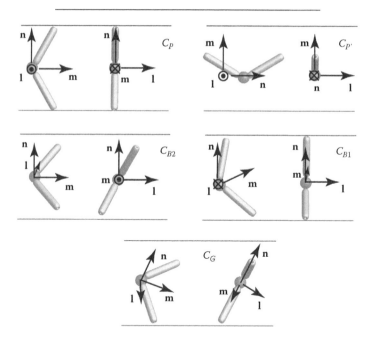

FIGURE 2.1 Minimal banana smectics: C_P (SmAP), $C_{P'}$ (not observed), C_{B2} (SmCP), C_{B1} (leaning phase, SmCLP), and C_G (SmC$_G$).

important to stabilize polar structures in smectics. That brings us to naturally introduce the order parameter capturing the *steric* nature of the interactions. The following paragraph outlines the theory of the nematic phases developed by Lubensky and Radzihovsky (2002). The order parameter can be constructed from the mass-moment tensors rather than from the electric charge distribution. A simple model capturing

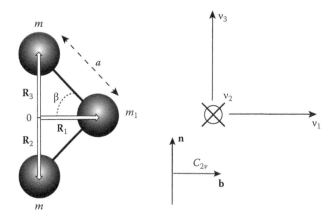

FIGURE 2.2 Simple three-atom model of a bent-shaped mesogen and a body fixed ortho-normal coordinate system.

the mass distribution in a bent-shaped mesogen is a rigid triplet of beads with a fixed bond between them (Figure 2.2). The central bead, the core of the "molecule," has a mass m_1, and the mass of each outer bead is m. We designate the nonpolar direction of this construction by υ_3 and the polar vector υ_1. The first vector contributes to the nematic director $\mathbf{n}$ and the second one contributes to the polarization $\mathbf{p}$. The third vector υ_2 is chosen to be orthogonal to υ_3 and υ_1. Using the angles introduced in Figure 2.2, the positions of the bead centers are given by the vectors

$$\mathbf{R}_{\alpha,1} = (a\cos\beta)\upsilon_{\alpha,1} \tag{2.1}$$

$$\mathbf{R}_{\alpha,2} = -(a\sin\beta)\upsilon_{\alpha,3} \tag{2.2}$$

$$\mathbf{R}_{\alpha,3} = (a\sin\beta)\upsilon_{\alpha,3} \tag{2.3}$$

where the index α designates the selected molecule.

Three mass moments can be constructed as follows: The zero-order moment is the center of mass of the molecule $\mathbf{R}_c^\alpha = \dfrac{1}{2m+m_1}\sum_{\mu=1}^{3} m_\mu \mathbf{R}_{\alpha\mu} = \left(\dfrac{m_1}{2m+m_1}a\cos\beta\right)\upsilon_{\alpha,1}$. Defining the positions of the beads relative to the mass center $\mathbf{r}_{\alpha,i} = \mathbf{R}_{\alpha,i} - \mathbf{R}_c^\alpha$, we obtain the second (quadrupolar) mass moment tensor:

$$C_{2,\alpha}^{ij} = \sum_{\mu=1}^{3} m_\mu \left(r_{\alpha,\mu}^i r_{\alpha,\mu}^j - \frac{1}{3} r_{\alpha,\mu}^2 \delta^{ij} \right) \tag{2.4}$$

The third-order moment can be decomposed into a vector and a third-rank symmetric traceless tensor:

$$C_{3,\alpha}^{ijk} = \sum_{\mu=1}^{3} m_\mu \left(r_{\alpha,\mu}^i r_{\alpha,\mu}^j r_{\alpha,\mu}^k - \frac{1}{5} r_{\alpha,\mu}^2 \left(\delta^{ij} r_{\alpha,\mu}^k + \delta^{jk} r_{\alpha,\mu}^i + \delta^{ki} r_{\alpha,\mu}^j \right) \right) \qquad (2.5)$$

These mass moments can be expanded in terms of a complete set of tensors formed by the unit vectors $\mathbf{v}_{\alpha,i}$:

$$C_{1,\alpha}^i = c_1 v_{\alpha,1}^i$$
$$C_{2,\alpha}^{ij} = c_{23} Q_{\alpha,3}^{ij} + c_{22} \left(Q_{\alpha,1}^{ij} - Q_{\alpha,2}^{ij} \right) \qquad (2.6)$$
$$C_{3,\alpha}^{ijk} = c_{31} T_{\alpha,1}^{ijk} + c_{32} T_{\alpha,2}^{ijk}$$

where

$$Q_{\alpha,a}^{ij} = v_{\alpha,a}^i v_{\alpha,a}^j - \frac{1}{3} \delta^{ij}, \quad a = 1,2,3 \qquad (2.7)$$

$$T_{\alpha,1}^{ijk} = v_{\alpha,1}^i v_{\alpha,1}^j v_{\alpha,1}^k - \frac{1}{5} \left(\delta^{ij} v_{\alpha,1}^k + \delta^{jk} v_{\alpha,1}^i + \delta^{ki} v_{\alpha,1}^j \right) \qquad (2.8)$$

$$T_{\alpha,2}^{ijk} = v_{\alpha,3}^i v_{\alpha,3}^j v_{\alpha,1}^k + v_{\alpha,3}^i v_{\alpha,1}^j v_{\alpha,3}^k + v_{\alpha,1}^i v_{\alpha,3}^j v_{\alpha,3}^k - \frac{1}{5} \left(\delta^{ij} v_{\alpha,1}^k + \delta^{jk} v_{\alpha,1}^i + \delta^{ki} v_{\alpha,1}^j \right) \quad (2.9)$$

and these coefficients are

$$c_1 = \frac{2mm_1 a^3 \cos\beta \left(-m_1 + 2m \cos 2\beta \right)}{\left(2m + m_1 \right)^2}$$

$$c_{23} = 2ma^2 \sin^2\beta - \frac{mm_1}{2m + m_1} a^2 \cos^2\beta$$

$$c_{22} = \frac{mm_1}{2m + m_1} a^2 \cos^2\beta \qquad (2.10)$$

$$c_{31} = \frac{2mm_1 \left(m_1^2 + 4m^2 \right)}{\left(2m + m_1 \right)^3} a^3 \cos^3\beta$$

$$c_{32} = \frac{4m^2 m_1}{2m + m_1} a^3 \sin^2\beta \cos\beta$$

The number of independent components of the tensors in the expansion (Equations 2.7 through 2.9) can be considerably reduced if one considers the symmetry and the completeness of the set of unit vectors $\mathbf{v}_\alpha$. There are only two independent molecular parameters in the set of second-rank tensors Q_α^{ij} and two nonvanishing third-rank tensors $T_{\alpha,a}^{ij}$ (Lubensky and Radzihovsky 2002). Coarse grained order parameters are introduced as

$$p^i(\mathbf{x}) = \frac{1}{\rho} \sum_\alpha v_{\alpha,1}^i \delta(\mathbf{x} - \mathbf{x}_\alpha)$$

$$Q_\alpha^{ij}(\mathbf{x}) = \frac{1}{\rho} \sum_\alpha Q_{\alpha,a}^{ij} \delta(\mathbf{x} - \mathbf{x}_\alpha) \qquad (2.11)$$

$$T_\alpha^{ijk}(\mathbf{x}) = \frac{1}{\rho} \sum_\alpha T_{\alpha,a}^{ijk} \delta(\mathbf{x} - \mathbf{x}_\alpha)$$

where the summation is done over all molecules at positions marked by $\mathbf{x}_\alpha$ in a microscopic volume at a coordinate $\mathbf{x}$, and ρ is the number density of the molecules. The full set of the order parameters describing a simple model of a bent-shaped molecule includes a vector order parameter, two second-rank tensors, and two third-rank tensors. A further simplification is made when only one tensor of each type is considered: Q^{ij} and T^{ijk}, respectively. This can be justified by integration of the two order parameters out of four (Lubensky and Radzihovsky 2002). We can choose a space-fixed orthonormal basis given by a set of three vectors $(\mathbf{m}, \mathbf{l}, \mathbf{n})$. A full set of tensors expressed in this basis $(\mathbf{m}, \mathbf{l}, \mathbf{n})$ is obtained with the help of a set of the symmetric-traceless orthonormal basis tensors J_μ^{ij} and I_μ^{ijk} (the expressions are given in [Lubensky and Radzihovsky 2002]), which transform under $L = 3$ and $L = 2$ representations of the rotation group in 3D. There are five second-rank tensors J_μ^{ij} and seven third-rank tensors I_μ^{ijk}. In this presentation, the three order parameters can be expressed as

$$p^i = \sum_\mu p_\mu n_\mu^i \qquad (2.12)$$

$$Q^{ij} = \sum_\mu Q_\mu J_\mu^{ij} \qquad (2.13)$$

$$T^{ijk} = \sum_\mu T_\mu J_\mu^{ijk} \qquad (2.14)$$

Landau mean-field theory can be developed by the expansion of the free energy density in rotationally invariant power-series in the order parameters p^i, Q^{ij}, and T^{ijk} and their cross-couplings:

$$f = f_p + f_Q + f_T + f_{pQ} + f_{pT} + f_{QT} + f_{pQT} + f_{QQTT} \qquad (2.15)$$

where the vector contribution f_p is similar to the energy of the paramagnetic-ferromagnetic transition:

$$f_p = \frac{1}{2}K_p\left(\partial_j p^i\right)\left(\partial_j p^i\right) + \frac{1}{2}r_p p^i p^i + u_p\left(p^i p^i\right)^2 \tag{2.16}$$

f_Q describes nematic–isotropic transition and K_Q is the corresponding Frank elastic constant (one constant approximation):

$$f_Q = \frac{1}{2}K_Q\left(\partial_k Q^{ij}\partial_k Q^{ij}\right) + \frac{1}{2}r_Q Q^{ij}Q^{ij} - w_Q Q^{ij}Q^{jk}Q^{ki} + u_Q\left(Q^{ij}Q^{ij}\right)^2 \tag{2.17}$$

Energy contribution due to the third-rank tensor is given by

$$f_T = \frac{1}{2}K_T\left(\partial_l T^{ijk}\partial_l T^{ijk}\right) + \frac{1}{2}r_T T^{ijk}T^{ijk} + u_T\left(T^{ijk}T^{ijk}\right)^2 + v_T T^{i_1 i_2 i_3}T^{i_1 i_4 i_5}T^{i_3 i_5 i_6} \tag{2.18}$$

The parameters are linear in temperature, $r_p \propto T - T_p, r_p \propto T - T_p, r_T \propto T - T_T$, and vanish at temperatures T_p, T_Q, and T_T. The lowest-order coupling terms are

$$
\begin{aligned}
f_{pQ} &= -w_{pQ}p^i p^j Q^{ij} \\
f_{pT} &= -w_{pQ}p^i p^j p^k T^{ijk} \\
f_{QT} &= -w_{QT}Q^{i_1 i_2}T^{i_1 jk}T^{i_2 jk} \\
f_{pQT} &= -w_{pQT}p^i Q^{jk}T^{ijk} \\
f^{(1)}_{Q^2 T^2} &= -w_1 Q^{i_1 l}Q^{i_2 l}T^{i_1 jk}T^{i_2 jk} \\
f^{(2)}_{Q^2 T^2} &= -w_2 Q^{i_1 j_1}Q^{i_2 j_2}T^{i_1 j_1 k}T^{i_2 j_2 k} \\
f^{(3)}_{Q^2 T^2} &= -w_3 Q^{i_1 i_2}Q^{j_1 j_2}T^{i_1 j_1 k}T^{i_2 j_2 k}
\end{aligned} \tag{2.19}
$$

The term f_{pQT} is particularly interesting. It is responsible for the development of polar order driven by T^{ijk} if the product $Q^{jk}T^{ijk} \neq 0$, and f_{pQT} will induce p^i (polar order). This model provides an unusually rich variety of phases listed in Table 2.2 and in Figure 2.3.

Some of these phases can be described by a single-order parameter such as p^i, Q^{ij}, and T^{ijk}. Properties of the low-symmetry phases, however, are functions of several order parameters that are coupled to each other. This means that the development of one type of order necessarily induces another type of order.

A symmetry lowering transition from the isotropic phase, for example, leads to a development of the uniaxial order described by Q^{ij} or even the polar order p^i (Figure 2.4). The latter case has not been observed in the bulk nematic phase yet. Isotropic-to-uniaxial nematic phase transition can be described purely by the f_Q term. Tetrahedratic order parameter T^{ijk} can lead to two distinct phases: the optically isotropic tetrahedral phase T and a $N + 3$ phase with uniaxial nematic and tetrahedratic types

TABLE 2.2

Rich Variety of Phases Predicted in Bent-Shaped Molecules, and Their Symmetry and Order Parameters

Phase	Symmetry	Order Parameters
V	$C_{\infty v}$	p_3, S, T_1
N	$D_{\infty h}$	S
$N + 2$	D_{2h}	$S, B_{1,2}$
$N + 3$	D_{3h}	$S, T_{2,3}$
T	T_d	$T_{6,7}$
N_T	D_{2d}	$S, T_{6,7}$
$(N_T + 2)^*$	D_2	S, B_1, T_6, T_7
$V + 2$	C_{2v}	p_3, S, B_1, T_1, T_6 or p_1, S, B_1, T_2, T_4
$V + 3$	C_{3v}	$p_3, S, T_1, T_{2,3}$
$(V_T + 2)^*$	C_2	$p_3, S, B_1, T_1, T_6, T_7$ or $p_1, S, B_1, T_2, T_4, T_5$
$N + V$	C_{1h}	$p_1, p_3, S, B_1, Q_3, T_1, T_2, T_4, T_6$

Source: Lubensky, T.C. and Radzihovsky, L., *Phys. Rev. E.*, 66, 031704-1-27, 2002.

Note: See also Figure 2.3.

of orders. The role of the tetrahedratic order has been intensively studied theoretically (Fel 1995, Cladis et al. 2003, Brand et al. 2005a,b, Radzihovsky and Lubensky 2007).

Experimental evidence of the tetrahedratic order was found in calorimetry studies, flow, and magneto-optical effects in the nematic phase (Pleiner et al. 2006, Ostapenko et al. 2008, Wiant et al. 2008). The transition driven by the T^{ijk} tensor is complex and may give rise to two distinct types of phases: the tetrahedratic (T) and the nematic ($N + 3$) phases with the D_{3h} symmetry. The latter phase shows a uniaxial nematic order where the f_{QT} coupling between the orientational and the tetrahedratic order parameters results in the triadic order transverse to the nematic axis. Such a phase, however, has not been discovered so far. Further symmetry breaking of the nematic phase, described by the theory based on the multipole expansion of the mass moments, gives rise to a variety of mesophases. Those phases do not have a positional order and can be described as nematics. Interestingly, the theory also predicts a spontaneous breaking of reflection symmetry in $(N_T + 2)^*$ and $(V_T + 2)^*$ phases without the need for smectic layers.

2.2 PHENOMENOLOGICAL THEORIES OF POLAR SMECTIC PHASES

Most of the mesophases discovered in bent-shaped mesogens are smectics. An exceptional property of the bent-shaped (or bent-core) liquid crystals (BLCs) is that they exhibit proper ferroelectric phases. Although the spontaneous formation of ferroelectric order in isotropic fluids and nematics is in principle possible, such transitions have never been observed experimentally (except for polymer nematics [Takezoe 2014]). Ferroelectric order in dipolar fluids would require very strong dipole moments.

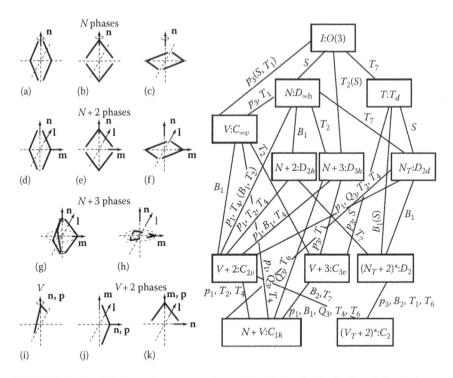

FIGURE 2.3 (Left) Schematic representations of the N, $N + 2$, $N + 3$, V, and $V + 2$ phases. Three versions, (a), (b), and (c), of the N phases are depicted with respective predominant alignment of ν_3, ν_1, and ν_2 along **n**, specifying the direction of the principal axis of Q_{ij} with the largest eigenvalue. The $N + 2$ phases, (d), (e), and (f), are obtained, respectively, from the N phases (a), (b), and (c) by restricting rotations in the plane perpendicular to **n** to have twofold symmetry, whereas the $N + 3$ phases, (g) and (h), are obtained by restricting these rotations to have a threefold symmetry. In the uniaxial V phase (i), the molecular ν_1 aligns on average along $\mathbf{p} \| \mathbf{n}$, sampling equally all orientations about the **p** axis. The $V + 2$ phase can be produced either by introducing biaxial order perpendicular to **p** and **n** into the V phase (j) or by introducing vector order into the $N + 2$ phase by aligning **p** along **m** (k). (Right) A flowchart of phase transitions between liquid-crystal phases illustrated in the left images. Order parameters, which become nonzero at each of the transitions, and their symmetry groups are indicated. For transitions that we have studied in detail, we have also indicated the secondary (explicitly induced by nonlinear couplings) order parameters by placing them in parentheses. (Reprinted with permission from the Lubensky, T.C. and Radzihovsky, L., Theory of bent-core liquid-crystal phases and phase transitions, *Phys. Rev. E.*, 66, 031704-1-27, 2002. Copyright 2002 by the American Physical Society.)

But this has a drawback that the dipoles rather tend to form rigid polar chains instead of the homogeneous bulk ferroelectric state (Terentjev et al. 1994).

The role of steric interactions in the formation of ordered phases has been demonstrated in a variety of molecular and colloidal systems (Meuer et al. 2007, Mourad et al. 2008, Vandenpol et al. 2010a,b). It dates back to the works of Onsager who developed the theory of the lyotropic isotropic–nematic phase transition of rod-shaped particles (Onsager 1949). In these models, pairs of hard particles tend to occupy the largest accessible volume to maximize their entropy. But the volume attainable by

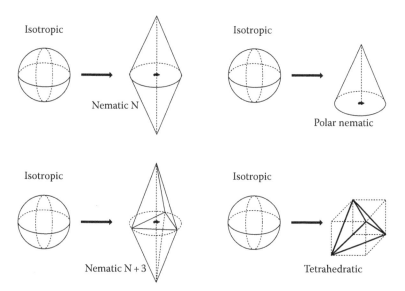

FIGURE 2.4 Schematics of the symmetry transformations on phase transition from the isotropic phase. (Redrawn from Lubensky, T.C. and Radzihovsky, L., *Phys. Rev. E*, 66, 031704-1-27, 2002.)

anisometric particles depends on their orientation (Figure 2.5). Some part of this volume, for a given configuration, becomes unattainable or *excluded*. This system then tries to minimize the excluded volume by allowing a particular orientation of the particles. At a sufficiently large concentration, it leads to the formation of ordered structures like the nematics or smectics.

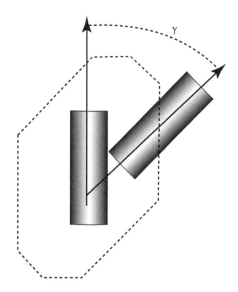

FIGURE 2.5 The projection of the excluded volume (marked by a dashed loop) formed by two rods with an angle γ between the symmetry axes.

In BLCs, the molecular shape does not have a center of symmetry but has the C_{2v} symmetry and possible polar effects on the packing need to be considered. These effects in two and three dimensions were considered by Bisi et al. (Bisi et al. 2008, 2010, Bisi and Rosso 2011). In the planar nematic-type structure composed of the rigid bent-shaped mesogens, the excluded area of the parallel configuration of the steric dipoles is nearly twice larger than that of the anti parallel one and it cannot minimize the excluded area (Bisi et al. 2008) (see Figure 2.6). The steric interactions favor molecules to group in antiparallel fashion in a biaxial but nonpolar structure.

Also the 3D analysis in (Bisi et al. 2008) shows that the antiparallel biaxial configuration of two bent-shaped molecules is favored and the polar order cannot be established (Figure 2.6). In the case of smectics, however, the molecules are restricted to the layers. Polar arrangement is sterically favored. Being driven by fluctuations, the interlayer penetrations of the molecules from one layer to another produce another force, which favors a parallel orientation of each arm between neighboring layers. Altogether, it favors the antiferroelectric order of molecular layers. However, this packing effect is not dominating. The intermolecular interactions yield a contribution that is about 5–10 times larger than the energy contribution of the packing entropy (Osipov and Pajak 2014). We shall discuss these interactions based on microscopic theories in Section 2.4.

To develop a macroscopic theory of such phases, it often suffices to consider a single smectic layer. Simple SmA and SmC-type phases can be described by the tilt order parameter and polarization. Designating the mean orientation of the mesogens in smectic layers by $\mathbf{n}$ and the smectic layer normal by $\mathbf{k}$, the tilt director can be introduced as a cross product $\boldsymbol{\xi} = [\mathbf{n} \times \mathbf{k}]$. The geometrical meaning of $\boldsymbol{\xi}$ is the

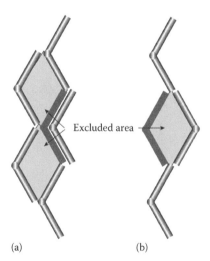

(a)   (b)

FIGURE 2.6 Two interacting bent-shaped particles in (a) parallel (ferroelectric) configuration and (b) antiparallel (antiferroelectric) configuration. The shaded region is inaccessible to the dark gray apex when this glides the molecule. The excluded area in the ferroelectric configuration (a) is twice as large as that in the antiferroelectric one (b).

rotation axis for the tilting, and the magnitude is equal to the projection of the direc-tor to the layer plane. Since both $\mathbf{n}$ and $\mathbf{k}$ are unit vectors, their product is a measure of the tilt. For small tilt angles θ, $[\mathbf{n} \times \mathbf{k}] \approx \theta$. Often $\left[\mathbf{k} \times \dfrac{\xi}{|\xi|} \right]$ is called the c-director.

Bent-shaped mesogens are strongly anisomeric molecules. To acknowledge that, we introduce another director that describes the mean orientation of molecular bows. For symmetric bent-shaped molecules, this orientation coincides with the spontane-ous polarization, which we shall call the p-director.

A variety of tilted polar smectic phases originating from polar order have been shown by Roy et al. (1999) using the Landau theory. The free energy expansion sug-gested by the authors reflected an achiral structure of the bent-shaped mesogens and contains no linear coupling terms specific for the chiral SmC* phase:

$$F_l = \frac{a}{2}\xi^2 + \frac{b}{24}\xi^4 + \frac{\alpha}{2}P^2 + \frac{\beta}{4}P^4 + \delta_1\xi^2 P^2 + c_1\left(\xi \cdot \mathbf{P}\right)^2 + \frac{1}{2\chi_0}\left(\mathbf{n} \cdot \mathbf{P}\right)^2$$

$$+ \frac{1}{2\chi_1}\left(\mathbf{k} \cdot \mathbf{P}\right)^2 + d_1\xi^2\left(\mathbf{n} \cdot \mathbf{P}\right)^2 + d_2\xi^2\left(\mathbf{k} \cdot \mathbf{P}\right)^2 + d_3\left(\mathbf{n} \cdot \mathbf{P}\right)\left(\mathbf{k} \cdot \mathbf{P}\right)\left(\mathbf{n} \cdot \mathbf{k}\right) - \mathbf{E} \cdot \mathbf{P} \quad (2.20)$$

Here, $\xi = |\xi|$. The first two terms, containing coefficients $a = a_0(T - T_\xi)$ and b, describe the SmA-SmC transition. The third and fourth terms describe the paraelectric–ferroelectric transition. $\alpha = \alpha_0(T - T_p)$ and β are expansion coefficients. The rest of the terms are couplings between the order parameters, the layer normal, and an electric field. Designating the angle between the vectors ξ and $\mathbf{P}$ by Ψ, the free energy can be expressed as

$$F_l = \frac{a}{2}\xi^2 + \frac{b}{24}\xi^4 + \frac{\alpha}{2}P^2 + \frac{\beta}{4}P^4 - \gamma_1\xi^2 P^2 \cos(2\Psi) + \delta\xi^2 P^2 - EP\xi\sin(\Psi) \quad (2.21)$$

where

$$\gamma_1 = \frac{1}{2}\left(\frac{1}{2\chi_1} - c_1\right) \quad \text{and} \quad \delta = \delta_1 + \frac{1}{2}\left(\frac{1}{2\chi_1} + c_1\right) \quad (2.22)$$

The interactions between the neighboring layers are described by the inhomoge-neous part of the free energy F_i

$$F_i = P\xi\cos(\theta)\sin(\Psi)\left(\mu_1\frac{\partial\varphi'}{\partial z} + \mu_2\frac{\partial\varphi}{\partial z}\right) + \frac{1}{2}K_{33}\xi^2\left(\frac{\partial\varphi'}{\partial z}\right)^2 + \frac{1}{2}KP_t^2\left(\frac{\partial\varphi}{\partial z}\right) \quad (2.23)$$

where
   P_t is the in-plane component of the polarization, φ and φ' are, respectively, the angles made by $\mathbf{P}_t$ and ξ
   μ_1 and μ_2 are the direct and inverse flexoelectric constants, respectively
   K_{33} is the bend elastic constant
   K is the bend elastic constant for the transversal polarization P_t

Minimization of F_l under the condition of $E = 0$ with respect to the order parameters ξ and P yields the equation for the equilibrium Ψ

$$2\gamma_1 \xi^2 P^2 \sin(2\Psi) = 0 \tag{2.24}$$

This equation has four solutions corresponding to four possible phases.

1. $\xi = \pm\sqrt{-a/b}$ and $P = 0$ correspond to a simple achiral and nonpolar SmC phase with C_{2h} symmetry.
2. When $\xi = 0$ and $P = \pm\sqrt{-\alpha/\beta}$, this is a biaxial polar SmA phase with a C_{2v} symmetry. The experimental evidence of this phase can be found in Reddy et al. (2011).
3. For $\Psi = 0$ or π, a transversely polarized SmC-like phase is stable. The polarization is perpendicular to the tilt plane. This is the typical B_2 or SmCP phase (Pelzl et al. 1999, Selinger 2003). The polarization is perpendicular to the tilt plane resulting in an overall chiral state, where the states with $\Psi = 0$ and $\Psi = \pi$ are of opposite handedness. For a fixed orientation of ξ, the layers can be combined in a synpolar or antipolar fashion. In the first case, the handedness of the layers remains equal, whereas in the latter case the handedness alternates and the overall structure is racemic.
4. $\Psi = \pm\pi/2$ and $\xi \neq 0, P \neq 0$ corresponds to a monoclinic C_s symmetry. The polarization P is in the tilt plane formed by the vectors $\mathbf{n}$ and $\mathbf{k}$. Each layer possesses an out-of-plane polarization P. However, the director remains $\mathbf{n} \rightarrow -\mathbf{n}$ invariant and the up–down invariance of the phase is broken. This is an example of a leaning (or tipping) phase. The experimental evidence of this phase can be found in Chattham et al. (2015) and Zhang et al. (2012). It is important to mention that lower symmetry phases such as the general SmC$_G$ with an arbitrary angle Ψ may occur in this model if the higher order terms are considered.

The coupling between the smectic layers is mostly governed by the flexoelectric terms μ_1 and μ_2 containing the gradient terms of the director azimuthal angle along the layer normal. The minimization of the full free energy also yields the phases with a heterogeneous director distribution. The flexoelectric terms do not affect the structure in the SmAP, SmC, and the leaning SmCPL phases. At the same time, heterogeneous structures can be expected in the case of the SmCP and SmC$_G$ phases. One example of a heterogeneous structure is the helical superstructure with a helix wave vector q. The equilibrium value of the wave vector is given by (see for details Roy et al. [1999])

$$q = -\frac{(\mu_1 + \mu_2) P_0 \xi_0 \cos(\theta_0) \cos(\Psi_0)}{K_{33}\xi_0^2 + K\left(1 - \xi_0^2 \sin(\Psi_0)\right)} \tag{2.25}$$

where P_0 and ξ_0 are equilibrium values of the polarization and the tilt, respectively. The pitch of the helix is $p = 2\pi/q$. The case of an antiferroelectric structure corresponds

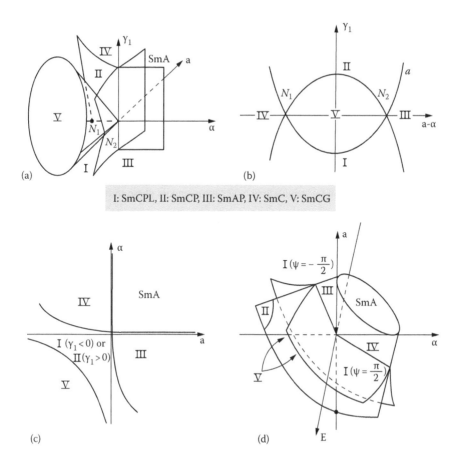

I: SmCPL, II: SmCP, III: SmAP, IV: SmC, V: SmCG

FIGURE 2.7 Phase diagrams of field-free states in the phenomenological parameter spaces (γ_1, α, a), $(\gamma_1, a, -\alpha)$, and (α, a). The phases are separated by the second-order transition surfaces (a) and lines ((b) and (c)). N_1 and N_2 are four-phase points. (d) Phase diagram under applied field E in the (α, a, E) space for $\gamma_1 > 0$. (Reprinted with permission from Roy, A., Madhusudana, N., Toledano, P., Longitudinal spontaneous polarization and longitudinal electroclinic effect in achiral smectic phases with bent-shaped molecules, *Phys. Rev. Lett.*, 82, 1466, 1999. Copyright 1999 by the American Physical Society.)

to the pitch equal to the double of the layer thickness. The regions of phase stability for various values of the phenomenological coefficients are given in Figure 2.7.

Although the phase diagram shows that the polar low-symmetry phases can be reached only through the SmA or SmC states, in the experiment, the SmCP phase often occurs right below the isotropic phase. The SmA-SmCP and SmC-SmCP transitions often occur in compounds with a large opening angle.

Except for ferro- and antiferroelectric SmCP and SmAP phases, phases with other periodicities can be encountered among the bent-shaped mesogens. This variety of structures can be described by a model where the free energy is expressed as a sum

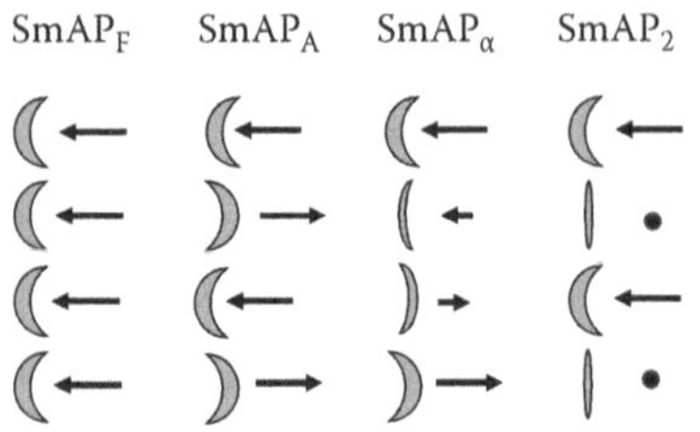

$$\text{SmAP}_F \quad \text{SmAP}_A \quad \text{SmAP}_\alpha \quad \text{SmAP}_2$$

FIGURE 2.8 Four possible arrangements and tentative structural names of non-tilted and polarly ordered layered systems (Pociecha et al. 2003). The arrows indicate the direction of the smectic layer polarization. The fifth arrangement is the smectic A random phase (SmAP_R), where the polarizations in the neighboring layers have random phase shifts (not shown here).

of Landau-type energy contributions of single smectic layers and the interlayer couplings:

$$F = \sum_j \left(\frac{1}{2} a_{0p} P_j^2 + \frac{1}{4} a_{0p} P_j^4 + \frac{1}{6} a_{0p} P_j^6 + \frac{1}{2} a_{0t} \xi_j^2 + \frac{1}{4} a_{0t} \xi_j^4 + \frac{1}{6} a_{0t} \xi_j^6 + \frac{\Omega}{2} \left(\xi_j \times P_j \right)^2 \right)$$
$$+ \sum_j \left(\frac{1}{2} a_{0p} \left(P_j \cdot P_{j+1} \right) + \frac{1}{4} a_{2p} \left(P_j \cdot P_{j+1} \right)^2 + \frac{1}{2} a_{0t} \left(\xi_j \cdot \xi_{j+1} \right) \right) \tag{2.26}$$

The first sum is the Landau expansion in terms of the tilt and polarization order parameters. The last term in this sum, $\frac{1}{2}\Omega\left(\xi_j \times P_j \right)^2$, describes the bilinear coupling between the tilt and polarization. For $\Omega<0$, the tilt of the molecules favors polar order within each layer. This term is responsible for the quadratic electroclinic effect, where an electric field applied in the SmA phase induces tilt above a certain threshold (Eremin et al. 2008). The linear tilt–polarization coupling term is symmetry forbidden when the molecules are achiral. The second sum contains interlayer coupling terms. The sum is taken over the nearest neighbors due to a short-range scale of the interactions and the electrostatic screening. The first term in the sum represents the dipole–dipole interactions, whereas the bilinear polarization term results from the van der Waals (vdW) interactions and accounts for a nonparallel orientation of the molecular parts. Even in an oversimplified case of the polar SmAP phase, where only a_{1p} and a_{2p} terms are considered, the minimization of the free energy with respect to varying polarization $p_j = P(\cos\varphi_j, \sin\varphi_j)$, where φ_j is the azimuthal angle of the polarization in the smectic layer, yields a rich variety of phases. Those phases include the ferro- (SmAP_F) and antiferroelectric (SmAP_A) phases, bilayer SmAP_2 phase, SmAP_α phase, where the polarization is helically modulated, and SmAP_R with randomly organized polar layers (see Figure 2.8).

2.3   ORDER PARAMETER CONDENSATION

Another theoretical approach to understand the development of the "banana" phases was suggested by Lorman and Mettout (1999, 2004). The attention of the authors was drawn by the fact that the majority of bent-core compounds shows a tilted *and* polar phase directly below the isotropic phase. This means that tilt and polarization result from a single symmetry breaking instability of the isotropic phase. A similar situation was observed in superconductors, which may exhibit several simultaneous symmetry breakings. This process cannot be accounted by a simultaneous onset of classical order parameters. Instead, a single order parameter, associated with *p*- or *d-anisotropic* wave functions, gathers all the relevant degrees of freedom involved in the transition in a single representation (Lorman and Mettout 1999).

This may be achieved in bent-core liquid crystals by a direct condensation in the isotropic liquid of the transverse vector wave order parameters: a wave of polarization and a wave of tilt. In this model, three types of phases can be encountered in the minimal phase diagram:

1. Rectilinear R-phase, which has a structure of linearly polarized wave and describes the SmAP$_A$ phase with D_{2h} symmetry.
2. EL phase with elliptic polarization belonging to the symmetry group D_2 is a chiral antiferroelectric structure and corresponds to chiral SmCP phases.
3. C-phase, where the wave vectors are circularly polarized, is tentatively attributed to the structure of the B$_7$ phase.

Further development of the wave-vector model showed its applicability to columnar and soft-crystal phases (Mettout 2007). Considering condensation of several non-parallel wave vectors, up to 24 different phases can be obtained in the framework of this model, including orthorhombic, monoclinic, and triclinic structures. A rich phase diagram resulting from this model describes nearly all known phases formed by bent-core mesogens. Other important theoretical findings including the models of such exotic phases as B$_7$, Dark Conglomerate, and others will be discussed in the following section.

2.4   MOLECULAR THEORIES AND SIMULATIONS

In this section, we review molecular theories of bent-core LCs and some results of numerical simulations. In particular, we will focus on the development of the polar order in achiral BLCs. Microscopic description considering interactions between the mesogens can help to shed light on the preference of the mesogens to form a particular type of mesophase such as polar or nonpolar, tilted or orthogonal.

Interestingly, ferroelectric order also occurs in the orthogonal smectic phases (see Section 2.2). This is another piece of evidence that BLC ferroelectrics are of the proper type. The primary order parameter is the polarization and not the tilt, unlike in the chiral SmC* phase. There are two major difficulties to achieve the ferroelectric order in a fluid phase. First, the dipoles tend to form chain-like structures that only weakly interact with each other. Second, the energy of the long-range dipole–dipole interactions between the molecules in general depends on the boundary conditions and the shape of the sample (Osipov and Teixiera 1987). This problem, however, can be circumvented in the quasi-2D geometry, such as freely suspended films. In 2D, the dipole–dipole interaction contributes to the free energy. Indeed, there are several examples where the ferroelectric order could not be established in bulk materials but the freely suspended films exhibited a spontaneous polar structure in the field-free state (Eremin et al. 2012).

In microscopic theories and numerical simulations, bent-core mesogens can be modeled as V-shaped assemblies of rigid spheres or rods with associated dipole moments. The dipole moments originate from the chemical structure and are particularly enhanced by introducing substituents. An example of such a model is analyzed in the works by Osipov, Emelyanenko, and Pajak (Emelyanenko and Osipov 2004, Osipov and Pajak 2014). The framework of these molecular models is based on the dipolar and vdW or dispersion interactions.

Since the interaction potentials rapidly decay with the distance, the interaction energy is calculated between the neighboring mesogens i and j

$$U(i,j) = V_{LJ}\left(r_{ij}, \mathbf{b}_i, \mathbf{b}_j\right) + V_{dd}\left(\mathbf{r}_{ij}, \mu_i, \mu_j\right), \tag{2.27}$$

where

$\mathbf{r}_{ij}$ is a vector connecting the mass centers of the neighboring mesogens
μ_i and μ_j are the respective dipole moments
$\mathbf{b}_i$ and $\mathbf{b}_j$ the unit vectors in the direction of the bows

This formula is valid in case of perfectly aligned long molecular axes. The mean field potential $U_{MF}(\mathbf{b})$ is determined in the self-consistent way by integration of the orientation of the mesogen:

$$U_{MF}\left(\hat{b}\right) = \rho \int f_1\left(\mathbf{b}_1\right) V\left(\mathbf{b}, \mathbf{b}_1\right) d\mathbf{b}_1, \tag{2.28}$$

where

ρ is the number density
f_1 is the orientation distribution function

The self-consistency condition for f_1 is

$$f_1\left(\mathbf{b}\right) = \frac{1}{Z} \exp\left(-\frac{U_{MF}\left(\mathbf{b}\right)}{k_b T}\right), \tag{2.29}$$

where Z is the one-particle partition function

$$Z = \int \exp\left(-\frac{U_{MF}\left(\mathbf{b}\right)}{k_b T}\right) d\mathbf{b}. \tag{2.30}$$

The preference for a particular mesophase structure can be found by minimizing the intermolecular interaction energy. In case of bent-core mesogens, this approach demonstrates the preference for tilt phases for different opening angles of the mesogen. An example of this calculation is given by Emelyanenko and Osipov (2004). They considered a model-mesogen consisting of two sphero-cylinders with an opening angle $\pi - 2\alpha$ and the dipole moments of each mesogenic arm μ (Figure 2.9a). The polar director is given by $\mathbf{b}_i$.

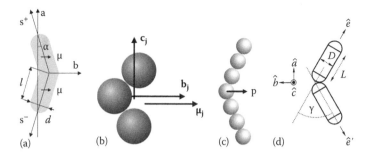

FIGURE 2.9 Different models of a bent-core mesogen used in microscopic theories and simulations: (a) a bent-rod model used in Emelyanenko and Osipov (2004). The mesogen is made of two units s^+ and s^-, the bow direction is given by **b**, and the dipole moment of each unit is µ; (b) is a three-bead model from Osipov and Pajak (2014). The beads represent Lennard-Jones centers. The electric dipole $\boldsymbol{\mu}_i$ is along the C_{2v} axis **b**. The "long molecular axis" is **c**. The third principle axis is given by the vector product $\mathbf{a}=\mathbf{b}\times\mathbf{c}$; (c) Soft-sphere model from Xu et al. (2001); and (d) Hard spherocylinder model from Lansac et al. (2003). The opening angle is given by γ.

The interaction energy of a pair of mesogens ($\mathbf{b}_1$ and $\mathbf{b}_2$) is given by the dipole–dipole (U_{dip}) and vdW (U_{disp}) interactions between their arms:

$$U_{12} = U_{disp} + U_{dip},$$

$$U_{disp} = -\int_{-l/2}^{l/2} dt_1 \int_{-l/2}^{l/2} dt_2 \frac{J_0 d^4}{\rho^6 (t_1, t_2)}, \qquad (2.31)$$

$$U_{dip} = \frac{\mu^2}{r_{12}^3}\left((\mathbf{b}_1 \cdot \mathbf{b}_2) - 3(\mathbf{u} \cdot \mathbf{b}_1)(\mathbf{u} \cdot \mathbf{b}_2)\right),$$

where
J_0 is the strength of the dispersion interactions
d is the diameter of the rod

The first term is the depletion interaction obtained by integration of the contribution from each volume element of the mesogens parameterized by t_1 and t_2. The vector ρ connects the corresponding points t_1 and t_2 on the neighboring molecules (for details see Emeleynenko and Osipov [2004]). Interestingly, the dispersion term as a function of the angle ψ between the vectors $\mathbf{b}_1$ and $\mathbf{b}_2$ exhibits a deep minimum as shown in Figure 2.10a.

This finding shows that it is the dispersion interactions that stabilize ordering of the short molecular axes and further stabilize the smectic phase for sufficiently large α. Accounting for the tilt of the mesogens results in more complicated expressions for the dispersion interactions: as shown in Figure 2.10b, the U_{disp} has a minimum at finite values of the tilt angle θ. In this phase, molecular bows are parallel to the tilt plane forming a leaning structure. Although the position of the minimum weakly depends on α, it requires a sufficiently large bend angle α for the phase to occur.

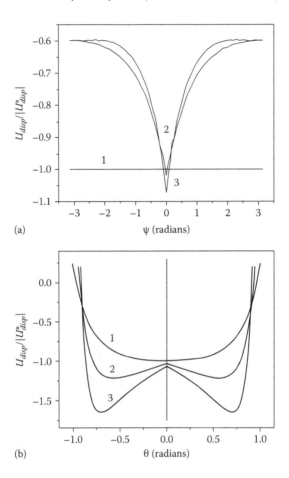

FIGURE 2.10 (a) Average dispersion interaction energy between non-tilted molecules located in the same smectic layer as a function of the angle between the short molecular axes $\mathbf{b}_1$ and $\mathbf{b}_2$ in the case $\langle l\cos(\alpha)\rangle/d = 1.5$: (1) $\alpha = 0$; (2) $\alpha = \pi/10$; (3) $\alpha = \pi/7$ and (b) average dispersion interaction energy between parallel banana molecules located in the same smectic layer as a function of the tilt angle for $\langle l\cos(\alpha)\rangle/d = 1.5$: (1) $\alpha = 0$; (2) $\alpha = \pi/15$; (3) $\alpha = \pi/10$. The energy is normalized by that for the pair of corresponding rod-like molecules $\langle U^*_{disp}\rangle$. (Reprinted with permission from Emelyanenko, A.V. and Osipov, M.A., Origin of spontaneous polarization, tilt, and chiral structure of smectic liquid-crystal phases composed of bent-core molecules: A molecular model, *Phys. Rev. E*, 70, 021704-1-021704-8, 2004. Copyright 2004 by the American Physical Society.)

The full energy including the dipolar term exhibits two equilibrium polar states: with the polarization perpendicular to the tilt plane like a common SmCP or B$_2$ phase and a phase where the polarization is parallel to the tilt plane (so-called leaning phase). The interplay between the two contributions selects the particular phase. The SmCP phase is primarily stabilized by the dipole moments and sufficiently large opening angles. Despite most of the polar SmC-type "banana" phases are of the

SmCP types, there are only a few examples of leaning phases given in Zhang et al. (2012), Chattham et al. (2015), and Westphal et al. (2016).

The microscopic theory of the transitions in the orthogonal SmA phases was developed by Osipov and Pajak (2014) using molecule models composed of triplets of spheres (Figure 2.9b). Restricting the description to the dispersion and the dipole–dipole interactions (see Equation 2.31), the mean-field interaction potential can be expressed as

$$U_{MF} = \rho \left(-3J\sigma^2\lambda^2 - \frac{\mu^2}{2\sigma} \right) \langle \cos\phi \rangle \cos\phi + \rho \frac{213}{4} \lambda^4 J\sigma^2 \left[1 + \langle \cos 2\phi \rangle \cos 2\phi \right], \quad (2.32)$$

where ϕ is the angle between the secondary m-director in the biaxial phase and the orientation of the molecular bow. $\langle \cos\phi \rangle$ and $\langle \cos 2\phi \rangle$ correspond to the dipolar and quadrupolar order parameters, respectively. The dimensionless mean-field free energy of the system deduced from potential Equation 2.32 is given by

$$\frac{F}{\rho\sigma^2 |J|} = \frac{1}{2} \left(\left[\frac{M}{2} - 3\lambda^2 \right] \langle \cos\phi \rangle^2 + \frac{213}{4} \lambda^4 \langle \cos 2\phi \rangle^2 \right) - t \ln Z, \quad (2.33)$$

where the control parameter $M = \dfrac{\mu^2}{\sigma^3 |J|}$ describes the relative strength of the dipole–dipole and dispersion interactions, $t = \dfrac{k_b T}{\sigma^2 |J|}$ is the dimensionless temperature, and Z is the one-particle partition function.

The phase diagram of the system is shown in Figure 2.11a. Depending on the strength of the dipole–dipole interaction, uniaxial SmA, biaxial SmA$_b$, and polar biaxial SmAP may occur. The biaxial-to-polar (SmA$_b$-SmAP) transition can be either of the first (dotted curve) or second (solid curve) order (Figure 2.11a and b), but the SmA-SmAP transition is always second order (Figure 2.11a and c). The latter is accompanied by the breaking of two symmetries and is different from a simple paraelectric–ferroelectric transition with a single order parameter (Figure 2.11b).

The mesophase behavior of bent-core molecules has been studied by computer simulations too. Atomistic simulations have an advantage that they can show how the macroscopic properties of the condensed phase depend on the chemical structure of the mesogens. Full-scale simulations are computationally very intensive and can be done for rather small ensembles. Therefore, various simplified models of the mesogens were used in simulations.

Steric models of mesogens are usually made up by pairs of hard cylinders or spherocylinders (Figure 2.9d) interacting through excluded-volume interactions (Camp et al. 1999, Memmer 2002, Lansac et al. 2003), and ellipsoids interacting through the Lennard-Jones or Gay-Berne potential. The molecules can also be modeled as sets of soft beads (Xu et al. 2001) (Figure 2.9c). Such molecules can be rigid or semiflexible.

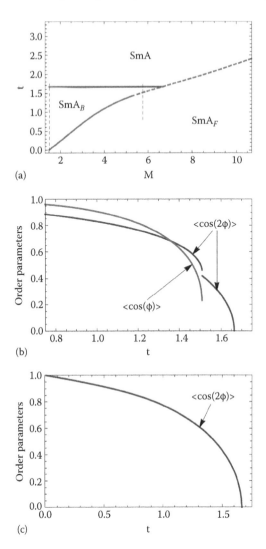

FIGURE 2.11   (a) Phase diagram for $\lambda = 1/2$ and the temperature dependence of the order parameter for selected values of (b) $M = 5.75$ (right vertical dotted line in (a)) and (c) $M = 1.5$ (left vertical dotted line in (a)). (With kind permission from Springer Science+Business Media: *Phys. Rev. Lett.*, Magnetic-field induced isotropic to nematic liquid crystal phase transition, 101, 2008, 247801-1-4, Ostapenko, T., Wiant, D.B., Sprunt, S.N., Jakli, A., and Gleeson, J.T.)

Structure formation in these simulations is characterized by various order parameters that are calculated by averaging over the whole molecular ensemble in the simulation. The degree of orientational order is described by the ensemble average over N molecules of the tensor formed by molecular long axes $\mathbf{n}_j$

$$Q_{\alpha\beta} = \left\langle \frac{1}{N} \left(\frac{3}{2} n_{j\alpha} n_{j\beta} - \frac{1}{2} \delta_{\alpha\beta} \right) \right\rangle. \tag{2.34}$$

The largest eigenvalue is assigned to the nematic order parameter S. The polar orientational order of molecular bows is given by

$$\mathbf{p} = \frac{1}{N} \sum_{i=1}^{N} \mathbf{p}_i. \tag{2.35}$$

The smectic order parameter is defined as

$$\rho = \frac{1}{N} \sum_{j=1}^{N} \exp\left(\frac{2\pi i}{d} \mathbf{k} \cdot \mathbf{r}_j \right). \tag{2.36}$$

The chiral order parameter in layered structures is given by a pseudo-scalar

$$\chi = \frac{1}{N} \sum_{i=1}^{N} \left(\left[\mathbf{k}_i \times \mathbf{n}_i \right] \cdot \mathbf{p}_i \right) \left(\mathbf{k}_i \cdot \mathbf{n}_i \right), \tag{2.37}$$

where
  $\mathbf{p}_j$ points the direction of the molecular bow
  $\mathbf{k}_j$ is the smectic layer normal
  $\mathbf{r}_j$ is the position of the center of mass of the molecule j

Important parameters in those simulations are opening angle ψ between 180° and 120° and the length-to-diameter ratio L/D (between 2 and 10). Monte Carlo (MC) simulations are usually made at various densities, which are converted to pressure using isobaric–isothermal conditions (Memmer 2002). MC simulations showed that steric interactions alone are capable of stabilizing polar order in smectic layers and antipolar arrangement of the layers (Memmer 2002, Lansac et al. 2003).

In MC simulations by Lansac et al. (2003), only excluded volume interactions between V-shaped spherocylinders with the aspect ratio $L/D = 5$ and varying opening angle ψ were considered. Although non-tilted phases occurred in this model, a transition between a nonpolar SmA and an antiferroelectric SmAP$_A$ phase was found when the opening angle was as large as 167°. Remarkably, this angle is much larger than the opening angle of mesogens experimentally observed in common bent-core compounds (145°–120°). The antiferroelectric SmAP$_A$ phase emerges directly from the isotropic or nematic phases at a reduced pressure (Figure 2.12). A direct SmA-SmAP$_A$ transition occurs in a very small parameter region of the opening angle ψ. Although the non-tilted phase was found in the pure steric model of the spherocylinders, tilted phases were observed in MC simulations of the molecules made of hard spheres making a mesogen with an opening angle of 140° confined in a box (Xu et al. 2001). This is accompanied by the spontaneous symmetry breaking measured by the chiral order parameter χ. Although the reported tilted phase in Xu et al. (2001) is crystalline, it is believed that the reflection symmetry breaking is independent of the crystallization and can take place in a noncrystal smectic phase.

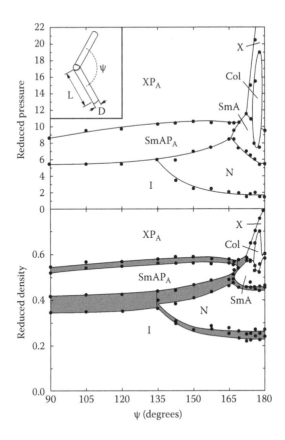

FIGURE 2.12   Phase diagram for various opening angles of the bent-core mesogens from the simulation. (Reprinted with permission from Lansac, Y., Maiti, P.K., Clark, N.A., and Glaser, M.A., Phase behavior of bent-core molecules, *Phys. Rev. A.*, 67, 011703-1-011703-6, 2003. Copyright 2003 by American Physical Society.)

The tendency for antiferroelectric order in the polar smectic phase occurring in various MC models suggests a steric stabilization mechanism. This was examined by Lansac et al. (2003) who put forward an idea that the fluctuation-driven gain of the translational entropy in case of the synclinic interface is responsible for the antiferroelectric order. This is demonstrated in Figure 2.12, where the free energy of the SmA phase was plotted for varying azimuthal angles between the bow directions in the adjacent layers. The energy was determined in MC simulations using a shifted periodic boundary condition. The antiferroelectric state (synclinic interlayer interface) is the global energy minimum. Stabilization of ferroelectric order requires accounting for electrostatic interactions. It was demonstrated by Pelaez and Wilson (2006) that switching off the electrostatic interactions leads to the disappearance of the ferroelectric order.

Another interesting simulation, where a spontaneous breaking of reflection symmetry takes place, was reported by Memmer (2002) (see Figure 4.26). The simulations were made with Gay-Berne bend-shaped particles with the fixed opening angle of 140° in the isobaric–isothermal ensemble (NpT). The molecular arms were modeled by a rotationally symmetric ellipsoid with the axis ratio of 1:1:3. This model exhibits an isotropic–nematic and a nematic–smectic transition. Both these transitions appear to be first order. The smectic phase has an orthogonal antiferroelectric SmAP$_A$ structure. No titled phase occurs in this model either. Remarkably, the nematic phase, close to the phase transition into the smectic phase, exhibits a helical superstructure making a reflection symmetry breaking. The existence of both directions of helicity suggests a spontaneous symmetry breaking. This nematic phase with helicity is now realized as the twist-bend nematic phase (N_{TB}).

REFERENCES

Bisi, F. and R. Rosso. Excluded-volume potential for rigid molecules endowed with C$_{2v}$ symmetry. *Eur. J. Appl. Math.* 23 (2011): 29–60.

Bisi, F., R. Rosso, E. G. Virga, and G. E. Durand. Polar steric interactions for V-shaped molecules. *Phys. Rev. E* 78 (2008): 011705-1-8.

Bisi, F., A. Sonnet, and E. Virga. Steric effects in a mean-field model for polar nematic liquid crystals. *Phys. Rev. E* 82 (2010): 041709-1-16.

Brand, H., P. Cladis, and H. Pleiner. Macroscopic properties of smectic C$_G$ liquid crystals. *Eur. Phys. J. B* 6 (1998): 347–353.

Brand, H., P. Cladis, and H. Pleiner. Polar biaxial liquid crystalline phases with fluidity in two and three spatial dimensions. *Int. J. Eng. Sci.* 38 (2000): 1099–1112.

Brand, H. R., P. E. Cladis, and H. Pleiner. Selected macroscopic consequences of tetrahedratic order. *Ferroelectrics* 315 (2005a): 165–172.

Brand, H. R., H. Pleiner, and P. Cladis. Tetrahedratic cross-couplings: Novel physics for banana liquid crystals. *Physica A* 351 (2005b): 189–197.

Camp, P., M. Allen, and A. Masters, Theory and computer simulation of bent-core molecules. *J. Chem. Phys.* 111 (1999): 9871–9881.

Chaikin, P. M. and T. C. Lubensky, *Principles of Condensed Matter Physics*. New York: Cambridge University Press, 1995.

Chattham, N., M. G. Tamba, R. Stannarius, E. Westphal, H. Gallardo, M. Prehm, C. Tschierske, H. Takezoe, and A. Eremin. Leaning-type polar smectic-C phase in a freely suspended bent-core liquid crystal film. *Phys. Rev. E* 91 (2015): 030502(R)-1-5.

Cladis, P., H. Brand, and H. Pleiner. Fluid biaxial banana phases: Symmetry at work. *Ferroelectrics* 243 (2000): 221–230.

Cladis, P. E., H. Pleiner, and H. R. Brand. Deformable tetrahedratic phases: The effects of external fields and flows. *Eur. Phys. J. E* 11 (2003): 283–291.

De Gennes, P. G. and J. Prost. *The Physics of Liquid Crystals*. Oxford, U.K.: Clarendon Press, 1995.

Emelyanenko, A. V. and M. A. Osipov, Origin of spontaneous polarization, tilt, and chiral structure of smectic liquid-crystal phases composed of bent-core molecules: A molecular model. *Phys. Rev. E* 70 (2004): 021704-1-021704-8.

Eremin, A. et al., Transitions between paraelectric and ferroelectric phases of bent-core smectic liquid crystals in the bulk and in thin freely suspended films. *Phys. Rev. E* 86 (2012): 051701-1-051701-10.

Eremin, A., S. Diele, G. Pelzl, H. Nádasi, W. Weissflog, J. Salfetnikova, and H. Kresse. Experimental evidence for an achiral orthogonal biaxial smectic phase without in-plane order exhibiting antiferroelectric switching behavior. *Phys. Rev. E* 64 (2001): 051707-1-6.

Eremin, A., S. Stern, and R. Stannarius. Electrically induced tilt in achiral bent-core liquid crystals. *Phys. Rev. Lett.* 101 (2008): 247802-1-4.

Fel, L. Tetrahedral symmetry in nematic liquid crystals. *Phys. Rev. E* 52 (1995): 702–717.

Lansac, Y., P. K. Maiti, N. A. Clark, and M. A. Glaser, Phase behavior of bent-core molecules. *Phys. Rev. E* 67 (2003): 011703-1-011703-6.

Lorman, V. and B. Mettout. Unconventional mesophases formed by condensed vector waves in a medium of achiral molecules. *Phys. Rev. Lett.* 82 (1999): 940–943.

Lorman, V. and B. Mettout. Theory of chiral periodic mesophases formed from an achiral liquid of bent-core molecules. *Phys. Rev. E* 69 (2004): 061710-1-16.

Lubensky, T. C. and L. Radzihovsky. Theory of bent-core liquid-crystal phases and phase transitions. *Phys. Rev. E* 66 (2002): 031704-1-27.

Memmer, R. Liquid crystal phases of achiral banana-shaped molecules: A computer simulation study. *Liq. Cryst.* 29 (2002): 483–496.

Mettout, B. Theory of two- and three-dimensional bent-core mesophases. *Phys. Rev. E* 75 (2007): 011706-1-18.

Meuer, S., P. Oberle, P. Theato, W. Tremel, and R. Zentel. Liquid crystalline phases from polymer-functionalized TiO_2 nanorods. *Adv. Mater.* 19 (2007): 2073–2078.

Mourad, M. C. D., E. J. Devid, M. M. van Schooneveld, C. Vonk, and H. N. W. Lekkerkerker. Formation of nematic liquid crystals of sterically stabilized layered double hydroxide platelets. *J. Phys. Chem. B* 112 (2008): 10142–10152.

Onsager, L. The Effects of shape on the interaction of colloidal particles. *Ann. NY Acad. Sci.* 51 (1949): 627–659.

Osipov, M. A. and G. Pajak. Molecular theory of proper ferroelectricity in bent-core liquid crystals. *Eur. Phys. J. E* 37 (2014): 79-1-7.

Osipov, M. A. and P. Teixeira. Density-functional approach to the theory of dipolar fluids. *J. Phys. A* 30 (1987): 1953–1965.

Ostapenko, T., D. B. Wiant, S. N. Sprunt, A. Jákli, and J. T. Gleeson. Magnetic-field induced isotropic to nematic liquid crystal phase transition. *Phys. Rev. Lett.* 101 (2008): 247801-1-4.

Peláez, J. and M. R. Wilson. Atomistic simulations of a thermotropic biaxial liquid crystal. *Phys. Rev. Lett.* 97 (2006): 267801-1-267801-4.

Pelzl, G., S. Diele, and W. Weissflog. Banana-shaped compounds: A new field of liquid crystals. *Adv. Mater.* 11 (1999) 707–724.

Pleiner, H., P. E. Cladis, and H. R. Brand. Splay-bend textures involving tetrahedratic order. *Eur. Phys. J. E* 20 (2006): 257–266.

Pociecha, D., M. Čepič, E. Gorecka, and J. Mieczkowski, Ferroelectric mesophase with randomized interlayer structure. *Phys. Rev. Lett.*, 91 (2003): 185501-1-185501-4.

Radzihovsky, L. and T. C. Lubensky. Fluctuation-driven 1st-order isotropic-to-tetrahedratic phase transition. *Europhys. Lett.* 54 (2007): 206–212.

Reddy, R. A., C. Zhu, R. Shao, E. Korblova, T. Gong, Y. Shen, E. Garcia, M. A. Glaser, J. E. Maclennan, D. M. Walba, and N. A. Clark. Spontaneous ferroelectric order in a bent-core smectic liquid crystal of fluid orthorhombic layers. *Science* 332 (2011): 72–77.

Roy, A., N. Madhusudana, P. Toledano, and A. Neto. Longitudinal spontaneous polarization and longitudinal electroclinic effect in achiral smectic phases with bent-shaped molecules. *Phys. Rev. Lett.* 82 (1999): 1466–1469.

Selinger, J. V. Chiral and antichiral order in bent-core liquid crystals. *Phys. Rev. Lett.* 90 (2003): 165501-1-4.

Takezoe, H. Historical overview of polar liquid crystals. *Ferroelectrics* 468 (2014): 1–14.

Terentjev, E. M., M. A. Osipov, and T. J. Sluckin. Ferroelectric instability in semiflexible liquid crystalline polymers of directed dipolar chains. *J. Phys. A* 27 (1994): 7047–7059.

Van den Pol, E., A. Lupascu, P. Davidson, and G. J. Vroege. The isotropic-nematic interface of colloidal goethite in an external magnetic field. *J. Chem. Phys.* 133 (2010a): 164504-1-8.

Van den Pol, E., A. V. Petukhov, D. M. E. Thies-Weesie, and G. J. Vroege. Simple rectangular columnar phase of goethite nanorods and its martensitic transition to the centered rectangular columnar phase. *Langmuir* 26 (2010b): 1579–1582.

Westphal, E., H. Gallardo, G. F. Caramori, N. Sebastian, M. G. Tamba, A. Eremin, S. Kawauchi, M. Prehm, and C. Tschierske. Polar order and symmetry breaking at the boundary between bent-core and rodlike molecular forms: When 4-cyanoresorcinol meets carbosilane end group. *Chem. Eur. J.* 22 (2016): 1–18.

Wiant, D., K. Neupane, S. Sharma, J. T. Gleeson, S. Sprunt, A. Jákli, N. Pradhan, and G. Iannacchione. Observation of a possible tetrahedratic phase in a bent-core liquid crystal. *Phys. Rev. E* 77 (2008): 617010-1-7.

Xu, J., R. Selinger, J. Selinger, and R. Shashidhar, Monte Carlo simulation of liquid-crystal alignment and chiral symmetry-breaking. *J. Chem. Phys.* 115 (2001): 4333–4338.

Zhang, C., N. Diorio, S. Radhika, B. K. Sadashiva, S. N. Sprunt, and A. Jákli. Two distinct modulated layer structures of an asymmetric bent-shape smectic liquid crystal. *Liq. Cryst.* 39 (2012): 1149–1157.

3 Phase Structures

In addition to conventional phases like N, SmA, and SmC phases, a variety of novel mesomorphic structures have been discovered in bent-core molecular systems. In this chapter, we give a brief summary of such phases followed by a detailed discussion in the forthcoming chapters.

Most of bent-core liquid crystalline compounds tend to form smectic phases. The tendency of molecular bows to order is strongly coupled to the breaking of the translational symmetry, giving rise to the smectic order. Therefore, compounds exhibiting both nematic and typical polar bent-core phases are rather rare. Usually they exhibit either a direct transition from the isotropic phase to a smectic phase or show only conventional nematic and smectic phases. Even if compounds seem to exhibit the conventional nematic phase, the physical properties are sometimes quite different. As will be discussed in Chapter 4, there are many distinct physical properties that cannot be seen in the conventional N phase consisting of rod-shaped molecules. They are as follows:

1. Flexoelectric coefficients are extremely large, although many opposing opinions do exist (see Section 4.1).
2. Characteristic electroconvection patterns are observed (see Section 4.2).
3. Bend elastic constant K_{33} is much smaller than K_{11} and K_{22}, although K_{33} is the largest elastic constant in rod nematogens (see Section 4.4).
4. Partly because of (3), chiral systems of bent-core molecules generally exhibit the stable blue phase (see Section 4.5).
5. In some nematic systems, a new nematic phase, the twist-bend nematic (N_{TB}) phase, appears below the conventional nematic phase (see Section 4.6).
6. Bend-core molecules had been expected to show the biaxial nematic phase. No conclusive evidence for the existence of the biaxial nematic phase has been reported, although in many papers the biaxial nematic phase is claimed to exist (see Section 4.3).

3.1 PECULIARITIES OF SmA AND SmC PHASES

Several kinds of bent-core mesogens form conventional nonpolar smectic phases typical for rod-shaped molecules. Besides the nematic phase, the SmA and SmC phases were found in various 6- and 4-ring mesogenic compounds such as 4,6-dichloro-1,3-phenylene bis[4-(4-n-alkyloxy-3-fluoro-phenyliminomethyl)benzoates] (Figure 3.1) (Eremin et al. 2004) and derivatives of N-benzoylpiperazine (Stern et al. 2009). Detailed structural studies performed on chloro-substituted mesogens in Eremin et al. (2004) showed that the opening angle between the wings of the bent-core molecules plays a crucial role in determining the occurrence of the SmA and SmC phases. The opening angle was determined experimentally by the NMR spectroscopy. An easy way to determine the opening angle α between two mesogenic

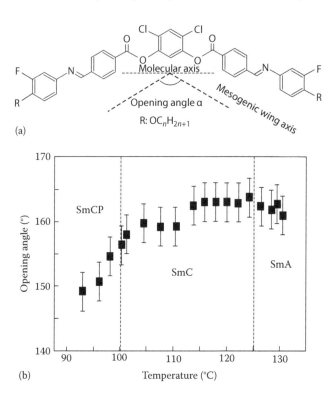

(a)

(b)

FIGURE 3.1 (a) Chemical structure of the compound used for NMR measurements and (b) temperature dependence of the opening angle α in the compound with $n = 12$. (Redrawn from Eremin, A. et al., *Phys. Chem. Chem. Phys.*, 6, 1290, 2004.)

wings (see Figure 3.1a) is to use mesogens with fluorine substitutions at the outer ring. A single splitting observed by ^{19}F NMR is proportional to the orientational order parameter S: $\Delta\nu = \Delta\nu_0 S(T)(3/2\cos^2(\varepsilon(T)) - 1/2)$, where T is the temperature, ε is the angle between the molecular long axis and the para-axis of the fluorinated ring, and $\Delta\nu_0 = 29.2$ ppm. The order parameter $S(T)$ was determined separately by ^{13}C NMR. Since the molecules undergo rapid conformational changes, the mean value of the angle can be determined by averaging over many conformational states. An example of the temperature dependence of the mean opening angle for 4,6-dichloro-1,3-phenylene bis[4-(4-12-alkyloxy-3-fluoro-phenyliminomethyl)benzoates] is shown in Figure 3.1b.

In the SmA phase, the opening angle is very large (160°–155°), which corresponds to a nearly stretched conformation. This state apparently results from the repulsion of the chlorine substituents in the central ring and the oxygen atoms in the ester groups. As the opening angle decreases with decreasing temperature, the steric moments of the mesogens increase. Collective dynamics slow down and polar correlations appear. This is also marked by an increase of the dielectric strength (Eremin et al. 2004). The polar SmCP phase already starts to develop when the opening angle reaches 155°, although in compounds with a direct isotropic-SmCP transition, the opening angle is about $120°$. The growth of polar correlations was demonstrated by the dielectric spectroscopy. The dielectric strength increases over the whole existence region of the SmA and SmC phases on cooling. Switching current measurements often show current peaks that are often interpreted as a manifestation of a polar structure. Some special orthogonal smectic phases are described in the following subsections.

3.1.1 RANDOM PHASES

Occurrence of highly electric-field-responsive uniaxial SmA phases is a striking feature of liquid crystals formed by bent-core liquid crystals. Apart from peaks in current transients reminiscent to the ferroelectric current response, those phases exhibit remarkable dielectric and nonlinear optical susceptibility. This suggests that strong polar correlations are already present in the phase. Yet, such SmA phases are uniaxial and have been shown to be completely miscible with the SmA phase of rodlike mesogens. In 2003, Pociecha et al. proposed an idea of a phase with a randomized interlayer structure (Pociecha et al. 2003). Currently, those phases are designated as $SmAP_R$ (or SmA random phase). Random phases were found in various bent-core compounds (Figure 3.2).

In homeotropically aligned sample cells, the $SmAP_R$ phase appears completely dark between crossed polarizers, suggesting that the phase is uniaxial. In planar cells, on the other hand, the texture is fan-shaped. The current response to a driving triangular-wave field exhibits a single current peak. This peak is similar to the polarization current peak in the ferroelectric SmC* phase. Switching polarization was found to be in the range of 150–300 nC/cm^2 (Pociecha et al. 2003). The electro-optical behavior is completely different from that in the SmC* phase. No residual polarization remains in the field-free state. In homeotropically aligned cells with in-plane electrodes, application of an electric field results in field-induced birefringence. This birefringence relaxes when the field is switched off. A new display mode was also demonstrated using the field-induced birefringence (Shimbo et al. 2006a). The details are described in Section 9.1. Another important argument in favor of a polar phase structure was reported by the Tokyo Group where the field-induced second harmonic generation (SHG) was studied in the $SmAP_R$ phase (Shimbo et al. 2006a,b). SHG is a useful tool to determine the polar structure without the help of switching (see Chapter 6). The measurements showed a very strong SHG in samples under an electric field. In agreement with the electro-optical studies, the SHG disappeared completely in the field-free state. This confirms that there is no spontaneous polarization in the $SmAP_R$ phase. These experimental observations altogether suggest that the $SmAP_R$ phase shows paraelectric behavior with an unusually strong response to an electric field. In analogy to ferrofluids, the term "super-paraelectric" would be a more appropriate designation of random phases.

What can be the reasons for a super-paraelectric behavior of the SmA_R phase formed by bent-core mesogens? The answer to this question has already been addressed in the work by Pociecha et al. (2003). The intralayer correlations between the molecules may lead to polar order within a single smectic layer. The interlayer coupling between the layers determines the global polar structure of the phase. Those interactions are usually of short range and can be described by a clock-type model.

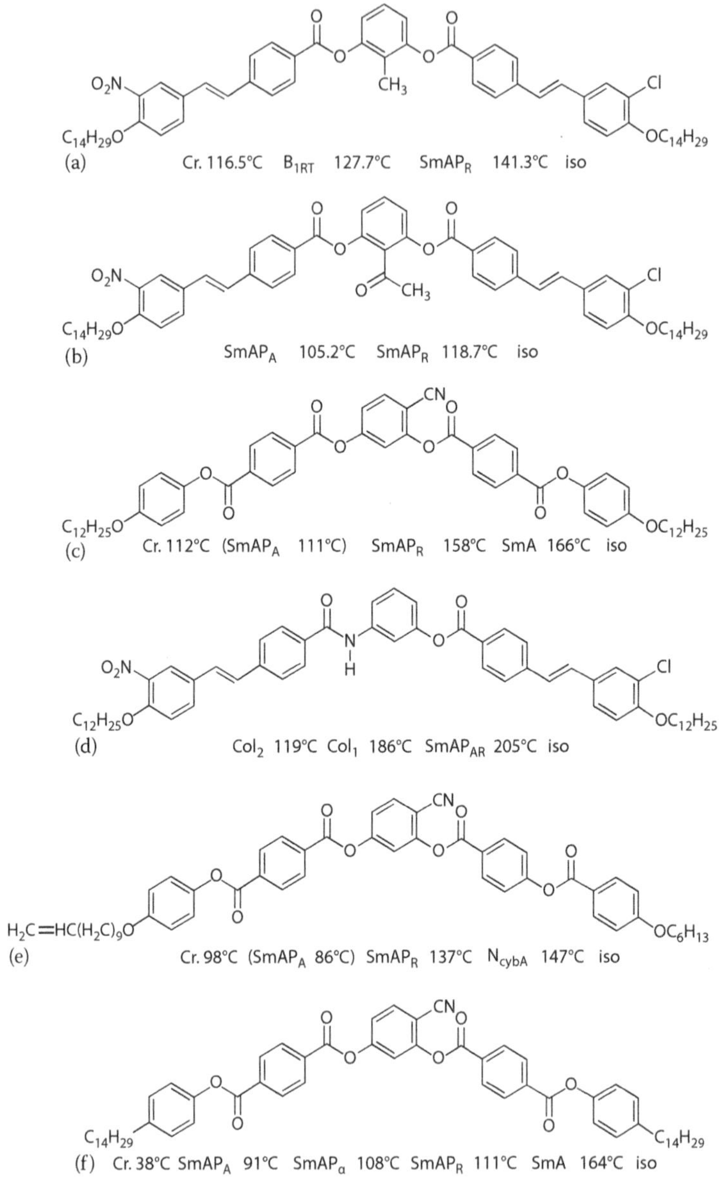

(a)   Cr. 116.5°C   B$_{1RT}$ 127.7°C   SmAP$_R$ 141.3°C   iso

(b)   SmAP$_A$ 105.2°C   SmAP$_R$ 118.7°C   iso

(c)   Cr. 112°C   (SmAP$_A$ 111°C)   SmAP$_R$ 158°C   SmA   166°C   iso

(d)   Col$_2$ 119°C   Col$_1$ 186°C   SmAP$_{AR}$ 205°C   iso

(e)   Cr. 98°C   (SmAP$_A$ 86°C)   SmAP$_R$ 137°C   N$_{cybA}$ 147°C   iso

(f)   Cr. 38°C   SmAP$_A$ 91°C   SmAP$_\alpha$ 108°C   SmAP$_R$ 111°C   SmA   164°C   iso

FIGURE 3.2   Chemical formulas of compounds exhibiting random phases: (a) first SmAP$_R$ phase (Pociecha et al. 2003, Shimbo et al. 2006a); (b) compound showing SmAP$_A$-SmAP$_R$ transition (Guo et al. 2011b); (c) compound with a three-phase sequence, SmAP$_A$-SmAP$_R$-SmA (Panarin et al. 2010); (d) compound with the first random phase with antiferroelectric clusters SmAP$_{AR}$ (Gomola et al. 2010); (e) compound showing a sequence of nematic and orthogonal random phases (Shanker et al. 2011); and (f) first compound showing the SmAPα phase (Panarin et al. 2011). (From Eremin, A. and Jákli, A., Polar bent-shape liquid crystals—From molecular bend to layer splay and chirality, *Soft Matter*, 9, 615–637, 2013. Reproduced by permission of The Royal Society of Chemistry.)

Polarization $p_j = P(\cos(\varphi_j), \sin(\varphi_j))$ in each layer is determined by the azimuthal angle φ_j and its amplitude $\mathbf{P}$. The free energy can be expressed as a function of the layer polarization p_j as a sum over all layers:

$$G_{NN} = \sum_j \left(\frac{1}{2} a_{1p} \left(p_j \cdot p_{j+1} \right) + \frac{1}{4} b_{1p} \left(p_j \cdot p_{j+1} \right)^2 \right) \tag{3.1}$$

Electrostatic interactions favor antiferroelectric order in neighboring layers, whereas the van der Waals interactions stabilize the ferroelectric order. Altogether, they determine the bilinear coefficient a_{1p}. The biquadratic term accounts for the interactions favoring a nonparallel orientation of molecular moieties. A whole range of new phases results after minimization of the free energy functional with respect to the azimuthal angle φ. Polar ferro- and antiferroelectric phases occur when $|a_{1p}| > b_{1p}P^2$. Additionally, a short-pitch helical phase $SmAP_\alpha$ and a bilayer $SmAP_2$ phase occur when $|a_{1p}| > b_{1p}P^2$ and a "random" phase $SmAP_R$ occurs where the free energy is degenerate with respect to the variation of the azimuthal angle φ. Both $SmAP_\alpha$ and $SmAP_2$ phases can be stabilized by long-range interactions. In particular, the $SmAP_\alpha$ phase can be stable in systems with competing next-nearest neighbor interactions between the layer polarizations. Random phase $SmAP_R$, on the other hand, should appear in systems where only interactions between the nearest neighboring layers are present. Several groups provided experimental studies of the phases with randomized layer structure (Shimbo et al. 2006a,b, Panarin et al. 2010, 2011, Gupta et al. 2011).

A randomized polar smectic phase exhibiting antiferroelectric-type switching ($SmAP_{AR}$) was observed in a compound with a 3-aminophenol central unit (Gomola et al. 2010). It was suggested that this phase consists of randomly aligned clusters with a local antiferroelectric order. The stability of the clusters is attributed to a nonsymmetric molecular architecture containing intermolecular hydrogen bonding (Figure 3.2d). The electro-optical switching in the $SmAP_{AR}$ phase occurs through a dielectric alignment of the clusters and a gradual formation of the $SmAP_A$ phase followed by the switching of the antiferroelectric clusters into the ferroelectric state (Figure 3.3). This behavior was confirmed by the SHG studies. In contrast to the $SmAP_R$ phase, where the signal is continuously increasing with increasing electric field, the SHG signal in the $SmAP_{AR}$ phase appears when the applied field exceeds a threshold value.

3.1.2 Electrically Induced Tilt

Another interesting phenomenon found in the SmA and SmC phases is an electrically induced tilt. This effect was observed in several bent-core and hockey-stick mesogens (Eremin et al. 2008). In the chiral SmA* and SmC* phases, the application of an electric field results in a tilt of the mesogens, known as the electroclinic effect (Garoff and Meyer 1977). This linear electro-optic response is a direct consequence of the chiral structure of the mesogens allowing a linear coupling between the tilt and polarization. This effect is forbidden in the achiral SmA and SmC phases. Yet a field-induced tilt has been observed in nonchiral bent-core and hockey-stick mesogens (Figure 3.4).

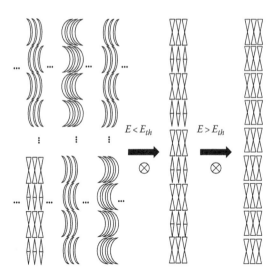

$E < E_{th}$

$E > E_{th}$

FIGURE 3.3 Cartoons showing switching in the SmAP$_{AR}$ phase. (From Gomola, K., Guo, L., Pociecha, D., Araoka, F., Ishikawa, K., and Takezoe, H., An optically uniaxial antiferroelectric smectic phase in asymmetrical bent-core compounds containing a 3-aminophenol central unit, *J. Mater. Chem.*, 20, 7944–7952, 2010. Reproduced by permission of The Royal Society of Chemistry.)

It should be noted that the properties of this field-induced tilt are different from the linear electroclinic effect. The tilt occurs at a threshold in the SmA phase (Figure 3.4a). The threshold field increases linearly with temperature. This phenomenon, designated as the quadratic electroclinic effect, is explained by a quadratic coupling between the tilt and polarization allowed in the nonchiral systems. Bent molecules favor polar alignment through a stronger hindrance of the molecular rotations about their long axes. Stronger hindrance occurs when the molecules are tilted with respect to the layer normal. In the SmA phase, it results in a field-induced transition from an orthogonal phase to a tilted one. In the SmC phase, the tilt continuously increases with the applied field.

3.2 ORTHOGONAL POLAR SMECTIC PHASES

Conventional smectic phases are constituted by stacks of layers, in which the molecules are aligned parallel (SmA) or obliquely (SmC) with respect to the layer normal. The SmA phase is optically uniaxial with the local symmetry $D_{\infty h}$; the SmC phase is biaxial with the local symmetry C_{2h}. A biaxial smectic phase may also appear, if one of the principal axes of the susceptibility (magnetic) tensor $Q_{\alpha\beta}$ is parallel to the layer normal but the transverse components of the tensor $Q_{\xi\xi}$ and $Q_{\eta\eta}$ are no longer equivalent to each other (de Gennes and Prost 1993). In this case, the molecules are aligned parallel to the layer normal but the properties with respect to the in-plane directions are different. de Gennes suggested to designate this phase as C_M, where M refers to McMillan (1971, 1973), who established the first microscopic theory of the SmA and SmC phases. In a theoretical investigation

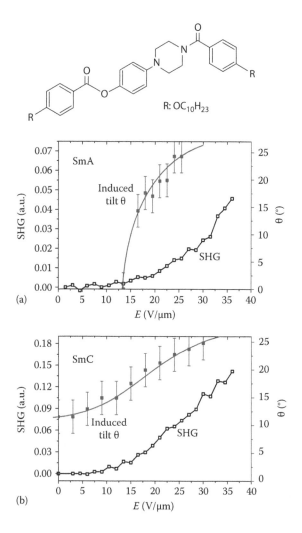

FIGURE 3.4 Dependence of optical tilt θ (E) and SHG intensity on the electric field in (a) the SmA phase (T = 115°C) and (b) the SmC phase (T = 110°C). Chemical structure of the molecule used is shown at the top. (Reprinted with permission from Eremin, A., Stern, S., and Stannarius, R., *Phys. Rev. Lett.*, 101, 247802-1. Copyright 2008 by the American Physical Society.)

made by Brand et al. (1992), the nonequivalence of the transversal components of Q has been taken into account by considering board-like molecules, where for steric reasons $Q_{\xi\xi} \neq Q_{\eta\eta}$, leading to a phase possessing the D_{2h} symmetry. The McMillan phase was first reported in non-bent-core systems (Leube and Finkelmann 1990, 1991, Hegmann et al. 2001). In the bent-core mesogens, the biaxial orthogonal phase was reported by Semmler et al. (1998). Later the phase was designated as SmA_b.

Another way to obtain $Q_{\xi\xi} \neq Q_{\eta\eta}$ can be realized in the mesophases with bent-core mesogens (Pratibha et al. 2000, Sadashiva et al. 2002). If the alignment of the molecules is orthogonal with respect to the layer planes, a local polar orthorhombic C_{2v} symmetry results (in contrast to the C_M phase). This phase is similar to the SmA phase, since one of the principal axes of Q is parallel to the layer normal. Therefore, it might be reasonable to designate such a phase as SmAP in analogy to the well-known tilted variant designated as SmCP (see Figure 3.5).

The first example of an antiferroelectric orthogonal polar $SmAP_A$ phase was found in a cyanoresorcinol derivative fluorinated on the outer rings (Eremin et al. 2001). This new phase was distinguished by the *Schlieren* texture in a homeotropically aligned SmA phase (Figure 3.6). The optical appearance of the $SmAP_A$ phase is similar to that of the SmC phase because of the biaxiality caused by the polar director aligned along the molecular bows. More strictly, the *Schlieren* texture with two-brush

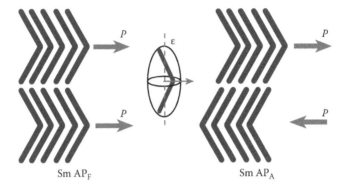

$Sm\,AP_F$ $\qquad\qquad\qquad\qquad$ $Sm\,AP_A$

FIGURE 3.5 Schemes of the ferroelectric $SmAP_F$ and antiferroelectric $SmAP_A$ phases. The dielectric ellipsoid (shown in the middle) is biaxial and has three unequal components. This biaxiality is responsible for the characteristic *Schlieren* textures (see Figure 3.6) observed in polarizing optical microscope (POM).

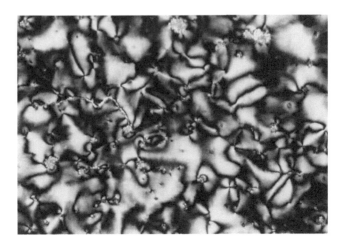

FIGURE 3.6 *Schlieren* texture in a homeotropically aligned antiferroelectric $SmAP_A$ phase of the compound. Two-brush defects can be seen due to the dispiration formation.

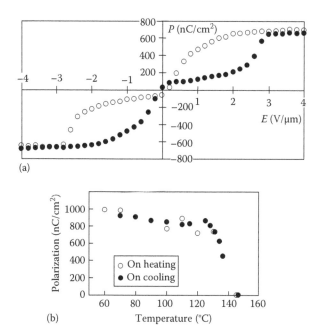

FIGURE 3.7 Antiferroelectric switching in the $SmAP_A$ phase: (a) double-loop hysteresis curve and (b) temperature dependence of the switched polarization. (Redrawn from Eremin, A. et al., *Phys. Rev. E*, 64, 051707-1, 2001.)

defects is more similar to the SmC_A phase. This is the dispiration, that is, a combined defect of an edge dislocation and a screw dislocation (Takanishi et al. 1992), which removes a plane defect inevitably introduced by the **p** director. One of the first SmAP materials showed antiferroelectric switching behavior (see Figure 3.7) and, thus, the phase was designated as the $SmAP_A$ (Eremin et al. 2001).

The orthogonal structure of the phase was established in x-ray studies, as shown in Figure 3.8. In most cases, no layer shrinkage upon the SmA–SmAP transition could be detected, but the layer thickness increased as shown in Figure 3.9. The mesogens remain parallel to the layer normal, as the wide-angle diffuse scattering maxima centered on the equator (Eremin et al. 2001, Pociecha et al. 2005).

Later, the orthogonal polar phases with the sequences SmA–$SmAP_A$ and SmA–$SmAP_R$–$SmAP_A$ were found in other compounds with 4-cyanoresorcinol central unit (Dunemann et al. 2005, Shanker et al. 2011) and with acetophenone unit (Guo et al. 2011b). The transition between the uniaxial random SmA ($SmAP_R$) phase and the biaxial antiferroelectric $SmAP_A$ phase exhibits a very small latent heat and it can be treated as a weak first-order transition (Eremin et al. 2001, Pociecha et al. 2005). The layer spacing in the $SmAP_A$ phase exhibits a weaker temperature dependence compared with the high-temperature SmA phase, and the layers become more defined as demonstrated by higher-order x-ray reflections. The system exhibits critical slowing down of polar fluctuations upon approaching the transition into the $SmAP_A$ phase, which is accompanied by a drastic increase of the dielectric response.

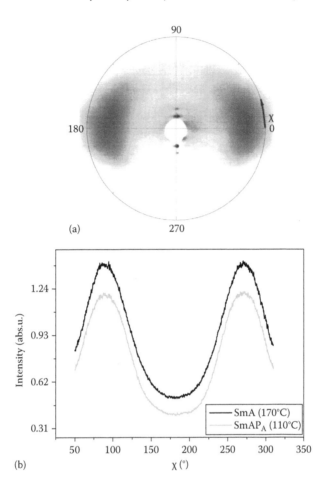

FIGURE 3.8 (a) X-ray pattern of the oriented sample and (b) χ-scans in the SmA and the SmAP phases of the compound. These figures clearly indicate orthogonal smectic phase. (Reprinted with permission from Eremin, A., Diele, S., Pelzl, G., Nadashi, H., Weissflog, W., Salfetnikova, J., and Kresse, H., *Phys. Rev. E*, 64, 051707-1. Copyright 2001 by the American Physical Society.)

Fast molecular rotations around the long axes in the SmA phase slow down as the temperature approaches the $SmAP_A$ phase. Collective character of the molecular dynamic becomes stronger pronounced, and the dielectric relaxation frequency is reduced below 200 Hz (Pociecha et al. 2005). The polar mode, observed by dielectric spectroscopy, is much stronger for the bent-core system than that observed at the SmA–SmC* transition in the chiral tilted smectics made of rodlike molecules. This can be attributed to a much weaker interlayer antiferroelectric coupling (Pociecha et al. 2005). The switching polarization developing in the antiferroelectric phase shows a strong temperature dependence.

Another type of biaxial SmAP phase was also found in compounds with a partial bilayer ordering (Sadashiva et al. 2001, 2002). The bent-core compounds are highly

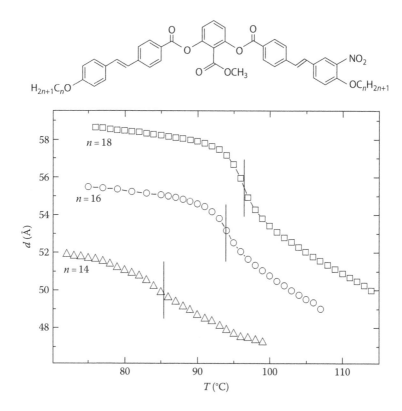

FIGURE 3.9 Layer spacing vs. temperature in the vicinity of the SmA–SmAP$_A$ phase transition of the compound shown at the top. Chemical structure of the molecule used is shown at the top. (Reprinted with permission from Pociecha, D., Gorecka, E., Čepič, M., Vaupotič, N., Gomola, K., and Mieczkowski, J., *Phys. Rev. E*, 72, 060701(R)-1. Copyright 2005 by the American Physical Society.)

asymmetric with a wing terminated by a highly polar cyano group and a wing with alkoxy chain (Figure 3.10 top). In most of the cases, the transition takes place from SmA$_d$ (uniaxial partially bilayered phase) to SmA$_{db}$ (biaxial SmA$_d$). The layer birefringence appears at the transition and steadily increases with decreasing temperature in SmA$_{db}$. Because of the absence of the electric-field-induced transition to the SmAP$_F$ phase, Sadashiva et al. (2002) suggested a phase structure shown in Figure 3.10a, where the polar order of pairs of interdigitated molecules is cancelled within each layer instead of forming antiferroelectric interlayer structures. But later, double peaks in switching current measurements were observed and the SmA$_{db}$ phase is reassigned to the phase structure shown in Figure 3.10b and designated as SmA$_d$P$_A$ (Reddy and Sadashiva 2004).

The first ferroelectric orthogonal SmAP$_F$ phase was reported by the Boulder group in 2011 (Reddy et al. 2011). In order to achieve the ferroelectric ground state, a particular molecular design was developed. To reduce the tendency for the tilt, mesogens with only single alkyl tail were synthesized. The tendency to form antiferroelectric

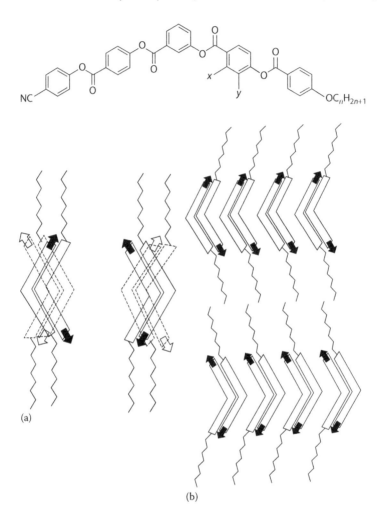

FIGURE 3.10 Two possible phase structures, (a) and (b), of the SmA_dP_A phase. Because of antiferroelectric switching behavior, (b) is the true phase structure. First synthesized molecule, which shows SmA_dP_A, is also shown at the top. (From Sadashiva, B.K., Reddy, R.A., Pratibha, R., and Madusudana, N.V., Biaxial smectic A phase in homologous series of compounds composed of highly polar unsymmetrically substituted bent-core molecules, *J. Mater. Chem.*, 12, 943–950, 2002. Reproduced by permission of The Royal Society of Chemistry.)

order was suppressed by using a terminating carbosilane group. This suppresses the interpenetration of tails in adjacent layers and lowers the in-plane entropic pressure on the mesogens. The chemical structure of such an asymmetric mesogen is given at the top of Figure 3.11. In-layer molecular orientation and the electric polarization in the $SmAP_F$ phase were studied in sandwich cells and freely suspended films by depolarized reflected light microscopy, as shown in Figure 3.11 (Reddy et al. 2011). *Schlieren* textures show topological defects with an integer strength of ±1 (Figure 3.11). No defects with a half-integer topological strength were found, which

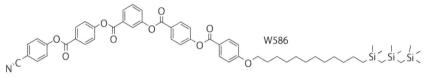

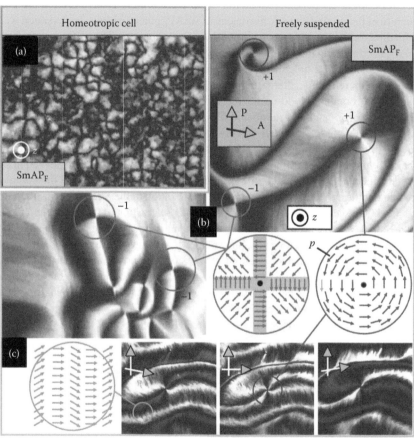

FIGURE 3.11 POM textures of a compound W586 (shown at the top) exhibiting the SmAP$_F$ phase. (a) Homeotropic cell showing four-brush defect patterns between crossed polarizers, indicating a 2π reorientation of an optic axis, and thus polar ordering in the SmAP$_F$ phase; (b) depolarized reflected light microscopy images of freely suspended films with oblique incidence and slightly decrossed polarizers. Characteristic brush patterns certify the biaxiality of the SmAP$_F$ phase; and (c) single-layer ($N = 1$) freely suspended film exhibiting equal intensity bright brushes emanating from a $+2\pi$ defect viewed with crossed polarizers, indicating an optical axis normal to the film. (From Reddy, R.A., Zhu, C., Shao, R., Korblova, E., Gong, T., Shen, Y., Garcia, E. et al., Spontaneous ferroelectric order in a bent-core smectic liquid crystal of fluid orthorhombic layers, *Science*, 332, 72–77, 2011. Reprinted with permission of AAAS.)

is indicative that the director order exhibits a polar symmetry. Another interesting feature of the optical textures is a suppression of the polarization splay. The director adopts a bend-only orientation. This is a consequence of the electrostatic interactions where a polarization splay is associated with a volume charge. π and 2π inversion walls in freely suspended films can be distinguished by leaning the films with respect to the observing direction because of different effective birefringence (Reddy et al. 2011, Eremin et al. 2012). Application of an electric field clearly reveals the polar structure of the ground state. A ferroelectric $SmAP_F$ phase exhibits 2π-inversion walls of the director, which is a fingerprint of the polar symmetry. The difference from the $SmAP_A$ becomes apparent from the electric response in freely suspended films. In the case of the $SmAP_A$ sample, the following odd–even effect can be observed. Namely, in even-numbered films, alternating layer polarizations cancel out, whereas residual polarization remains in odd-numbered films (see Chapter 7). Hence, the switching behaviors in even- and odd-numbered films are different.

The $SmAP_F$ phase shows an electro-optical response in sandwich cells too. The analysis of the optical transmission response and current transients shows a switching behavior typical for the thresholdless V-shaped switching occasionally observed in chiral SmC materials (Reddy et al. 2011). This occurs as a consequence of the complete screening of the applied electric field by the polarization of the $SmAP_F$ material. Reddy et al. (2011) suggested that the bending plane and polarization of the molecules are parallel to surfaces in the absence of an electric field and they become perpendicular upon the application of a field. However, Guo et al. (2011a) suggested a polarization splayed structure, in which polarizations at both surfaces point opposite directions and splayed in between. In this model structure, the in-plane component of the polarization remains finite in the absence of a field. The existence of the polar order in the absence of an electric field was also evidenced by a strong dielectric response (dielectric strength of more than 1000) and the SHG activity (Figure 3.12) (Guo et al. 2011a).

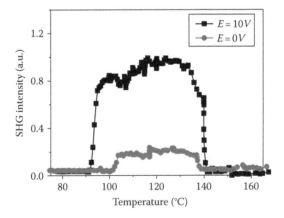

FIGURE 3.12 Temperature dependence of SHG in the presence (10 V) and absence of an electric field. (Reprinted with permission from Guo, L., Gorecka, E., Pociecha, D., Vaupotič, N., Čepič, M., Reddy, R.A., Gornik, K. et al., *Phys. Rev. E*, 84, 031706-1. Copyright 2011a by the American Physical Society.)

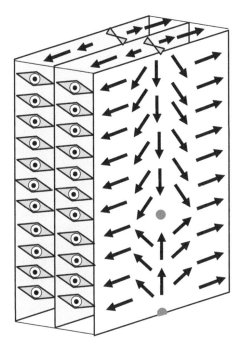

FIGURE 3.13   Phase structure of the SmAP$_{Fmod}$ phase.

Another ferroelectric phase, polarization modulated SmAP$_{Fmod}$, was discovered above the SmAP$_F$ phase (Zhu et al. 2012). As shown in Figure 3.13, the SmAP$_{Fmod}$ phase has a periodic undulation of the polarization. This splay modulation leads to a bistable response to electric-field application.

3.3   TILTED SMECTIC PHASES

A majority of phases formed by bent-core mesogens are tilted smectic phases. In 1996, Niori et al. reported the electro-optic behavior of 5-ring 1,3-phenylenebis[4-(4-n-alkyloxyphenyliminomethyl)benzoate] exhibiting new mesophases. In their first paper, all the phases were assumed to be orthogonal smectics with the C$_{2v}$ symmetry and to be a ferroelectric phase (Niori et al. 1996) due to their electro-optic switching behavior. These assumptions were preserved even after publishing a homologous series of this compound with alkyl (P-n-PIMB) and alkoxy (P-n-OPIMB) tails in 1997 (Sekine et al. 1997a,b) (see Figure 1.8). At the end of 1997, Link et al. published an important work based on texture observations of P-8-PIMB (original compound by Niori et al.) and P-9-OPIMB under an electric field (Link et al. 1997). They confirmed that the switchable phase below the isotropic phase is a tilted anti-ferroelectric phase, and designated this phase as SmCP$_A$, where **P** stands for polar and A antiferroelectric, and later also the B$_2$ phase. More importantly, they showed two spontaneous symmetry breaking instabilities, that is, layer polarity and chirality by achiral bent-core molecules, leading to the existence of four different structures.

The details will be described in this section later (see Figure 3.16). It is important to note here that the first two homologues, P-n-PIMB and P-n-O-PIMB, show four important lamellar phases, B_1, B_2, B_3, and B_4 (Sekine et al. 1997b, Watanabe et al. 1998).

X-ray diagrams of the B_2 phase revealed several orders of layer reflections and broad diffuse scattering in the wide-angle region (Figure 3.14a). These findings indicate a well-defined layer structure and liquid-like in-plane order. A large number of layer reflections distinguish bent-core liquid crystal (LC) phases from the ones of small rod-shaped molecules, where only the first- and, sometimes, the second-order layer reflections are usually present along the $q_{\parallel}$ (parallel to the layer normal) direction. Another important feature of the x-ray scattering from aligned samples is that the maxima of the diffuse scattering at the wave vectors q_0 are situated off the equator $q_{\perp}$ (perpendicular to the layer normal) of the diagram (Figure 3.14c and d). The direction of q_0 shows the lateral order of the mesogens in a smectic layer. The angle between the wave vectors q_0 and $q_{\perp}$ describes the tilt of the mesogens θ, which is in the range from a few degrees to $50° - 60°$. In the case of a monodomain sample with the same sense of tilt in all layers, only two diffuse maxima emerge, as shown in Figure 3.14c. In the case of a so-called fiber sample, where the monodomains have the same longitudinal orientation and a random transversal distribution, the diagram features four symmetric maxima, as shown in Figure 3.14d. The latter case is also observed when the tilt sense alternates from one layer to another (Figure 3.14b right). No cross-reflections occur in the scattering diagram, indicating no correlation between the transversal positions

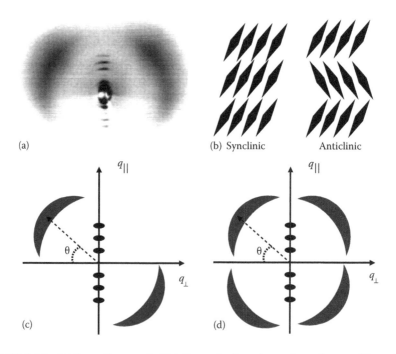

(a) (b) Synclinic   Anticlinic

(c) (d)

FIGURE 3.14 (a) X-ray diagram of a B_2 LC sample aligned on a glass substrate; (b) synclinic and anticlinic structures of the B_2 phase; (c) a schematic of an oriented x-ray pattern in the synclinic state; and (d) the pattern in the anticlinic state or a fiber pattern in the synclinic state.

of the mesogens in the neighboring layers. These findings suggest the structure of the B_2 phase equivalent to that of the tilted SmC phase.

The configuration of a tilted polar smectic layer can be defined by the three directions **n**, **p**, and the layer normal **k** (Figure 3.15). Two vectors **k** and **n** define the tilt plane. The polar director **p** defines the mean direction of the molecular bows and the polarization. However, as it will be shown later, different modifications of the SmCP phase exist, which share in common the fluidity of layers and the polar order perpendicular to the tilt plane. In a nonpolar SmC phase, spontaneous polarization does not exist and, thus, **p** is not defined. In the SmCP phase, the polarization is perpendicular to the tilt plane as was demonstrated in the experiments on freely suspended films (Link et al. 1997) (see Chapter 7 for details).

An important consequence of the polarization, which emerges in each layer due to packed bent-core molecules, is the chirality of a single SmCP layer. Namely, **k**, **n**, and **p** form a system with these three vectors, giving a particular handedness (Figure 3.15). Chiral order parameter can be introduced as $q = ([\mathbf{k} \times \mathbf{n}] \cdot \mathbf{p})$. It is positive for one handedness and negative for another. Thus, the polar order of the bent-core mesogens spontaneously breaks the reflection symmetry in a single layer of the SmCP phase even in case of nonchiral molecules. This type of chirality in liquid crystals is designated as structural chirality. Despite the fact that reflection symmetry breaking instabilities are often observed in solids, SmCP phase is the first example of the reflection symmetry breaking occurring in a bulk state of a liquid (Selinger 2003).

Two types of tilt correlations between the layers can be distinguished. Let us start with rod-shaped molecular systems. When the molecules are tilted in the same direction, it is called synclinic structure (Figure 3.14b left). Alternating tilt is designated as anticlinic (Figure 3.14b right). Both syn- and anticlinic structures like SmC and

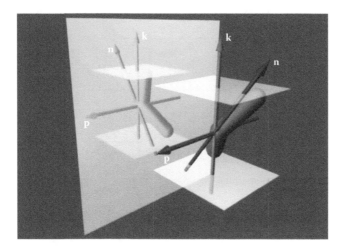

FIGURE 3.15 (**See color insert.**) Mirror image of a bent-core molecule in the B_2 phase. A chiral structure formed by polar packed tilted bent-core molecules. The handedness is determined by three vectors **p**, **n**, **k**. The state of opposite handedness can be constructed by inversion of the tilt or polarization.

SmC_A, respectively, have comparable energies and may occur even in the same compound. The synclinic state appears to be slightly more favorable than the anticlinic one because of packing entropy or excluded volume effect. Namely, the molecules can easily interpenetrate from one layer to another in the synclinic case. This freedom makes the synclinic interface more favorable over the anticlinic one. The situation is quite different in bent-core molecules. As shown in Figure 3.16, four types of basic structures can be distinguished: two antiferroelectric synclinic SmC_SP_A and anticlinic SmC_AP_A and two ferroelectric synclinic SmC_SP_F and anticlinic SmC_AP_F. It is easy to understand from Figure 3.16 that the parallel orientation of the terminal chains can be realized only in the antiferroelectric synclinic structure. Hence, the SmC_SP_A structure is the most stable configuration and is an often encountered banana phase.

Electro-optic studies on samples in bulk and freely suspended films showed that, in most cases, the macroscopic polarization is absent in the field-free state. Two peaks per half a period of the driving triangular-wave voltage distinguish the current

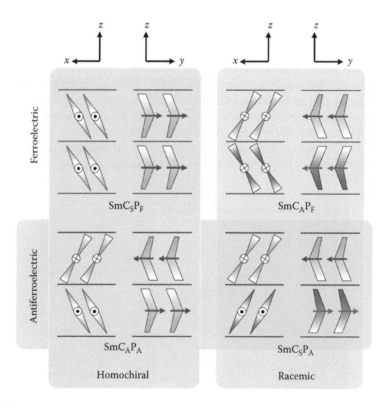

FIGURE 3.16 (See color insert.) Two directional views of SmCP phase of nonchiral banana-shaped molecules, that is, views in the planes perpendicular (x–z plane) and parallel (y–z plane) to the layer polarization. The molecular plane is tilted with respect to the layer normal. Ferroelectric and antiferroelectric states (upper and lower columns, respectively) exist in homochiral and racemic layer chirality (left and right columns, respectively). (Reprinted from Takezoe, H. and Takanishi, Y., *Jpn. J. Appl. Phys.*, 45, 597, 2006. With permission from the Japan Society of Applied Physics.)

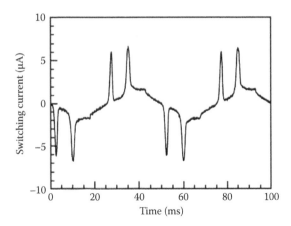

FIGURE 3.17 A typical current response of an antiferroelectric $SmCP_A$ liquid crystal subjected to a triangular-wave voltage. The peaks correspond to the switching from ferroelectric (FE) to antiferroelectric (AFE) and AFE to FE. (From Zennyoji, M., Takanishi, Y., Ishikawa, K., Thisayukta, J., Watanabe, J., and Takezoe, H., Partial mixing of opposite chirality in a bent-core liquid crystal molecular system, *J. Mater. Chem.*, 9, 2775–2778, 1999. Reproduced by permission of The Royal Society of Chemistry.)

response of an antiferroelectric phase in a sandwich cell (see Figure 3.17) (Pelzl et al. 1999a, Zennyoji et al. 2000). This occurs in the antiferroelectric-type SmCP (or $SmCP_A$) phase.

As already mentioned in the preceding text, among the four structures in the SmCP phase, SmC_SP_A is the most stable, since mesogenic groups (arms of bent-core molecules) in adjacent layers are parallel to each other so the interlayer penetrations are favorable. Hence, the ferroelectric ground-state SmCP phase ($SmCP_F$) is less stable compared to the $SmCP_A$ phase, because of both steric effect and polar interactions. Actually, most of the B_2 phases exhibit a $SmCP_A$ structure. Hence, ingenious molecular designs are necessary to realize the ferroelectric $SmCP_F$ phase. The first $SmCP_F$ phase was realized by mimicking interlayer clinicity of the SmC_A phase (Walba et al. 2000); asymmetric bent-core molecules with one racemic 1-methyl-heptyloxycarbonyl tail and one nonyloxy tail (Figure 3.18a) produce ferroelectric B_7 (though not B_2) with the local structure of SmC_SP_F. Note that the synclinicity of bent-core molecules is equivalent to anticlinic layer interfaces, as expected by the SmC_A structure.

A similar attempt also brought about SmC_AP_F (Gorecka et al. 2000). The molecule is a symmetric P-8-O-PIMB chiral analogue, P-8-O-PIMB6*, where chiral carbons are located in molecular tails, far from the mesogenic core (Figure 3.18b). The ferroelectricity was confirmed by a single switching current peak under a triangular voltage wave, large dielectric strength, and SHG activity (see Figure 6.3) in the absence of an electric field (Gorecka et al. 2000). The racemic layer structure (SmC_AP_F) was established by the texture observation and SHG interferometry (see Figure 6.3) (Nakata et al. 2001). The importance of the interlayer steric interaction was clearly shown by the odd–even effect for the emergence of $SmCP_F$ and $SmCP_A$

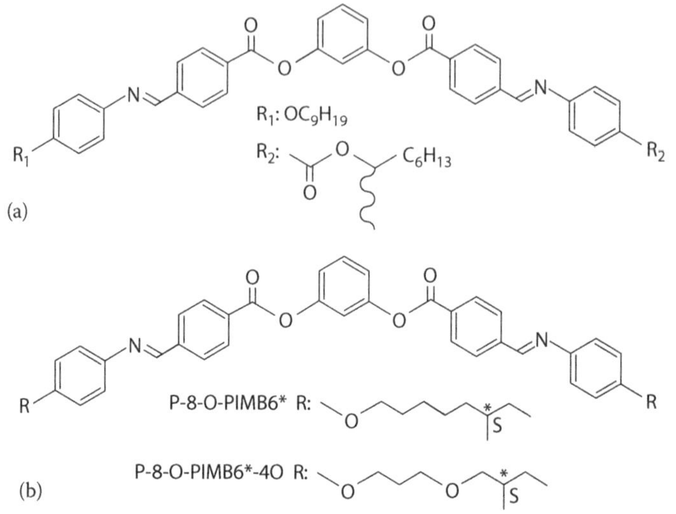

(a)

(b)

FIGURE 3.18   (a and b) Compounds showing SmCP$_F$ phase.

(Kumazawa et al. 2004, Lee et al. 2005); using eight molecules in two homologues, P-n-O-PIMB(n − 2)* and P-n-O-PIMB(n − 2)*(n − 4)O (Figure 3.18b), ferroelectric for even n and antiferroelectric for odd n were confirmed. Theoretical background was also given for the odd–even effect (Nishida et al. 2006, Niigawa et al. 2007). The importance of the structure of the tail chain was also confirmed using 1:1 mixtures of neighboring homologues of symmetric bent-core mesogens and nonsymmetric molecules with odd and even tail groups. Longer terminal chains play an important role in determining the polar structure. Namely, the ferroelectric state emerges when the carbon number n of the longer chain is even, whereas the antiferroelectric state emerges for odd n (Nishida et al. 2006). In case of nonsymmetric molecules with odd and even tail groups, longer tails with even n give SmCP$_F$ and those with odd n give SmCP$_A$ (Nishida et al. 2006).

Another attempt to realize SmCP$_F$ was also made by controlling layer interfaces, that is, the introduction of oligosiloxane molecular tails (Dantlgraber et al. 2002a,c). This is the same concept to realize SmAP$_F$ (Reddy et al. 2011). The first SmCP$_F$ phase using such sublayers was found in 5-ring siloxane derivatives (Figure 3.19a) (Dantlgraber et al. 2002c). These compounds exhibit a bistable switching (single switching current peak) typical for ferroelectrics (Figure 3.19b). The origin of ferro-electricity in this system is attributed to layer decoupling. This decoupling is achieved by a microphase separation into three subphases: (1) a layer of aromatic cores, (2) a layer of aliphatic chains, and (3) an oligosiloxane sublayer (Figure 3.19c). The bulkiness of the oligosiloxane moieties suppresses the interlayer penetration of the mesogens preventing the entropic driving force for the synclinic interlayer interface and the antiferroelectric order.

Fluid order in smectic layers and therefore the absence of the positional cross-correlations of the molecules in the neighboring layers forbid Bragg cross-reflections (k n) with both $k \neq 0$ and $n \neq 0$. As a result, the syn- or anticlinic structure

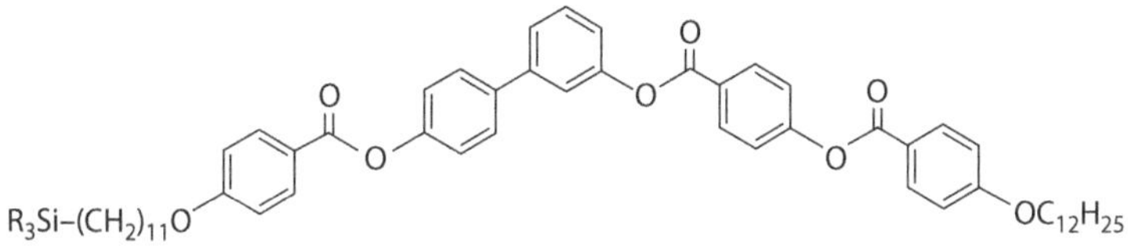

Compound	R_3Si–	T (°C)
2	$Me_3SiOSiMe_2$–	Cr 77 $SmCP_A$ 118 Iso
3	$Me_3Si(OSiMe_2)_2$–	Cr 70 $SmCP_F$ 115 Iso
4	$(Me_3SiO)_2SiMe$–	Cr 63 $SmCP_F$ 116 Iso

(a)

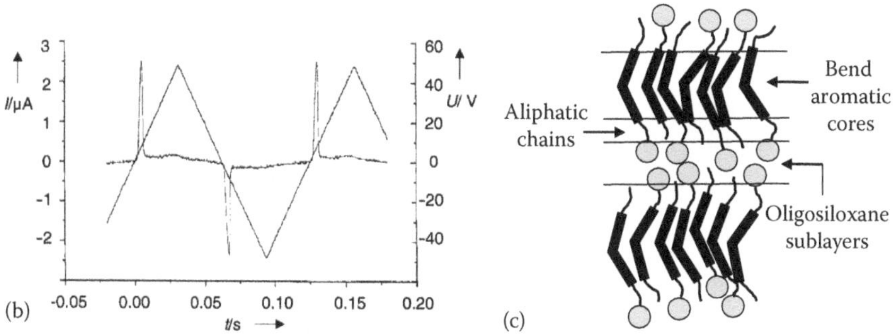

(b) (c)

FIGURE 3.19 (a) Chemical formula and phase scheme of the siloxane compounds exhibiting ferroelectric $SmCP_F$ phase; (b) a typical for ferroelectric switching one-peak response to a triangular-wave voltage; and (c) the structure of sublayers formed by aromatic, aliphatic, and oligosiloxane moieties. (From Dantlgraber, G., Eremin, A., Diele, S., Hauser, A., Kresse, H., Pelzl, G., and Tschierske, C.: Chirality and macroscopic polar order in a ferroelectric smectic liquid-crystalline phase formed by achiral polyphilic bent-core molecules. *Angew. Chem. Int. Ed.* 2002c. 41. 2408–2412. Copyright Wiley-VCH Verlag GmbH & Co. KGaA. Reproduced with permission.)

of the phase cannot be verified using conventional x-ray techniques. A solution to this problem is given by the resonant x-ray scattering, which reveals the anisotropy of the tensorial structure factor (Levelut and Pansu 1999). This technique was employed to study incommensurate chiral SmC* phases of rod-shaped molecules and the polar SmCP phases of bent-core molecules (Cady et al. 2002, Fernandes et al. 2007, Takanishi et al. 2010, Barois et al. 2012). This technique requires the presence of an absorbing substituent such as Br, Cl, S, or Se. The polarization of the resonant Bragg peak was recorded as a function of the sample azimuthal position and later compared with the theoretically predicted dependence. Using a microbeam x-ray source (Takanishi et al. 2010), the scattering on glass cell sample could be recorded with and without an applied electric field. Experiments performed on different compounds confirmed the existence of SmC_SP_A, SmC_SP_F, and SmC_AP_A phases. In some materials, a coexistence of the SmC_AP_A and SmC_SP_A was demonstrated (Barois et al. 2012). These results suggest that the spontaneous reflection symmetry breaking instability is similar to the breaking of the up–down symmetry in Ising-type systems. The global symmetry is unchanged due to the formation of the domains of opposite handedness.

Single-layer ferroelectric and double-layer antiferroelectric structures are not the only combinations expected to occur in bent-core LCs. Other periodicities of tilt and polarization can be also expected, as in the ferroelectric subphases (Takezoe et al. 2010). One example of such behavior was found in a bent-core mesogen with salicylidene moieties. The SmCP phase in these compounds exhibits multistage switching (Figure 3.20) (Findeisen-Tandel et al. 2008). No residual polarization is in the

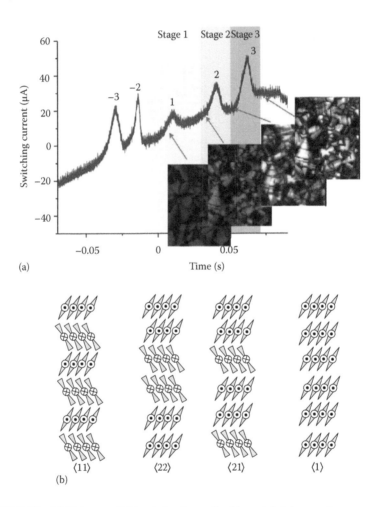

FIGURE 3.20 Multistage switching in bent-core liquid crystal: (a) current–response curve showing switching between three distinct states and (b) proposed structure for three states: antiferroelectric ground state, ferroelectric intermediate state, and ferroelectric end state. (With kind permission from Springer Science+Business Media: *Eur. Phys. J. E*, Multistage polar switching in bent-core mesogens, 25, 2008, 395, Findeisen-Tandel, S., Schroder, M.W., Pelzl, G., Baumeister, U., Weissflog, W., Stern, S., Nemeş, A., Stannarius, R., and Eremin, A.)

ground (field-free) state. In an electric field, there is a two-step switching of the polarization and the birefringence. The analysis of the electro-optical switching and SHG revealed a ferrielectric intermediate state with the polar periodicity given by a unit cell consisting of a pair of synpolar layers and a single layer polarized in the opposite direction. The saturated state under a sufficient electric field is synpolar. The ground antiferroelectric state, however, cannot be unambiguously determined from optical measurements. It has either a single- or a double-layer antipolar structure. This is a phenomenon observed in a homochiral domain.

Coexistence of domains with mixed chiralities was also found (Zennyoji et al. 1999) in P12-PIMB. While the extinction direction rotates by 34° in a homochiral domain, in different domains, the extinction direction does not rotate or rotates by 26°. The former is a 1:1 mixture of two homochiral domains, and the latter a 1:3 mixture of two homochiral domains.

In most cases, the polarization is orthogonal to the tilt plane in polar SmCP-type phases. This seems to be natural, if one considers steric interactions in idealized bent mesogens. There are, however, more exotic cases where $\mathbf{p}$ is not perpendicular to the tilt plane. In a "leaning" or "tipping" phase, the polarization is in the tilt plane. The individual layer polarizations are not parallel to the smectic layers, and the symmetry of the layer is C_S. Different combinations of tilt and polarization in the adjacent layers result in four types of structures: SmC_SLP_F, SmC_SLP_A, SmC_ALP_F, and SmC_ALP_A (Figure 3.21). The out-of-layer polarization in the SmC_ALP_F, SmC_SLP_A phases is averaged out. The experimental evidence for the leaning phases was reported in Chattham et al. (2015) and Zhang et al. (2012).

In a general case, the polarization can be neither perpendicular nor parallel to the tilt plane. This is the so-called general SmC_G phase with a triclinic symmetry C_1. In this optically biaxial phase, none of the principal axes of the macroscopic dielectric tensor lie either parallel or perpendicular to the layer normal. The structure of the SmC_G phase is sketched in Figure 3.22 (Chattham et al. 2015). Until now, no indication has been found for the existence of the SmC_G phase in calamitic LCs. Optical studies of freely suspended films of bent-core LCs provided a direct

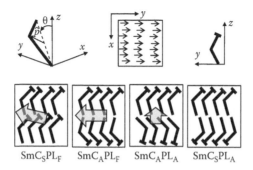

SmC_SPL_F $\quad$ SmC_APL_F $\quad$ SmC_APL_A $\quad$ SmC_SPL_A

FIGURE 3.21 Leaning phase: two-dimensional illustration of the four possible situations when only the long axis is tilted (leaned) with respect to the layer normal. (Redrawn from Eremin, A. and Jákli, A., *Soft Matter*, 9, 615, 2013.)

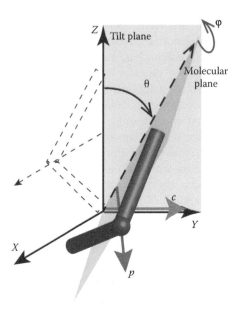

FIGURE 3.22   (See color insert.) A schematic of a bent-core mesogen in the general polar SmC$_G$ phase. (Reprinted with permission from Chattham, N., Tamba, M.-G., Stannarius, R., Westphal, E., Gallardo, H., Prehm, M., Tschierske, C., Takezoe, H., and Eremin, A., *Phys. Rev. E*, 91, 030502(R)-1. Copyright 2015 by the American Physical Society.)

evidence for the existence of the polar general SmCP phase in NORABOW mesogen (Chattham et al. 2010). In addition to the experiments with freely suspended films, other indirect evidence for the triclinic symmetry was found in other bent-core compounds. These studies were based on the analysis of the behavior of the microscopic textures under electric fields and x-ray scattering (Jákli et al. 2001a, Eremin et al. 2003, Chen et al. 2011b). A SmC$_G$ structure superimposed on B$_{1Rev}$-type structure (see Section 3.5) was also reported (Gorecka et al. 2008, Vaupotič et al. 2009a).

A rare intercalated SmC-type phase was designated as B$_6$. The first B$_6$ phase was found in a compound without Schiff's base units (Figure 3.23), exhibiting fan-shaped textures in polarizing microscopy (Pelzl et al. 1999b, Shen et al. 1999). The layer thickness is significantly smaller than the projection of the molecular length on the layer normal. This indicates a nonpolar intercalated biaxial structure shown in Figure 3.23c.

An example of a higher-ordered smectic phase is the B$_5$ phase. This was found in methyl-substituted five-ring compounds. They usually form the SmCP phase on cooling and the transition into the B$_5$ phase is accompanied by a small transition enthalpy and a small texture change. (Figure 3.24). The phase may exhibit antiferro- and ferroelectric switching (Nadasi et al. 2002). A distinctive feature of this phase is the presence of additional x-ray reflections in the wide-angle region (Figure 3.24b). The in-plane structure of the phase can be described by a rectangular 2D lattice, as shown in Figure 3.24c.

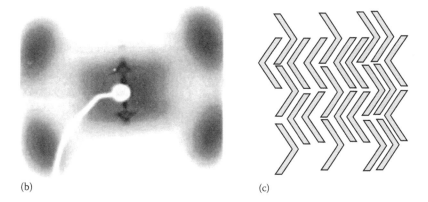

No	R	Y	X	X'	Y'	G	Cr		B$_6$		B$_1$		Iso
1a	C$_4$H$_9$	COO	—	—	OOC	CH	•	195	•	203	—		•
1b	C$_6$H$_{13}$	COO	—	—	OOC	CH	•	209	•	222	—		•
1c	C$_8$H$_{17}$	COO	—	—	OOC	CH	•	178	•	206	—		•
1d	C$_6$H$_{13}$	COO	—	—	OOC	N	•	224	—		•	242	•
1e	C$_8$H$_{17}$	COO	—	—	OOC	N	•	231	—		•	(230)	•
1f	C$_8$H$_{17}$	COO	C≡C	—	OOC	CH	•	183	•	220	—		•
1g	C$_8$H$_{17}$	COO	C≡C	C≡C	OOC	CH	•	204	•	236	—		•

(a)

(b)　　　　　　　　　　　　　　　　　　　　　(c)

FIGURE 3.23 (a) Chemical structure showing the B$_6$ phase; (b) X-ray pattern of an oriented liquid crystal in the B$_6$ phase; and (c) a schematic of the phase structure. (From Pelzl, G., Diele, S., and Weissflog, W.: Banana-shaped compounds—A new field to liquid crystals. *Adv. Mater.* 1999. 11. 707–724. Copyright Wiley-VCH Verlag GmbH & Co. KGaA. Reproduced with permission.)

3.4 FRUSTRATED PHASES

3.4.1 POLARIZATION-MODULATED PHASE

In bent-core mesogens, the ground state with a uniform director alignment is often prone to instability toward a state with a modulated structure. One of the brightest examples of such phases with a modulated polarization is the family of B$_7$ phases. Originally, B$_7$ was attributed to a particular phase that exhibits one of the most beautiful polarizing microscopy textures featuring a variety of complex structures such as ribbons, telephone wires and screw filaments, etc. The textures are

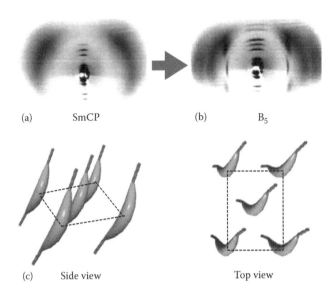

(a)   SmCP   (b) B_5

(c)   Side view   Top view

FIGURE 3.24 Comparison of x-ray patterns of (a) an oriented liquid crystal in the SmCP phase and (b) the same sample in the ferroelectric B_5 phase. Molecular arrangement in a unit cell of the B_5 phase is also shown in (c). (a and b: From Nadasi, H., Weissflog, W., Eremin, A., Pelzl, G., Diele, S., Das, B., and Grande, S., Ferroelectric and antiferroelectric "banana phases" of new fluorinated five-ring bent-core mesogens, *J. Mater. Chem.*, 12, 1316–1324, 2002. Reproduced by permission of The Royal Society of Chemistry; c: Reprinted from Takezoe, H. and Takanishi, Y., Jpn. *J. Appl. Phys.*, 45, 597–625, 2006. With permission from the Japan Society of Applied Physics.)

often irresponsive to an electric field but in some cases, they can be transformed into a fan-shaped texture typical of the SmCP phase. Further studies concluded that these textures are not unique for a single B_7 phase but occur in several different mesophases. The compounds studied are shown in Figure 3.25. The standard resolution x-ray diffraction (XRD) revealed two types of structures. One type is very similar to the structure of the SmCP phase: smectic layers, tilted mesogens, and no in-plane order, but not uniform layers such as in MHOBOW and H87 compounds (Figure 3.25) (Coleman et al. 2003). The second type of structure exhibits multiple reflections in wide-angle and small-angle XRD, which suggest a columnar structure similar to the B_1 phase (see Section 3.5), in which a layer interdigitation provides effective molecular packing (Coleman et al. 2003).

The polarization-modulated SmC (PM-SmCP) phase is distinguished from the phase with a simple layer structure SmCP by a spontaneous splay of the polarization. Since the splay cannot fill an infinite plane, periodic structures of defects separating the splay domains appear. The domains exhibit stripes with a width in the range of 50–1000 nm depending on the material and temperature. These stripes were directly visualized using freeze-fracture TEM studies (Coleman et al. 2003). They revealed coherent periodic structures with a similar layer modulation wavelength obtained by XRD studies.

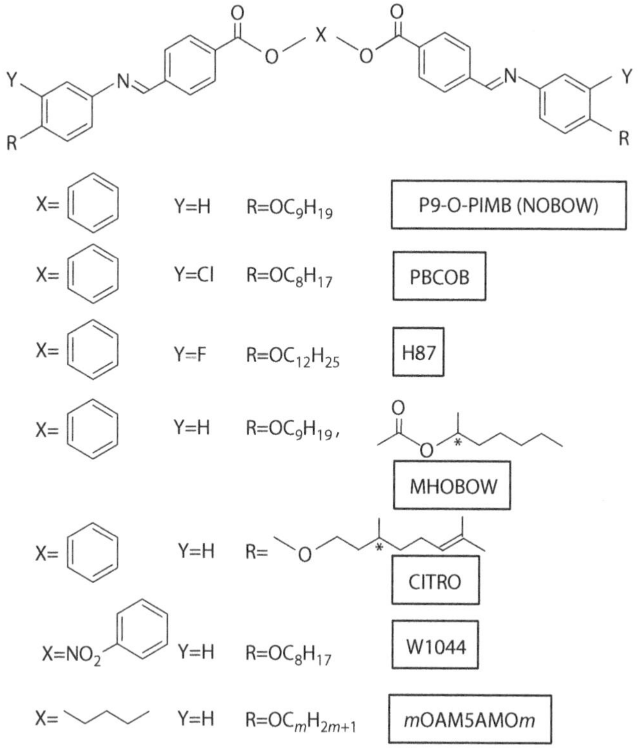

FIGURE 3.25 Compounds that exhibit the B_7 (polarization modulated) phase.

The modulation develops mostly in the layer plane. But the detailed analysis of different materials showed several possible variations (Coleman et al. 2008). The main unit of the structure is a stripe with splayed polarization. The splay of polarization couples to the layer curvature. At the edge of the splay domains, the mesogens experience a different environment compared to the middle part. As a result, the mesogens tend to accommodate a tilt variation across the stripe. This creates a dilative pressure, which deforms the layers as shown in Figure 3.26. Such deformed tiles can be arranged in different structures. In case of MHOBOW and H87 compounds (see Figure 3.25), the XRD can be well described by a rectangular lattice. This phase is designated as SmCPU$_A$. The tiles are arranged so that the layer curvature alternates from one tile to another. In compounds like UD50 and 2299 (Coleman et al. 2008), the tiles are arranged in an oblique lattice with an angle typically about 80°–85°. This phase is designated as SmCPU$_S$.

Local polar structure of the polarization-modulated phase was established by the observations of freely suspended films under in-plane electric fields and SHG studies (Coleman et al. 2003, Eremin and Jákli 2013). In the case of a columnar polarization-modulated B_1 phase, there is a half-layer discontinuity between the polarization splay stripes. Thus, the tiles are shifted by half a layer. Yet both rectangular and oblique structures are possible.

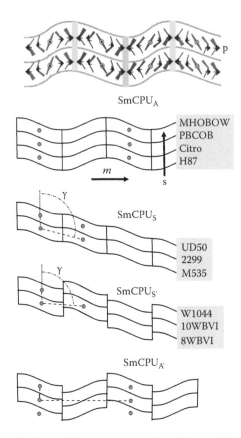

FIGURE 3.26   (See color insert.) Layer structures in various modifications of the B_7 phase. (Partly reprinted from Kim, H. et al., *Liq. Cryst.*, 41, 2014a, 328. With permission from Taylor and Francis; Redrawn from Coleman, D.A. et al., *Phys. Rev. E*, 77, 021703-1, 2008.)

A deeper insight into the structure of the layer-type B_7 phase was obtained using high-resolution XRD at a synchrotron facility. In addition to the layer reflections, satellite reflections were observed in LC cells with a bookshelf geometry suggesting a modulated layer structure (Coleman et al. 2003, 2008, Nakata et al. 2005). Upon slowly cooling a compound, CYTRO (Figure 3.26), which has chiral citronellyl tails (Nakata et al. 2005) from the Iso phase, a polarization-modulated SmCP (PM-SmCP) state exhibiting the B_7 texture emerges, while a metastable $SmC_SP_F^*$ is obtained by quenching from Iso. These two structures were characterized by XRD (Figure 3.27), where multiple satellite peaks around the first-order layer reflection occur, indicating undulated layers in PM-SmCP. It is also interesting that the $SmC_SP_F^*$ can also be obtained from the PM-SmCP by applying a field above the threshold value, but it thermally relaxes to the PM-SmCP state (Nakata et al. 2005).

3.4.2   HELICAL NANOFILAMENT PHASE

B_4 or helical nanofilament (HNF) phase was one of the first new "banana" phases to be observed in bent-core LCs (Pelzl et al. 1999b). It appeared in various compounds

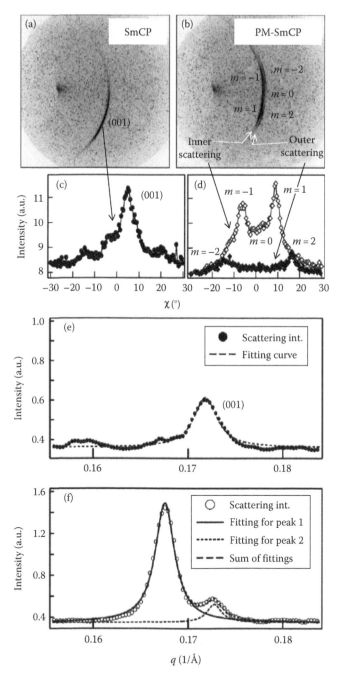

FIGURE 3.27 Microbeam XRD scattering patterns. 2D scattering patterns from (a) the field-induced SmC$_S$P$_F$* and (b) the ground-state PM-SmCP phases. (c) and (d) χ scan of (a) and (b). (e) and (f) Scattering intensity of (a) and (b) as a function of scattering angle. (Reprinted with permission from Nakata, M., Link, D.R., Takanishi, Y., Takahashi, Y., Thisayukta, J., Niwano, H., Coleman, D.A. et al., *Phys. Rev. E*, 71, 011705-1. Copyright 2005 by the American Physical Society.)

exhibiting the B_2 phase at a higher temperature. This phase was studied by various experimental techniques including XRD, optical microscopy, SHG, nuclear magnetic resonance, and atomic force microscopy.

The first x-ray study on B_4 was conducted using two beam directions parallel (geometry A) and perpendicular (geometry B) to a substrate (Figure 3.28), on which molecules form layers parallel to the substrate in the B_2 phase (Thisayukta et al. 2001). The layer reflection appears only in geometry A in B_2, whereas the layer reflection can be seen in both geometries in B_4. In addition, the reflection spots in B_4 change to circular reflection in geometry A, as shown in Figure 3.28. Based on these XRD results, the B_4 structure was suggested to be the same as that of the twist grain boundary (TGB) phase, in which a periodic organization of screw dislocations results in a helical winding of smectic layers. Later, the structure was shown to be a helical nanofilament with a continuous layer twist different from the TGB structure, as described in the following text.

One of the distinctive features of this phase was relatively low birefringence and the appearance of chiral domains, as shown in Figure 3.29a. Four bent-core molecules, Pn-O-PIMB (n = 8, 9, 12) (Figure 1.8), and a chiral analogue of P-9-O-PIMB (MHOBOW) (Figure 3.25) were carefully studied (Hough et al. 2009a). With crossed polarizer (P) and analyzer (A), the sample transmits only weakly because of its low birefringence. Uncrossing the analyzer by 5° reveals areas of uniform but opposite optical activity, several hundred micrometers in width, indicating the formation of chiral conglomerate domains (Figure 3.29a), as also previously shown by Earl et al. (2005). High-resolution powder XRD of P9-O-PIMB shows that the resolution-limited peaks of the high-temperature fluid smectic B_2 phase give way to a diffuse reflection at the B_2–B_4 transition. The broadening of the peaks is due to the limited

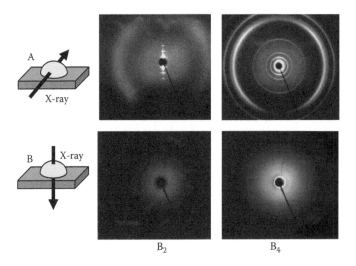

FIGURE 3.28   XRD patterns in B_2 and B_4 using two geometries, A and B. In a drop sample, the layers are parallel to a substrate surface. (Partly reprinted from Thisayukta, J. et al., *Jpn. J. Appl. Phys.*, 40, 3277, 2001. With permission from the Japan Society of Applied Physics.)

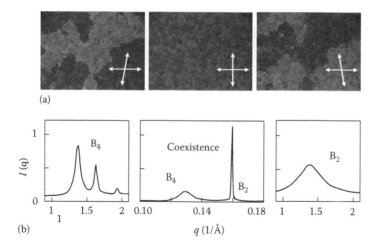

(a)

(b)

FIGURE 3.29 (a) Microphotographs of the B_4 phase of P-8-O-PIMB and (b) high-resolution powder XRD of P-9-O-PIMB through the B_4–B_2 phase transition. (a: Reprinted with permission from Earl, D.J., Osipov, M.A., Takezoe, H., Takanishi, Y., and Wilson, M.R., *Phys. Rev. E*, 71, 021706-1. Copyright 2005 by the American Physical Society; b: Redrawn from Hough, L.E. et al., *Science*, 325, 456, 2009a.)

crystallite size effect. In the present case, the diffuse reflection is due to the limited width of the membranes forming filaments. At the same time, the broad wide-angle peak indicative of liquid-like in-plane order of the B_2 phase is replaced by several diffuse peaks, indicating hexagonal in-plane order within the layers in the B_4 phase (Figure 3.29b) (Hough et al. 2009a).

The formation of large chiral domains is a dramatic demonstration of the spontaneous breaking of the reflection symmetry. Sometimes, the chiral domain size extends as large as a few millimeters. It must be pointed out that this symmetry breaking is secondary if the B_4 phase is formed from the B_2 phase, since the single layers of the B_2 phase are already chiral. Actually, cyclic temperature variation between B_2 and B_4 does not change the shapes of chiral domains (Niwano et al. 2004). The detailed chirality-related topics are described in Chapter 5.

Local layer structure of the B_4 phase was studied by the atomic force microscopy (AFM) and freeze-fracture transmission electron microscopy (FFTEM) (Hough et al. 2009a). As observed using these techniques, the layers are strongly distorted to form a saddle-splay local curvature, resulting in a helical nanofilament, as shown in Figure 3.30 (Chen et al. 2010, 2011). Molecular packing in a layer is discussed in the caption of Figure 3.30. The typical width w and pitch p of the filament are 20–40 and 80–200 nm, respectively. The polarity of the structure was confirmed by the SHG (Choi et al. 1998). More detailed analysis of chiral SHG measurements (see Section 6.3) confirmed that the polar direction is along the helical axis (Araoka et al. 2005).

It is believed that the formation of macroscopic chiral domains proceeds through the nucleation and chirality preserving growth of the nanofilaments. Single filaments bifurcate when they exceed a critical width (Chen et al. 2012). The chirality of the B_4 phase can also be controlled by different techniques, as will be described in

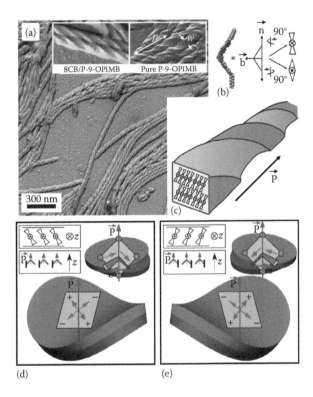

FIGURE 3.30 (**See color insert.**) B_4 helical nanofilaments and their corresponding molecular organization. (a) FFTEM images of a mixture (8CB/P-9-OPIMB = 75/25) quenched at 37°C reveal large homochiral regions with either left- or right-handed helices. In an inset, an FFTEM image of pure P-9-OPIMB is also shown. Different morphologies in the pure and mixture samples can be noticed; Layer edges are visible in the pure P-9-OPIMB, whereas they are not visible because of the coverage by 8CB in the mixture. (b) The convention for the molecular director **n** and the molecular bow direction **b** of bent-core molecules, with the polarization **P** along **b**. The handedness of the helices is determined by the corresponding layer chirality. (c) The structure of a helical nanofilament. (d) and (e) The formation of helical nanofilaments. Each molecular arm can be viewed as an elastically isotropic slab that dilates parallel to and compresses perpendicular to the molecular tilt direction as a result of the hexagonal in-plane ordering. Because the tilt directions of the top and bottom molecular arms are orthogonal, the two elastic slabs adopt a saddle-splay curvature to relieve the intralayer mismatch. The local layer chirality, which is determined by the polarization and molecular tilt, results in distinct regions with orthogonal saddle-splay and opposite signs of filament twist. (a: Reprinted with permission from Chen, D., Zhu, C., Shoemaker, R.K., Korblova, E., Walba, D.M., Glaser, M.A., Maclennan, J.E., and Clark, N.A., Pretransitional orientational ordering of a calamitic liquid crystal by helical nanofilaments of a bent-core mesogen, *Langmuir*, 26, 15541. Copyright 2010 American Chemical Society; b, d, and e: Reprinted with permission from Chen, D., Maclennan, J.E., Shao, R., Yoon, D.K., Wang, H., Korblova, E., Walba, D.M., Glaser, M.A., and Clark, N.A., Chirality-preserving growth of helical filaments in the B4 phase of bent-core liquid crystals, *J. Am. Chem. Soc.*, 133, 12656. Copyright 2011a American Chemical Society.)

Section 5.3. On the contrary, alignment control of helical nanofilaments is not easy. Successful results were reported using two techniques (Kim et al. 2014a): (1) restricted geometries (YoonAdMat 2011, Kim et al. 2014b) and physicochemical confinement (Lee et al. 2015b) and (2) nematic alignment field (Araoka et al. 2013). For method (1), Yoon et al. used two different restricted geometries, microchannels (YoonAdMat 2011) and chiral nanopore arrays (Kim et al. 2014b), which were both effective. In the case of microchannels, however, air flow is required to continuously apply a shear force. In the latter case, a chirality-controlled single helical nanofilament is grown in each pore (Figure 3.31). As for the method (2), a high-temperature nematic compound was mixed with a bent-core compound, so that helical nanofilaments appear in an oriented nematic field. Together with a temperature gradient, well-aligned millimeter-sized homochiral domains were obtained (Figure 3.32).

There are two other important questions: (1) Why does the B_4 phase look blue? (2) How does the transition to the B_4 phase occur? We will explain these questions in the following.

The B_4 phase was first called the smectic blue (SmBlue) phase (Sekine et al. 1997a) because of its transparent blue appearance. The blue color was first attributed to the selective reflection as in the cholesteric phase or to the scattering of light. Both suggestions have been rejected at present. According to Hough et al. (2009a), the half pitch of the helical nanofilament is about 100 nm, which is too short to exhibit blue reflection. The explanation due to stronger Rayleigh scattering at shorter wavelength is not applicable because of the fact that the blue color is observable only in short homologues in case of P-n-OPIMB. Alternative explanation due to molecular aggregation was proposed (Araoka et al. 2010). It is known that aggregations of molecules make their absorption spectra shift to blue (H-aggregate) or red (J-aggregate). In liquid crystals, however, such absorption band shift was not well

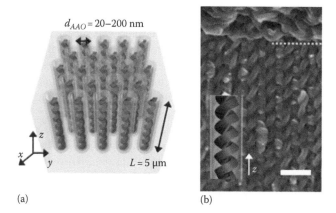

(a)   (b)

FIGURE 3.31 (a) Cartoon of porous column array of anodic aluminum oxide films and a single helical nanofilament in each pore and (b) SEM image of helical nanofilaments aligned in pores. (From Kim, H., Lee, S., Shin, T.J., Korblova, E., Walba, D.M., Clark, N.A., Lee, S.B., and Yoon, D.K., Multistep hierarchical self-assembly of chiral nanopore arrays, *Proc. Natl. Acad. Sci. USA*, 111, 14342–14347. Copyright 2014b National Academy of Sciences, U.S.A.)

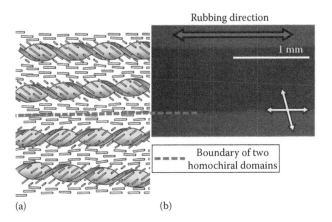

FIGURE 3.32 (a) Cartoon of aligned helical nanofilaments in a nematic field and (b) microphotograph of homochiral domains consisting of well-aligned helical nanofilaments. Arrows represent a rubbing direction, and a broken line is a boundary of two homochiral domains. (From Araoka, F., Sugiyama, G., Ishikawa, K., and Takezoe, H.: Highly ordered helical nanofilament assembly aligned by a nematic director field. *Adv. Funct. Mater.* 2013. 23. 2701–2707. Copyright Wiley-VCH Verlag GmbH & Co. KGaA. Reproduced with permission.)

recognized, because absorption peaks cannot be seen in the experiments using a cell of a few microns thick and, in most cases, there is no 2D/3D positional order. Araoka et al. (2010) carefully measured the absorption spectra of 12O-PIMB exhibiting the Iso–B_2–B_4 phase transition using 400 nm thick cells. They found that the absorption spectral shape in the B_2 phase is essentially the same as that in the Iso phase with a small blue shift (0.15 eV), suggesting that the spectra are attributable to isolated molecules (Figure 3.33). But at the B_2–B_4 phase transition, a large blue shift of the absorption band by about 0.74 eV was observed. In addition, peak broadening and small periodic fringes appear in B_4 (Figure 3.33). All of these results strongly suggest some intermolecular interactions. The blue shift was successfully reproduced in a simulation by considering an intralayer molecular arrangement based on Kasha's molecular exciton model (Kasha 1976). The small periodic fringes with a period of 1700 cm^{-1} are ascribed to the vibronic structure due to the aromatic stretching mode. Strong absorption shifted to the near-UV region may result in strong blue reflection from transparent background.

More recently, a more plausible explanation for the blue color was made based on the observation of doubly twisted structure. Zhang et al. (2014) found that nanofilaments can stack themselves not in a parallel fashion, but preferentially rotated by an angle of 35°–40° with respect to each other as shown in Figure 3.34 (doubly twisted structure). They observed selective reflection peaks at 475 and 460 nm for P-7-OPIMB and P-8-OPIMB, respectively. Taking the narrowest thickness of an individual filament $w \approx 30$ nm and the rotation angle $\beta \approx 37°$, the structural pitch would be ~300 nm. The estimated optical pitch is ~480 nm, which agrees quite well with the selective reflection wavelength.

Another interesting observation was made by differential scanning calorimetry (DSC) at the phase transition to the B_4 phase. Sasaki et al. (2011) studied the

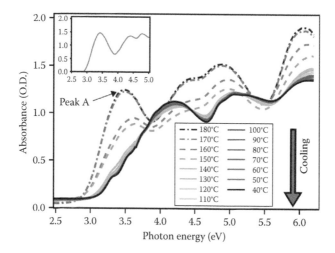

FIGURE 3.33 **(See color insert.)** Temperature dependence of absorption spectra of P-12-O-PIMB on cooling. The broken, dashed, and solid lines represent spectra in the isotropic, B_2, and B_4 phases, respectively. An inserted spectrum is from P-12-O-PIMB dissolved in chloroform. (Reprinted with permission from Araoka, F., Otani, T., Ishikawa, K., and Takezoe, H., *Phys. Rev. E*, 82, 041708-1. Copyright 2010 by the American Physical Society.)

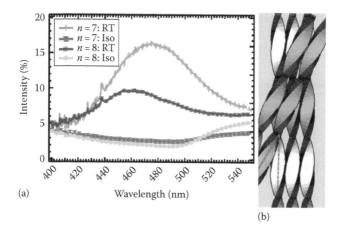

FIGURE 3.34 (a) Selective reflection spectra of P-7-O-PIMB and P-8-O-PIMB at RT and Iso and (b) schematic image of a twisted structure of two HNFs. (Reprinted by permission from Macmillan Publishers Ltd. *Nat. Commun.*, Zhang, C., Diorio, N., Lavrentovich, O.D., and Jákli, A., Helical nanofilaments of bent-core liquid crystals with a second twist, 5, 3302–3308. Copyright 2014.)

transition behavior using high-resolution DSC (HR-DSC) with a very slow scan rate (0.05 K/min) using pure P8-O-PIMB and its mixtures with rod-shaped molecules. In both cases, a bent-core compound only and mixtures, distinct formations of helical nanofilaments were thermally observed. One example is shown in Figure 3.35a, where the B_2–B_4 transition appears as many discrete DSC peaks on cooling over wide

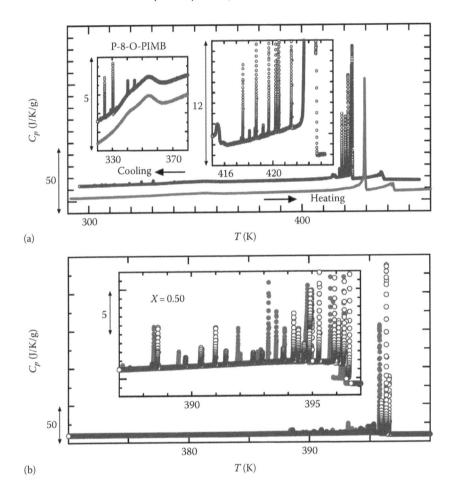

FIGURE 3.35 (a) HR-DSC charts of P8-O-PIMB and (b) DSC charts for different scans (open and closed symbols) near the Iso–B$_4$ phase transition in a 1:1 mixture of P-8-O-PIMB and 5CB. (Reprinted with permission from Sasaki, Y., Nagayama, H., Araoka, F., Yao, H., Takezoe, H., and Ema, K., *Phys. Rev. Lett.*, 107, 237802-1. Copyright 2011 by the American Physical Society.)

temperature ranges. In the mixture with 5CB, the Iso–B$_4$ direct transition appears also as many discrete DSC peaks, suggesting microphase separation of P8-O-PIMB and 5CB. Examples of two independent scans are given in Figure 3.35b, which shows that the formation of helical nanofilaments is independent in each measurement (open and closed symbols). In both cases, just simple normal transitions were observed on heating.

3.4.3 Dark Conglomerate Phase

Dark conglomerate (DC) phase is distinguished by little or no birefringence and a conglomerate texture of right- and left-handed domains, which becomes visible between slightly decrossed polarizers (Hough et al. 2009b). An example exhibiting chiral domains under decrossed polarizers is shown in Figure 3.36 (Findeisen-Tandel et al. 2008). The DC phase often emerges in compounds with the ferroelectric $SmCP_F$ phase or the B_7 phase (Keith et al. 2007c, Hough et al. 2009b, Chen et al. 2011b). It often coexists with the B_7 phase. In some cases, the DC phase can be induced by an electric field in the ferroelectric $SmCP_F$ state and in other phases (see later in this subsection). The size of the macroscopic chiral domains varies from a few microns up to a millimeter. The domains exhibit strong optical activity reaching as much as 1°/micron (Hough et al. 2009b). A racemic DC phase may also occur on cooling from the isotropic phase without smectic correlations (Sicilia et al. 2008). The chiral deracemization occurs under an applied electric field (Sicilia et al. 2008, Deepa and Pratibha 2014) (Figure 3.37). This process is driven by the chiral segregation and is believed to occur due to a spontaneous choice of the layer chirality when the

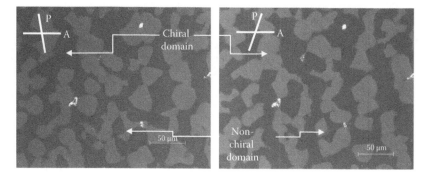

FIGURE 3.36 Optical microscopy texture of a DC between slightly uncrossed polarizers. White arrows mark nonchiral domains and chiral domains that change the brightness when the polarizers are uncrossed in the opposite sense. (With kind permission from Springer Science+Business Media: *Eur. Phys. J. E*, Multistage polar switching in bent-core mesogens, 25, 2008, 395, Findeisen-Tandel, S., Schroder, M.W., Pelzl, G., Baumeister, U., Weissflog, W., Stern, S., Nemeş, A., Stannarius, R., and Eremin, A.)

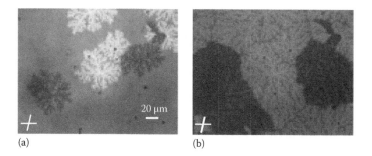

FIGURE 3.37 Formation of the DC_{dc} phase from the isotropic phase under a dc electric field of 10 V/μm in a sample cooled at 1°C/min. (a) Conglomerate fractal domains (126.6°C) and (b) further growth of fractal domains (126.3°C). Analyzer rotated by 2° in the clockwise direction from the crossed position. (Reprinted with permission from Deepa, G.B. and Pratibha, R., *Phys. Rev. E*, 89, 042504-1. Copyright 2014 by the American Physical Society.)

polar direction is fixed by an applied electric field. The chiral domains grow in a fractal fashion with the fractal dimension of about $D = 1.8968$ (Deepa and Pratibha 2014). Such a high value of D suggests the percolation mechanism of the chiral deracemization.

The chiral domains are not usually responsive to an applied electric field. However, in some materials, the DC phase can be induced by an electric field or even destroyed by an external field (Eremin et al. 2003, Ortega et al. 2011). One of the fascinating features of some DC materials is the chirality switching (see more details in Section 5.3.3). Under an external field, the chiral domains exhibited bistable switching of the handedness showing a current response typical for ferroelectrics (Pelzl et al. 2006, Zhang et al. 2010) (Figure 3.38). This switching process occurs

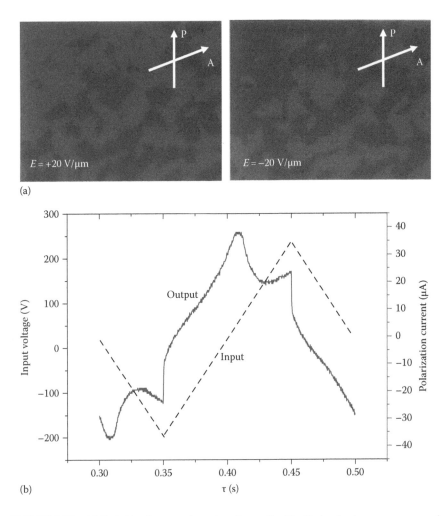

(a)

(b)

FIGURE 3.38 (a) Switching between the states of opposite chirality in a bent-core mesogen and (b) the current response to a triangular-wave voltage applied to the cell $E_{pp} = 40$ V/μm, $f = 2$ Hz.

without any detectable birefringence and is accompanied by a current response with switching polarization as high as 300 nC/cm^2.

Ortega et al. proposed a structure of the DC phase consisting of randomly distributed SmC$_A$P$_A$ clusters (Ortega et al. 2003) with a high tilt angle. Strong optical rotation is attributed to the helical arrangement of molecular wings of bent-core molecules under the anticlinic SmC$_A$P$_A$ structure with a two-molecular-length pitch. The clusters having such a structure behave as uniaxial crystals with optical axes aligned along helical axes. They claimed that the gyrotropic tensor component is proportional to the ratio of the pitch to the wavelength p/λ for the light propagating perpendicular to the helical axis, but to the third power of p/λ for light propagating parallel to the helical axis and is essentially negligible.

More detailed studies of the structure of the DC phase lead to a different model (Hough et al. 2009b, Chen et al. 2011b, Ortega et al. 2011). The structure of the phase was elucidated using FFTEM and x-ray techniques. Layer reflections in the x-ray scattering become diffuse in the DC phase reflecting short-range layer correlations (Chen et al. 2011b), the same as in the B$_4$ phase (Hough et al. 2009a). In smectics, this occurs when the layers become strongly frustrated as in LCs in restricted confinement. The x-ray reflection profile has a Lorentzian shape indicating an exponential decay of the density–density correlations with a typical correlation length $\xi \approx 40$ nm. Yet the higher harmonics of the layer reflections have wider diffraction width, indicating that phase fluctuation effects are the principal contribution to the disordering in the DC phase (Chen et al. 2011b). Figure 3.39a shows an FFTEM image of the DC phase of top-12-OPIMB (Chen et al. 2014). The image displays two important features of the DC structure: (1) saddle topography of individual layers and (2) the layer continuity on the length scales far above the layer correlation length (Hough et al. 2009b). This structure is similar to that of the lyotropic sponge phase where the empty volume is filled with the smectic layers with a fixed layer spacing. Layers that appear above a flat surface resemble saddles. At an LC/air interface, the structure is distinguished by a well-ordered array of toroidal focal conic domains as shown in Figure 3.39b (compound shown at the bottom) (Chen et al. 2011b). These findings demonstrate that the overall structure of the DC phase is analogue of the lyotropic sponge phase. Smectic layers remain flat only on a length scale of approximately 100 nm, and they adopt a nested curved configuration with defect lines where the curvature radius approaches the smectic layer spacing (Figure 3.39c). At the substrate surface, focal conic domains form a hexagonal structure with the periodicity of the order of 400 nm (compound shown at the bottom) (Figure 3.39b). The saddle curvature of the layers allows them to be continuous over long distances despite even the short layer correlation length.

A phase with a saddle-splay layer deformation is known in lyotropic systems as the sponge phase. A model of a sponge-like smectic phase with a cubic or hexagonal long-range order in the thermotropic system was suggested by DiDonna and Kamien (2002). They demonstrated that the energy cost of strongly deformed smectic layers can be compensated by the gain of the Gaussian curvature of the surfaces when the saddle-splay elastic constant K_{24} is negative enough. The preference for the saddle splay of the layers that resulted from the reduction of the elastic energy was suggested to be responsible for the large degree of disorder in the DC phase.

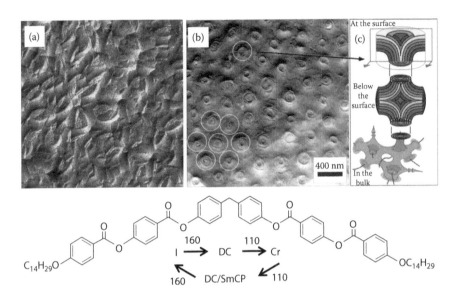

FIGURE 3.39 (a) Freeze-fracture TEM image of the DC phase. (b) and (c) TEM image of the DC phase at the free surface of a drop of liquid crystal material placed on a glass cover slip that was quenched after cooling from isotropic to 140°C. Samples used are shown at the top and the bottom for (a) and (b), respectively. (a: From Chen, D., Shen, Y., Aguero, J., Korblova, E., Walba, D.M., Kapernaum, N., Giesselmann, F. et al.: Chiral isotropic sponge phase of hexatic smectic layers of achiral molecules. *ChemPhysChem*. 2014. 15. 1502–1507. Copyright Wiley-VCH Verlag GmbH & Co. KGaA. Reproduced with permission; b and c: From Chen, D., Shen, Y., Zhu, C., Hough, L.E., Gimeno, N., Glaser, M.A., Maclennan, J.E., Ros, M.B., and Clark, N.A., Interface structure of the dark conglomerate liquid crystal phase, *Soft Matter*, 7, 1879–1883, 2011. Reproduced by permission of The Royal Society of Chemistry.)

The free energy for a symmetric bilayer to the lowest order is

$$F_c = \int \left(2K_1 H^2 + K_{24} K_G \right) d^3 x, \tag{3.2}$$

where

K_1 is the bending modulus
K_{24} is the Frank constant associated with saddle splay of the layers
K_G is the Gaussian curvature
H is the mean curvature

The saddle-splay curvature ($K_G < 0$) lowers the total energy. However, when the Gaussian curvature term is nonzero, defects must be present. Thus, this contribution to the free energy can be ignored in ordinary smectics.

In contrast to the lyotropic sponge phase, where the disorder arises from the thermally generated fluctuations, the lamellar structure of the thermotropic smectic phase suppresses the fluctuations of the layer thickness by the layer compressional elasticity. In case of mesogens with flexible tails, the mismatch of the

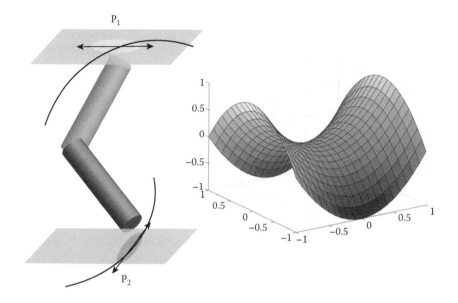

FIGURE 3.40 (See color insert.) Formation of the saddle-splay structure in the DC phase: frustration between molecular fragments of a bent-core mesogen can be relieved by saddle-splay curvature of the layers. The projections P_1 and P_2 of the molecular wings on the smectic layer are nearly orthogonal in a tilted SmCP phase. In case of tilted phase of rod-shaped mesogens, the in-plane correlation function shows an elongation in the direction of the molecular tilt. This leads to frustration in case of tilted bent-core mesogens. This frustration can be relieved by layer curvature along the mutually orthogonal principal directions.

in-plane area of the tails and cores can be relieved by a tilt (Figure 3.40). The tilt, however, creates a mismatch between the tilt planes of the two half-arms of the mesogens, making their projections on the layer plane nearly perpendicular to each other. This situation produces a strong tendency for the negative Gaussian curvature of the layers in the SmCP phase via the coupling of the polar order and the tilt of the molecular arms to the positional order within the smectic layers. This tendency results in a structure dominated by focal conic layer curvature motifs forming arrays of saddle splays of a characteristic size of the order of 200–400 nm. The tendency to form saddle-splay structures is inherently chiral and is expected to occur in homochiral local SmC_AP_A and SmC_SP_F states. The racemic subphases tend to form flat layers since the handedness alternates from one layer to another. Yet, the racemic structures with longer pitch may exhibit homogeneous saddle-splay structure, since the layer curvature tensor $\mathbf{L} = -\mathbf{e}_i \cdot [\mathbf{e}_j \cdot \nabla]\mathbf{n}$ is symmetric under combined layer reflection and 90° rotation about the layer normal (Hough 2002). This can explain occasional occurrence of the optically inactive DC phases reported in Reddy et al. (2007) and Findeisen-Tandel et al. (2008). On the other hand, SHG studies made by Martínez-Perdiguero et al. (2006) on both optically active and inactive types of DC phases suggest that the ground state of the DC phase is not polar (antiferroelectric) and the polarity is not responsible for the formation of the DC phase.

The behavior of the DC phase was studied in detail by several groups (Eremin et al. 2003, Martínez-Perdiguero et al. 2006, Zhang et al. 2010, Ortega et al. 2011, Deepa and Pratibha 2014, Milton et al. 2014, Nagaraj et al. 2014a,b, 2015). Various types of behaviors in an electric field have been reported. DC phase can be induced by an electric field (Eremin et al. 2003) or by slow cooling from the isotropic phase under a square-wave AC field (Ortega et al. 2011). In other materials, electric-field application can suppress the DC phase and induce birefringence (Findeisen-Tandel et al. 2008). In many cases, the texture does not respond to electric fields at all. Electric field can stimulate the chiral segregation resulting in larger chiral domains or the dominance of one handedness over the other one (Nagaraj et al. 2015). This process depends on the field amplitude as well as the frequency. As mentioned earlier, bistable switching between two chiral domains can be found in some compounds. This is accompanied by a single-peak current response (Figure 3.38) (Eremin et al. 2003). In other cases, no chirality switching is observed, but the current response shows an antiferroelectric two-peak structure (Nagaraj et al. 2014b). In dielectric spectroscopy, a transition into the DC phase is accompanied by an increase of the dielectric constant suggesting the formation of polar structure of the phase.

The electric-field-induced B_{1rev}-DC transition was also observed (Deepa et al. 2013). Two different DC phases named DC-B_{1rev} and DC-$B_{1revtiltM}$ are induced depending on the B_1 phases, at which a square wave field (1 kHz, up to 10 V/µm) is applied. The B_{1rev} and $B_{1revtiltM}$ phases appear at higher and lower temperatures and have a *cmm* rectangular lattice and a noncentered *pmm* rectangular lattice, respectively. The different induction processes of the two DC phases are shown in Figure 3.41 (left column for DC-B_{1rev} and right column for DC-$B_{1revtiltM}$). Once formed, the DC phase persists even if the field is switched off. The B_1 phase is recovered only if the sample is heated back into the Iso phase and cooled again in the absence of an electric field. When the applied field to the B_{1rev} phase increases up to ~9 V/µm, molecular rotation about its long axis results in the antiferroelectric–ferroelectric transition associated with further segregation of the aromatic and aliphatic parts in the columnar structure. After all, instead of the transformation to a SmC_SP_F phase with flat layers, the DC domains with opposite chirality develop. This is because in the DC phase the difference in the in-plane areas of the alkyl chains and the aromatic cores gives rise to a mismatch between the tilt planes of the two half-arms of the bent-core molecule, leading to layer frustration (Hough et al. 2009b).

Both structures of the B_4 and DC phases are driven by the same mechanism: intralayer mismatch of the two arms of molecules and the spontaneous tendency for saddle-splay curvature formation. There are many similar appearances such as spontaneously segregated chiral domain structures, low birefringence, short coherence length of the layer structure, etc. and in both phases they are similar. Because of these reasons, we sometimes encountered confusion in the assignment to these phases. But it is also true that both phases coexist and transfer to each other by small external stimuli. One example is P-12-OPIMB; both sponges (DC) and HNFs (B_4) are stable (Chen et al. 2014). Another example is twin dimers, consisting of two rod-shaped mesogens linked by a flexible chain. A phase of a compound 12OAz5AzO12 exhibiting chiral resolution was assigned to the B_4 phase in the literature (Choi et al. 2006), but it seems to be the DC phase, according to our recent studies based on scanning

FIGURE 3.41 **(See color insert.)** Textures of two different DC phases formed by the molecule shown at the top. Left (a, c, e) DC-B$_{1rev}$ process and right (b, d, f) DC-B$_{1revtiltM}$ process. (Reprinted with permission from Deepa, G.B., Radhika, S., Sadashiva, B.K., and Pratibha, R., *Phys. Rev. E*, 87, 062508-1. Copyright 2013 by the American Physical Society.)

electron microscopy (SEM), AFM, and XRD results (Le et al. 2016). But, by adding some amount of solvents such as *n*-dodecane, the sponge structure was transformed into the HNF structure.

3.5   COLUMNAR PHASE

At least two types of columnar phases can be recognized in bent-core liquid crystals. One is a columnar phase formed from broken-layer fragments. The other one is a columnar phase consisting of the stacking of disc-like assemblies of bent-core molecules. The former is historically called B_1 phase. The existence of the B_1 phase was first recognized by Sekine et al. (1997b) in the shortest homologue ($n = 6$) of P-*n*-O-PIMB. Based on XRD, they suggested a frustrated antiphase for the structure. Watanabe et al. (1998b) performed detailed XRD studies and suggested a frustrated smectic structure shown in Figure 3.42a. The terminology of frustrated smectics is mainly used for smectic phases (SmA_1, SmA_2, SmA_d, etc.), which require two order parameters to describe their properties, the center of mass density of the constituent molecules, and long-range head-to-tail correlations of polar molecules such as cyanobiphenyl (Barois et al. 1985, de Gennes and Prost 1993). The physical situation is the same for the structure shown in Figure 3.42. But the phase is more often called columnar, modulated, or broken-layer phase. Actually, columns are formed, with two-dimensional positional order, along the molecular bending (polar) direction (Figure 3.42b). Here, we can also recognize that infinite layer structures are broken or otherwise flat layers are modulated. Because of bent-core molecules, each block of layers has polarization and even chirality if the constituent molecules are tilted from the layer normal as in the B_2 phase. The formation of such columnar structures is driven by the tendency for forming splayed polarizations (see Figure 3.26) and by the escape from macroscopic polarization.

In the course of structure studies of main-chain polymeric liquid crystals having mesogenic groups linked by flexible alkylene chains with odd-numbered carbons, Nakata and Watanabe (1997) found a modulated smectic phase and suggested two possible models. One is the same as Figure 3.42a of *p2mg* symmetry and the other is a structure shown in Figure 3.42b of *pm2₁n* symmetry. The identification of the B_1

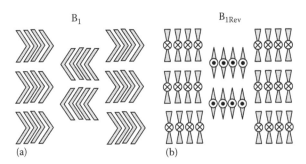

FIGURE 3.42   Structures of (a) B_1 and (b) B_{1Rev} phases. The B_1 phase has not been identified yet.

structure to Figure 3.42a was made without any experimental evidence, excluding the possible model shown in Figure 3.42b (Watanabe et al. 1998). In contrast, the structure of Figure 3.42b was identified by Takanishi et al. (1999) in a bent dimer (a bent molecule with two mesogens linked by an alkylene spacer; see Figure 1.7) by x-ray microbeam diffraction from monodomains and a birefringence change by applying an electric field. In the absence of an electric field, birefringence was smaller and only (0 2) XRD peaks were observed. In contrast, under an electric field, birefringence was larger and four additional XRD peaks, (1 1) and (−1 1), emerged. These results are consistent with the model structure of Figure 3.42b; in the absence and the presence of an electric field, polarization (and the molecular bend plane) is parallel and perpendicular to the sample plane, respectively. The same model structure of Figure 3.42b was first confirmed in the bent-core B_1 phase, in which electric-field switching was possible, by Szydlowska et al. (2003) using birefringence measurements. They designated this phase as the B_{1Rev} phase. Moreover, they observed two B_{1Rev} phases, where molecules are non-tilted or tilted from the layer normal direction. They also designated the phase with tilted molecules as $B_{1RevTilted}$. The confirmation of the model structure of Figure 3.42b was also made by Pelz et al. (2003) and Takanishi et al. (2006a) using XRD. Pelz et al. also identified the $B_{1RevTilted}$ phase by XRD. An important difference between B_1 and B_{1Rev} structures exists in their unit cells. In the B_1 phase, the body-centered unit cell is always rectangular, while in the B_{1Rev} phase, the body-centered unit cell could be oblique. Actually, in many cases, the B_{1Rev}^{obl} structure was observed. As far as we are aware of, no one has experimentally confirmed the B_1 structure shown in Figure 3.42a.

Furthermore, the B_{1Rev} phase is further classified into two subgroups based on the lattice types (rectangular and oblique lattices) and electron density modulation (with and without density modulation along the direction parallel to the layer) (Vaupotič et al. 2012, Gorecka et al. 2014). Such a variety of phase structures can be studied using XRD. Figure 3.43 summarizes the structures of these subgroups together with their diffraction spots. The difference with and without density modulation arises in the diffraction spot in the equatorial line (parallel to the layer). Note that the indexing is made based on body-centered unit cells for rectangular lattices and primitive cells for oblique lattices, as illustrated by dotted squares. For most of the bent-core LCs, the B_{1Rev}^{obl} phase was observed. It is easily identified as B_{1Rev}^{rec} and B_{1Rev}^{obl} by the (1 1) and (−1 1) XRD peaks. In the former case, two peaks are overlapped, and in the latter case, two peaks split generally with different intensities. Whether the unit cell is rectangular or oblique is irrelevant to the molecular tilt, so there are 8 (=2^3) different B_{1Rev} phases. In addition, the alternation of polarization (polarization modulation) is possible along two directions, parallel or perpendicular to the smectic layer. Some examples are shown in Figure 3.43, where the alternations parallel and perpendicular to the layer are shown in B_{1Rev} and $B_{1Rev,mod}$ phases, respectively. Such different types of polarization alternations cannot be detected using normal XRD measurements, because electron density modulation does not depend on the polarization direction. In order to differentiate the two types of polarization modulations along two crystallographic directions (parallel and perpendicular to the smectic layer), resonant XRD studies have to be used. There is only one resonant XRD applied to the B_1-type phase (Folcia et al. 2011). According to their results, the polarization modulation

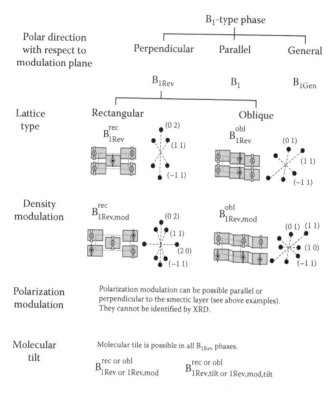

FIGURE 3.43 Summary of the B_1-type phase structures.

exists perpendicular to the smectic layer in the B_1-type phase, as shown by $B_{1Rev,mod}^{obl}$ in Figure 3.43.

Because of such a variety of subphases mentioned in the preceding text, sequential transitions among several B_1-type phases are possible. So far, however, only one B_1-type phase has been observed in most of single compounds. Only a few reports suggested compounds with multiple B_1-type phases based on XRD data showing a layer modulation (Pelzl et al. 1999b, Szydlowska et al. 2003, Zhang et al. 2012, Gorecka et al. 2014). For the full understanding of the structures, detailed XRD measurements are necessary.

The B_{1Rev} phase is normally quite irresponsive to an electric field. In some cases, however, electric-field-induced changes occur. As mentioned in the second paragraph of this section, the first observation was made in a bent dimer by Takanishi et al. (1999). Based on birefringence and XRD measurements in the absence and presence of a DC field, they suggest the orientation change such that the bending plane parallel to the substrate switches to perpendicular as shown in Figure 3.44. The switching occurs dielectrically and the model is consistent to the birefringence change. When the applied field was removed, original orientation reappeared. A similar (but not quite) type of switching was also observed in bent-core mesogens (Szydlowska et al. 2003). By applying higher fields such as 15 V/μm, the texture change was irreversible.

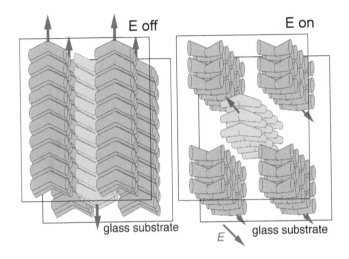

FIGURE 3.44 Orientation change in the B_{1Rev} phase by applying an electric field. XRD peaks due to columnar structures can be observed only under a field. (Redrawn from Takanishi, Y. et al., *J. Mater. Chem.*, 9, 2771, 1999.)

At this stage, switching current was observed without changing the texture in the B_{1Rev} phase. Another type of switching in the B_{1Rev} phase was observed in bent-core molecules by Ortega et al. (2004). By applying a field of 15 V/μm, only slight bire-fringence increase was observed within the B_{1Rev} phase, as same as the switching demonstrated by Takanishi et al. (1999). At 25 V/μm, the B_{1Rev}–B_2 phase transition occurred and the homochiral B_2 phase was established above this field. Upon the removal of the field, the B_{1Rev} structure was recovered. The established B_2 phase is homochiral antiferroelectric phase. Since the tilt angle is 45°, no texture change was observed between two ferroelectric states (homochiral synclinic, SmC_SP_F) with opposite polarizations by applying a square wave field. This tilt angle also gives a dark texture in the ground state (antiferroelectric homochiral anticlinic, SmC_AP_A) during the switching under a 50 Hz triangular field.

The electric-field-induced switching in the B_{1Rev} phase seems to be more compli-cated. Zhang et al. (2008) reported that the electric-field-induced phase transition to B_2 (SmC_SP_F) occurs through the rotation of molecules along their long axis associated with the chirality change of the layer fragments. However, the switching to another SmC_SP_F with the opposite polarization occurs by the rotation of molecules around cones keeping the chirality unchanged. By removing a field, B_2 changes to B_{1Rev} through the former process. These processes are schematically shown in Figure 3.45.

Let us turn to description of another columnar structure consisting of bent-core molecules: hexagonal columnar phase. Each column consists of discs, which are formed by molecular assemblies either of a few polycatenar bent-core molecules (Gorecka et al. 2004, 2006) or of several bent-core molecules with an acute bent angle (Li et al. 2010, Kang et al. 2012).

The first example is a hexagonal columnar stack of cones ("umbrellas") consist-ing of three or four bent-core polycatenar molecules shown in Figure 3.46, where

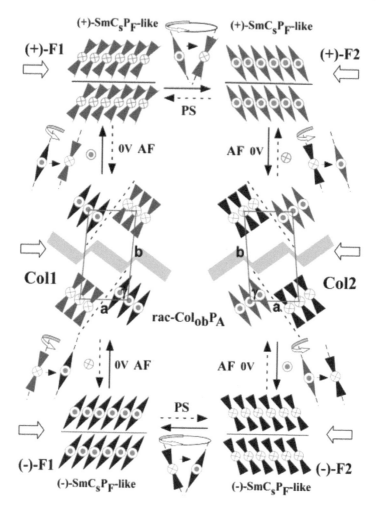

FIGURE 3.45 Electric-field-induced phase transition between SmC_SP_F and B_{1Rev}. (From Zhang, Y., O'Callaghan, M.J., Baumeister, U., and Tschierske, C.: Bent-core mesogens with branched carbosilane termini: Flipping suprastructural chirality without reversing polarity. *Angew. Chem. Int. Ed.* 2008. 47. 6892–6896. Copyright Wiley-VCH Verlag GmbH & Co. KGaA. Reproduced with permission.)

three end chains at both ends ensure the density uniformity within each disc (cone) (Gorecka et al. 2006). The compound shows the Col_h–Col_hP_A phase transition. Based on XRD and birefringence measurements, each disc is flat in the Col_h phase, whereas it is conical in the Col_hP_A phase, and the transition is of the first order (Gorecka et al. 2004). If the disc is flat, polar order cannot be formed. In the Col_hP_A phase, the conical object possesses dipoles, so that polar columns could be constructed, if the cones are closely packed. Gorecka et al. observed a switching current peak in the Col_hP_A phase. One of the most conventional and powerful methods to demonstrate the polar order is the optical SHG, which is a second-order nonlinear optical effect and is active only in

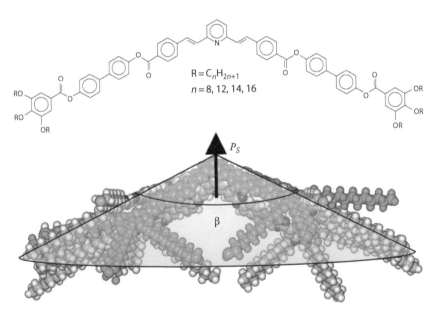

FIGURE 3.46 **(See color insert.)** Polycatenar molecule and its organization to an umbrella assembly. (Reprinted with permission from Gorecka, E., Pociecha, D., Matraszek, J., Mieczkowski, J., Shimbo, Y., Takanishi, Y., and Takezoe, H., *Phys. Rev. E*, 73, 031704-1. Copyright 2006 by the American Physical Society.)

noncentrosymmetric media in the framework of electric dipole transitions (see details in Chapter 6). SHG measurements were conducted in this polycatenar compound. SHG appeared only when an electric field was applied in the vicinity of the Col_h–Col_hP_A phase transition (Gorecka et al. 2006). In the Col_h phase, the dielectric strength increased and the relaxation frequency became slower when approaching the Col_hP_A phase. This is the typical behavior of the soft mode, in the present case, umbrella fluctuation inducing the polar order along the column axis. By applying an electric field, discs are deformed into cones, resulting in the Col_hP_F phase.

Another interesting behavior is in the dielectric response of a long homologue ($n = 16$), which shows a continuous (second-order) Col_h–Col_hP_A phase transition, whereas the transition is of the first order in short homologues ($n = 8, 12$). The dielectric relaxation frequency does not follow a Vogel–Fulcher dependence (Line and Glass 2001), but a superposition of Vogel–Fulcher and Curie–Weiss dependence. This means the energy barrier for dipole reorientation under an electric field depends on temperature. Such a coexistence of the critical slowing down and glassy freezing has often been considered for solid relaxors (Vugmeister and Rabitz 2000). The temperature dependence of the dielectric response measured at different frequencies (Figure 3.47) is characteristic of relaxors, that is, shifts of the maximum position and value of the dielectric susceptibility as a function of frequency.

The detailed SHG measurements including SHG interferometry were first made in a different Col_hP_A phase (Okada et al. 2005). Among various compounds (Gorecka et al. 2004, Kishikawa et al. 2005, Fitie et al. 2010, Miyajima et al. 2011), which

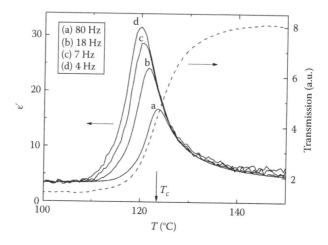

FIGURE 3.47 Temperature dependence of the dielectric response measured at different frequencies. (Reprinted with permission from Gorecka, E., Pociecha, D., Matraszek, J., Mieczkowski, J., Shimbo, Y., Takanishi, Y., and Takezoe, H., *Phys. Rev. E*, 73, 031704-1. Copyright 2006 by the American Physical Society.)

show switchable columnar phase (Takezoe et al. 2006, Araoka and Takezoe 2014), the ferroelectric columnar phase Col_hP_F, which has switchable and remnant polarization, was just recently discovered (Miyajima et al. 2012).

Another interesting columnar phase was reported in a system composed of bent-core V-shaped molecules with an acute bend angle. The details of such bent-core V-shaped mesogens are described in Section 3.8. One such mesogenic molecule (see R3 in Figure 3.60) exhibits a unique Col_h phase, the model structure of which was proposed, as shown in Figure 3.48 (Li et al. 2010, Kang et al. 2012). Li et al. (2010) refuted the structure shown in the previous example (Figure 3.46) (Gorecka et al. 2004) based on XRD data. Electric-field-induced switching was observed, and the structural transformation was proposed. In the absence and presence of the field, bent-direction or bent-plane is perpendicular and parallel to the column axis, respectively (Figure 3.48). Further evidence, however, is needed to confirm the model structure (Gorecka et al. 2016).

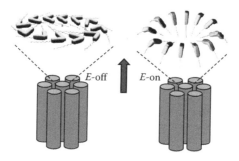

FIGURE 3.48 Characteristic columnar structure formed by bent-core molecules with an acute bend angle in the absence and under an electric field.

3.6 BENT DIMERS

By bent dimers, we designate here dimeric mesogens consisting of two mesogenic units connected via a spacer so that the overall shape of the mesogen is bent. This means that the two mesogens, which are of rod shape, are linked by a flexible chain at both mesogenic tails to form a bent-core molecule. There are other types of dimers, that are, bend–rod dimers and bend–bend dimers with end-to-end, side-to-end, and side-by-side linkages. These types of dimers are described in Section 3.7. Readers may also refer to a review article by Imrie and Henderson (2007) for various types of dimers and oligomers consisting of different units.

Dimeric liquid crystals were first synthesized to expand the variety of chemical structures for low molar mass liquid crystals from simple and conventional rod-shaped molecules (Emsley et al. 1984, Date et al. 1992, Barnes et al. 1993). The molecular structures of selected dimers are shown in Figure 3.49a and b. It is interesting to mention that the studies on bent dimers were started not only to explore their properties compared to conventional low molar mass LCs but also as a model system for main-chain liquid crystalline polymers (Griffin and Britt 1981, Emsley et al. 1984, Date et al. 1992, Watanabe et al. 1993). In these papers, an odd–even effect in the phase structures was observed. Particularly interesting phase behavior in bent-core mesogenic systems is that the SmA and SmC_A phases emerge depending on the number of carbons in the flexible methylene spacer: SmA for even and SmC_A for odd carbons. The SmC_A here is equivalent to the B_6 phase (see Section 3.3), since the periodicity along the layer normal is equal to half the molecular length, that is, interdigitated layer structure (see Figure 1.7). Soon after these works, Niori et al. (1995) made a systematic study of the phase structures by changing both the lengths of the methylene spacer ($n = 3$–9) and the end chains ($m = 2, 4, 8, 12$) of compounds

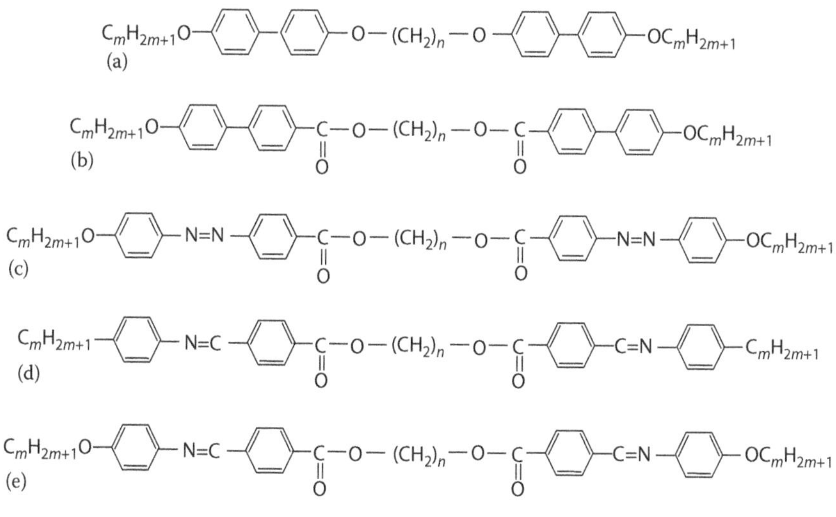

FIGURE 3.49 Bent dimers studied. (a) $mOnOm$, $m = 0$–10, $n = 1$–12; (b) $m = 4$, $n = 4$–9; (c) $m = 2, 4, 8, 12$, $n = 3$–9; (d) 12AM5AM12, $m = 12$, $n = 5$; and (e) $mOAM5AMOm$, $m = 6, 8, 10, 12, 14$, and 16, $n = 5$.

n \ m	2	4	8	12
3		B_6		
4	N	N SmA	N	N
5	B_6	B_6	B_6	B_6
6	N	N SmA	N SmA	N
7		B_6	B_6	B_6
8	N	N SmA	SmA	N
9		B_6	B_6	B_6

FIGURE 3.50 Phases shown by various lengths of end chains (m), a methylene spacer (n) in compounds shown in Figure 3.47c.

bearing azo linkages (Figure 3.49c). As summarized in Figure 3.50, the distinct odd–even behavior is seen; N and SmA appear only for even n and B_6 only for odd n. It is noted that the emergence of the SmA is restricted in compounds with a similar number of n and m. All the compounds with odd n studied in this work seem to be the interdigitated B_6 phase because of the comparison between the layer spacing d and the molecular length L: $d \sim L/2$. The existence of columnar (modulated or frustrated) phases will be discussed later with the compounds shown in Figure 3.49e.

Soon after the discovery of polar switching in bend-core LCs in 1996 (Niori et al. 1996), electro-optic measurements on bent dimers were started. Actually, the antiferroelectric switching was observed in 12AM5AM12 (see Figure 3.49d) (Watanabe et al. 1998). During the switching, the extinction direction did not alter and only the birefringence changed. This indicates that the bend plane does not tilt from the smectic layer normal, that is, the SmAP$_A$ phase. The switching occurs between an antiferroelectric state, where the bend plane and the polarization direction are parallel to substrate surfaces, and two ferroelectric states, where they are perpendicular to the surfaces, as illustrated in Figure 3.51.

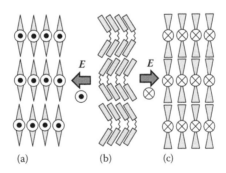

(a)   (b)   (c)

FIGURE 3.51 Electric-field-induced switching between (b) SmAP$_A$ and (a) and (c) SmAP$_F$. (Redrawn from Watanabe et al., 1998.)

One of the interesting aspects in bent dimers is the emergence of various phase structures depending on the combination of the lengths of both end chains and a linkage connecting two mesogenic groups. Experimental results using micro-beam XRD measurements are reported by Takanishi et al. (2006b). They worked on the bent dimers mOAM5AMOm (Figure 3.49e) with five carbons in their spacers, which make the molecules a bent shape, and various end chain lengths ($m = 6$, 8, 10, 12, 14, and 16). An interdigitated structure appears when m is small ($m = 4$ and 6) (Figure 3.52a). This phase has a periodicity of half the molecular length and is identical to the B_6 phase in bent-core mesogens. By increasing the end chain length, the mesogenic groups in neighboring molecules become more separated from each other, destabilizing the phase structure. Molecules favor the formation of a simple non-interdigitated layer structure with a sufficient mesogenic interaction. However, because of the tendency of polarization (along the bent direction)

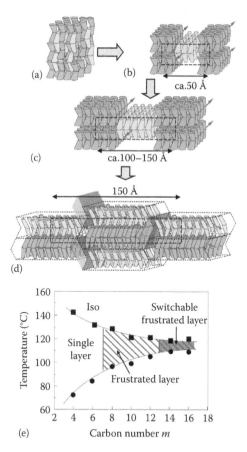

FIGURE 3.52 Phase structures formed by mOAM5AMOm with different m. (a) $m = 6$; (b) $m = 8$; (c) $m = 10$, 12; (d) $m = 14$, 16; (e) mesogenic phase range as a function of carbon number. (Reprinted with permission from Takanishi, Y., Toshimitsu, M., Nakata, M., Takada, N., Izumi, T., Ishikawa, K., Takezoe, H., Watanabe, J., Takahashi, Y., and Iida, A., *Phys. Rev. E*, 74, 051703-1. Copyright 2006 by the American Physical Society.)

splay and of avoidance of macroscopic polarization, molecules form a columnar structure, as shown in Figure 3.52b through d. This phase structure is sometimes called a frustrated or modulated phase and is identical to the B_1-type phase. There are many B_1-type phase structures (see Section 3.5 and Figure 3.43), and the structures in the bent-core dimers mOAM5AMOm are B_{1Rev}^{rec} (m = 8, 10, and 12) and $B_{1Rev,mod,tilt}^{rec}$ (m = 14 and 16) phases, as shown in Figure 3.52. The overall temperature range of these different phases as a function of m is shown in Figure 3.52e. Two-step electric-field-induced structure change was also found, as shown in Figure 3.53. Izumi et al. (2006a) also reported the phase structure of the same homologues. The results are essentially the same as Takanishi et al. (2006b) except for long homologues (m = 14 and 16). Izumi et al. assigned them to the B_2 (SmCP$_A$) phase. Since the layer modulation period (lattice constant a) is quite long, they may not detect small-angle diffractions.

Bent dimers with chemically different spacers and mesogenic units were also reported by other authors (Prasad et al. 2001b, Yelamaggad et al. 2002, Sepelj et al. 2006, 2007, Imrie and Henderson 2007, Bialecka-Florjanczyk et al. 2008). Some of them show the conventional phases such as N, SmA and SmC (Prasad et al. 2001b, Sepelj et al. 2007), Col$_r$ (Yelamaggad et al. 2002, Sepelj et al. 2006, 2007), B$_6$ (Sepelj et al. 2007), and B$_4$ (Bialecka-Florjanczyk et al. 2008) phases in various dimers with spacers of odd numbers of carbons.

The mixtures of homologues mOAM5AMOm with short (m = 4) and long (m = 16) end chains (Izumi et al. 2006b) were also studied. In a mixture

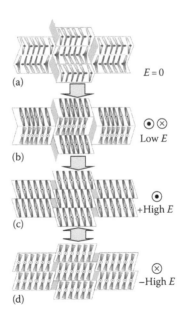

FIGURE 3.53 Electric-field-induced structure change in mOAM5AMOm from (a) zero to (b) low, (c) high, and (d) high opposite fields. (Reprinted with permission from Takanishi, Y., Toshimitsu, M., Nakata, M., Takada, N., Izumi, T., Ishikawa, K., Takezoe, H., Watanabe, J., Takahashi, Y., and Iida, A., *Phys. Rev. E*, 74, 051703-1. Copyright 2006 by the American Physical Society.)

(4OAM5AMO4/16OAM5AMO16 = 65/35), a sequential transition Iso–SmAP$_F$–SmAP$_A$–B$_4$ was confirmed. In the mixtures with high 16OAM5AMO16 content, an electric-field-induced structure change, which is essentially the same as that reported by Takanishi et al. (2006b) in a pure 16OAM5AMO16, was observed.

One of the most important aspects in bent dimers is the emergence of a new nematic phase, sometimes called the twist-bend nematic N$_{TB}$. As will be described in Section 4.6, some bent dimers show the N–N transition, where the upper-temperature N phase is the normal N phase and the lower-temperature N phase is a phase with a local chiral helical order of arbitrary handedness. See Section 4.6 for details.

3.7 PHASES FORMED BY MULTIPLE BENT-MESOGENIC UNITS: DIMER, TRIMER, OLIGOMER, POLYMER, AND DENDRIMER

In the previous section, we described symmetric dimers consisting of rod-shaped mesogens linked by flexible chains containing odd numbers of carbons, which make the dimer bend. There are other types of dimers combining rod- and bent-core units (bent–rod dimer) and two bent-core units (bent–bent dimer) by flexible chains. Moreover, various kinds of chemical structures with more than one bent-mesogenic unit have been designed, as summarized in Figure 3.54.

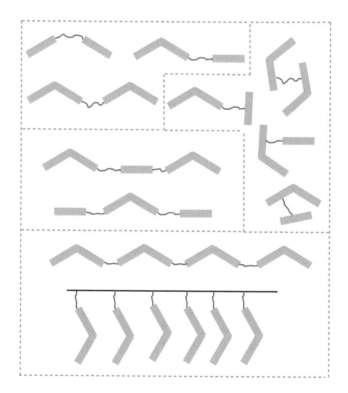

FIGURE 3.54 Structure of different kinds of multiple mesogenic units.

3.7.1 Bend–Rod Dimer

The first example of terminally connected bend–rod dimers was reported by Yelamaggad et al. (2004a). Because of a strong shape biaxiality of bent-core molecules, biaxial phases can be excpected in such a system. Several authors reported the biaxial nematic phase (see Section 4.3), although the biaxiality is possibly attributed to the existence of biaxial cybotactic smectic clusters (Le et al. 2009, Vaupotič et al. 2009b). Yelamaggad et al. also suggested biaxial nematic (N_b) phase followed by the biaxial smectic A (SmA_b) phase based on polarizing microscopy observations: exclusively two-brush disclinations in the textures and well-separated isogyres in conoscopy. Anchoring transition between homeotropic, tilted, and planar orientations by changing temperature was observed in similar compounds (Lee et al. 2010). This is important, since tilted director orientation with respect to the substrate surfaces possibly leads to a misjudgment about the anchoring transition as of a biaxial nature.

More systematic studies of bend–rod dimers were made by Tamba et al. (2006, 2007, 2010, 2011). A wide variety of mesophases were found: conventional N, SmA and SmC, simple and undulated SmCP, and Col_r phases. It is interesting that photochromic azobenzene functionalized bend–rod dimers exhibit an unusual phase sequence, Iso–SmC_AP_A–SmC_SP_F, on cooling (Tamba et al. 2011). Moreover, in the SmC_AP_A phase, a characteristic switching behavior was observed; double switching current peaks consistent with the antiferroelectric phase were observed but an extinction cross parallel to polarizer and analyzer directions remained unchanged upon switching. This indicates the switching of layer chirality (see Section 5.3) through the polarization reversal by the rotation about the molecular long axis. The same kind of switching process was observed in the SmC_SP_F phase. A photo-orientation process by polarized UV and Vis irradiation in a perpendicular direction to the polarization direction was also observed in this compound. Unusual defect walls and the electro-optic characteristics different from those in usual calamitic nematics were also investigated using a bend–rod dimer (Tamba et al. 2007).

Another type of dimer, bend–bend dimer, was first reported by Dantlgraber et al. (2002b). They showed a distinct influence of the number of dimethylsiloxane units used for the spacer on the mesophase structures, that is, $SmCP_F$ for the dimer with three (odd) siloxane units and $SmCP_A$ for that with four (even) units. This is consistent with the simple correlation between mesogenic arms connected to the siloxane spacer; two arms are straight for even numbers of units, whereas they are bent for odd, resulting in antiferroelectric and ferroelectric structures, respectively. A complementary work on the same bent-core structures was also reported later (Achten et al. 2007).

Laterally connected dimers have been studied by Shanker et al. (2012). In their work, bend–bend dimers and bend–rod dimers connected in lateral–lateral or lateral–terminal manners were studied (see Figure 3.54). All compounds except one with the lateral–lateral connection exhibit wide N phases with cybotactic SmC clusters. Bend–rod dimers with short end chains and short spacer units can provide wide N phases down to −20°C. However, no evidence of biaxiality could be found in these N phases. A bend–rod dimer with a lateral–terminal linkage shows the Col_r phase (B_{1rev}) in addition to the N phase.

3.7.2 Dendrimers

The first dendrimer with bent-core mesogenic units was reported in 2002 (Dantlgraber et al. 2002a), in which a tilted polar SmCP phase with random tilt and polarization (SmCP$_R$) was proposed based on a single switching current peak and high dielectric constants. It is noted that the report was published earlier than the report of the SmAP$_R$ phase (Pociecha et al. 2003, Shimbo et al. 2006a). Later, some reports on dendrimers followed (Kardas et al. 2005, Hahn et al. 2007). A cartoon of a dendrimer with terminally attached bent-core units and possible organization of this dendrimer are shown in Figure 3.55 (Kardas et al. 2005). The first- and second-generation diaminobutane (DAB) dendrimers with some different bent-core molecules were synthesized and studied. The most interesting mesomorphic behavior is the two sequential columnar phases, B$_{1Rev}$ and B$_{1RevTilt}$, in a dendrimer with pentafluorophenyl ester derivatives. Third-generation carbosilane dendrimers, which show electro-optic switching, were also synthesized by Hahn et al. (2007) and Keith et al. (2007a). They systematically studied the influence of the shape and the size of silyl units on the mesomorphic properties of bent-core liquid crystals, from dimers via oligomers and dendrimers to polymers. They synthesized a variety of molecules with silicon-containing linking units, as shown in Figure 3.56. All of them exhibit the tilted SmCP phase. The phase structures are strongly influenced by the ratio of silicon atoms to bent-core units; for ratios smaller than 1:1, antiferroelectric switching is observed. Surprisingly, only single polarization peak accompanied the tristable optical switching. This phase was designated as SmCP$_A'$. For ratios larger than 1:1, ferroelectrically switching dark texture emerges. In the most part of the SmCP$_A'$ phase existence range, chiral and achiral dark domains coexist. A randomly tilted phase with a random polar direction (SmC$_R$P$_{FE}$) was also observed in one of the compounds (1/4b in Figure 3.56). Besides the number of silyl groups, the shape of these units also has a significant influence. Linear or branched oligosiloxane or carbosilane units are particularly efficient in inducing optically isotropic mesophases with ferroelectric switching.

Supramolecular chiral architectures from bent-core dendritic molecules were reported by Cano et al. (2014). Being inspired by the work of Lin et al. (2008, 2012),

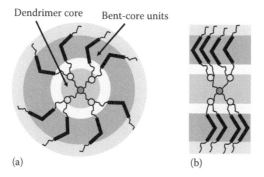

FIGURE 3.55 Structure of dendrimers with terminally attached bent-core units (a) and possible smectic structure (b). (Redrawn from Kardas, D. et al., *J. Mater. Chem.*, 15, 1722, 2005.)

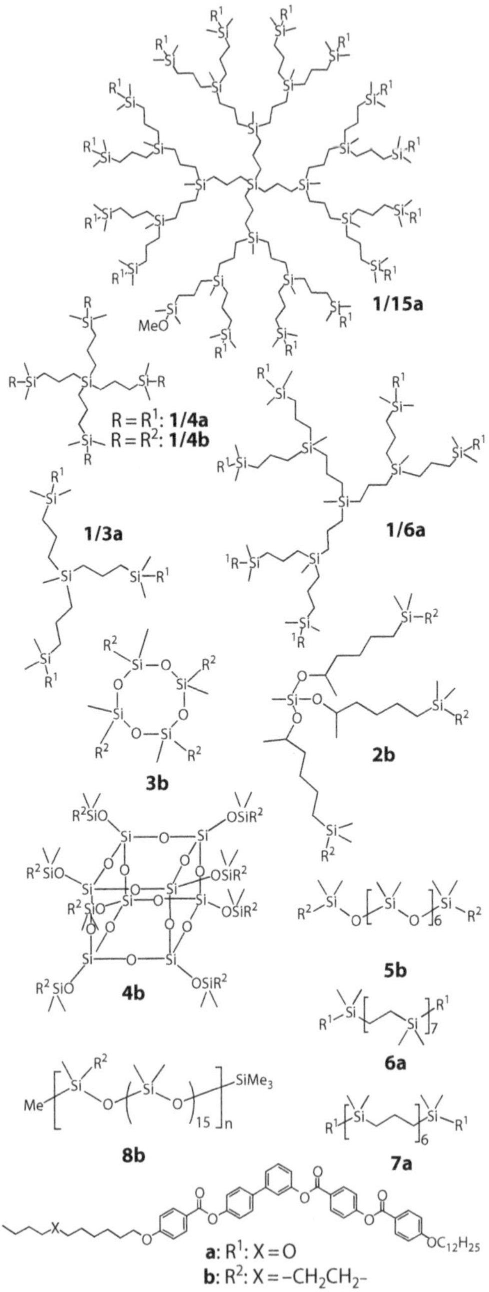

FIGURE 3.56 Molecules with silicon-containing linking units synthesized. (From Keith, C., Reddy, R.A., Baumeister, U., Hahn, H., Lang, H., and Tschierske, C., Continuous transition from antiferroelectric to ferroelectric switching liquid crystalline phases in two homologous series of bent-core mesogenic dimers based on carbosilane spacer units, *J. Mater. Chem.*, 16, 3444–3447, 2006. Reproduced by permission of The Royal Society of Chemistry.)

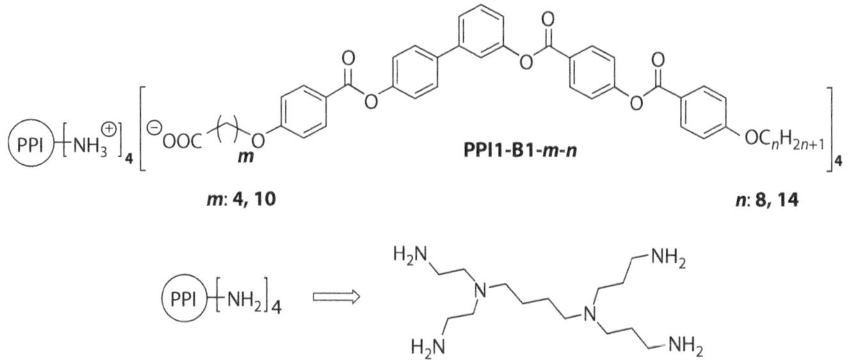

FIGURE 3.57 Compound of ionic first-generation dendrimers bearing achiral or chiral bent-core mesogenic units. (From Cano, M., Sanchez-Ferrer, A., Serrano, J.L., Gimeno, N., and Ros, M.B.: Supramolecular architectures from bent-core dendritic molecules. *Angew. Chem. Int. Ed.* 2014. 53. 13667–13671. Copyright Wiley-VCH Verlag GmbH & Co. KGaA. Reproduced with permission.)

that is, hierarchical helical superstructure made by achiral bent-core molecules in a solvent, they synthesized ionic first-generation dendrimers bearing achiral or chiral bent-core mesogenic units, or calamitic units at four ends (Figure 3.57). The compound with a long terminal chain and a flexible spacer (PPI1-B$_1$-10-14) exhibits the SmCP phase, whereas PPI1-B$_1$-4-8 exhibits the Col$_r$ phase. Different aggregates were observed by SEM and TEM in the ionic dendrimers in solution (tetrahydrofuran [THF] and water), that is, rods, spheres, nontwisted or twisted fibers, helical ribbons, and tubules (empty nanotubes). Depending on the lengths of the spacer (*m*) and the terminal chain (*n*), the aggregates show a variety of morphologies. The compounds with a short spacer (*m* = 4) are particularly interesting, that is, chiral organization formed by achiral units. The structure of the helical strands also depends on the end chain lengths (*n* = 8 and 14). Not only the chiral architectures, but also the gelation ability are different; *n* = 14 is reminiscent of organogels, but *n* = 8 does not show gel formation. It was also confirmed that the use of calamitic units instead of bent-core units does not induce chiral structure. Chirality control by using chiral bent-core mesogens was not successful, in contrast to the success of Lin et al. who used bent-core mesogens in solutions (Lin et al. 2012).

3.7.3 POLYMERS

Both side-chain and main-chain polymer liquid crystals with bent-core mesogenic units have been reported. There are two types of side-chain polymers; bent-core mesogenic units are linked to a polymer main chain at one end or in the middle of the bent-core mesogen. The former was first reported by Keith et al. (2005). The compound has a polysiloxane backbone and exhibits an optically isotropic SmCP$_F$ phase consisting of chiral domains with opposite handedness. A polysiloxane backbone was also used by several groups (Achten et al. 2006, Bubnov et al. 2011).

Block co-polymers with bent-core moieties and polystyrene blocks were synthesized and were shown to give lamellar (SmCP) or columnar (Col or B_1-like) monomorphisms depending on the content of the bent-core block (Tenneti et al. 2009). Side-chain bent-core liquid crystalline polymers have also been obtained by hydrogen-bonding interaction (Barbera et al. 2006, Wang et al. 2010, Yang et al. 2010, Chen et al. 2011). The polymers also show either B_1-like or SmCP phases depending on the structure of bent-core mesogenic units and their number (Barbera et al. 2006), and they show the SmC_G phase (Chen et al. 2011). The second example, in which bent-core units are side-on attached to the polymer backbone, was synthesized (Chen et al. 2006, Xu et al. 2009) and found to show columnar mesophases. In 2014, side-chain homo- and copolymers with bent-core mesogens bearing an azo linkage were synthesized (Gimeno et al. 2014). Although there are many azo-containing bent-core molecules, only a very few molecules were polymerized (Srinivasan et al. 2013). It is interesting to know whether a monomer and its own side-chain polymer exhibit the same phase or not. All the side-chain homo- and copolymers show the SmCP phase, whereas one of the azo-monomers only shows a B_1-like (Col_r) phase, although other monomers exhibit the SmCP phase, the same as in polymers.

Only a few main-chain polymers have been synthesized. The first example was synthesized by Choi et al. (2004). When the conventional Pn-O-PIMB (see Figure 1.8) monomers were polymerized, the $SmCP_A$ phase was obtained. With some substitutions, N or B_1-like phase emerged instead of $SmCP_A$. A polymer consisting of naphthalene central core with partially fluorinated two wings at an acute angle (about 60°) even shows the ferroelectric $SmCP_F$ phase (Choi et al. 2010b). Later, Gimeno et al. (2011) synthesized the first bent-core main-chain polymer with oligosiloxane spacer units and found the DC mesophase over a wide temperature range. This DC phase can be converted into the SmCP phase by applying strong electric fields.

3.8 PHASES FORMED BY UNCONVENTIONAL BENT-CORE MOLECULAR STRUCTURES

A majority of bent-core molecules have their bend angle of about 120°. However, many bent-core molecules with a bend angle of about 60° have also been reported. The first such bent-core V-shaped mesogens were synthesized by Vorländer and Apel (1932). Since then, even before going into an active research phase of bent-core mesogens, bent-core V-shaped molecules were synthesized and characterized (Kuboshita et al. 1991, Matsuzaki and Matsunaga 1993). Most of the studied compounds have a 1,2-phenylene core, as shown in Figure 3.58a (Vorländer and Apel

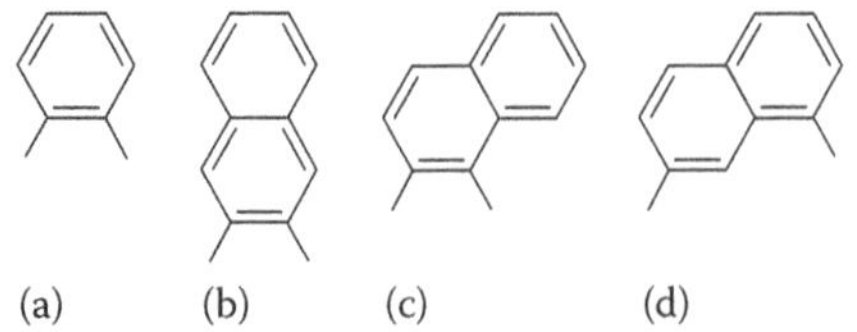

(a)   (b)   (c)   (d)

FIGURE 3.58   (a–d) Central groups for bent-core molecules with an acute bend angle.

1932, Kuboshita et al. 1991, Matsuzaki and Matsunaga 1993, Attard and Douglass 1997, Prasad 2001a, Yoshizawa and Yamaguchi 2002, Yelamaggad et al. 2004b). But later, 1,2- and 2,3-naphthalene core was also used (Figure 3.58b and c) (Matsuzaki and Matsunaga 1993, Lee et al. 2007a, Alonso et al. 2010, Choi et al. 2010a). In most of these compounds, N, SmA, and sometimes highly ordered smectic phases such as SmB appeared. In a homologous series, the N and SmA phases are stabilized in shorter and longer homologues, respectively. In the middle homologues, the two phases coexist. In further homologues, SmA appears together with N, and finally N is destabilized and only SmA remains (Kuboshita et al. 1991, Matsuzaki and Matsunaga 1993, Yelamaggad et al. 2004b). One of the important and naïve questions is how bent-core V-shaped molecules are accommodated in the mesophase structure; whether the bending direction (symmetry axis) is parallel or perpendicular to the director (layer normal in case of SmA). This problem was discussed in several papers (Yelamaggad et al. 2004b, Lee et al. 2009, Alonso et al. 2010, Choi et al. 2010a). Based on the comparison between the layer spacing and calculated molecular configuration, Yalamaggad et al. suggested an interdigitation of the molecules between neighboring layers, partially bilayered structure, where the bending direction is perpendicular to the layer (Figure 3.59b). A similar model was also shown by Attard and Douglass (1997) and Alonso et al. (2010) based on XRD measurements,

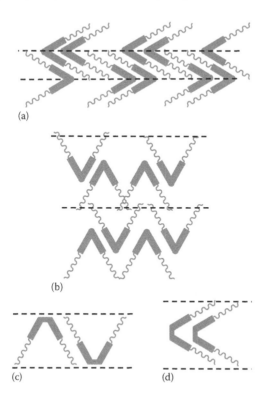

FIGURE 3.59 (a-d) Possible smectic layer formation by bent-core mesogens with an acute bent angle.

although the former and the latter suggested bilayered and single-layered structures, respectively, contrary to a partially interdigitated layer structure by Yalamaggad et al. Birefringence and optical second-harmonic generation as a function of electric field also support this model (Figure 3.59c) (Alonso et al. 2010).

In contrast, Lee et al. (2007a, 2009) used a compound with a 1,7-naphthylene core (Figure 3.58d), which shows a variety of phases (see later), although compounds with a 1,2- or 2,3-naphthylene core show only the SmA phase (Lee et al. 2007a). According to XRD studies, the inner and outer reflections were invariably observed on the meridian and equator, respectively, in the N, SmA, and $SmAP_A$ phases. This means that the molecular orientation does not essentially change in these phases. They observed completely dark texture in the N and SmA phases for homeotropically aligned samples and the texture changed to be birefringent in the $SmAP_A$ phase. This observation clearly indicates that the molecules are packed with their bent direction parallel to the layer (Figure 3.59d) but not perpendicular to the layer (Figure 3.59c). This also means that, even in such an arrangement of bent-core V-shaped molecules, the rotation about the molecular long axis is not hindered and there is no sign of biaxiality in N and SmA at least in an optical scale. Another important observation supporting this model is the dependence of the layer spacing d on the number of carbons of the alkyl tail n. The increasing rate $\Delta d/\Delta n$ of the layer spacing differs by factor two for the two models. The experimentally obtained value, about 1.4 Å/carbon (Lee et al. 2009), is consistent with the model shown in Figure 3.59d. The similar result was also obtained by Matsuzaki and Matsunaga (1993). The electro-optic switching (electric-field-induced antiferroelectric-to-ferroelectric transition) in the $SmAP_A$ phase is also consistent with the model. The molecular arrangement shown in Figure 3.59d may be a consequence of the relatively long distance between the linking positions of two side wings. Different molecular packing structures in the smectic layer, that is, bend direction is perpendicular or parallel to the layer, may be attributed to different core structures, 1,2- or 2,3-phenylene and 1,7-naphthylene cores. It is also important to note that Choi et al. (2010a) suggested the polar SmA model similar to Figure 3.59d using a compound with a 2,3-naphthylene core. The conclusion was made based on birefringence, switching current response, IR dichroism, and SHG measurements.

While bent-core V-shaped molecules with a 1,2- or 2,3-phenylene core show only conventional N and SmA phases, those with a 1,7-naphthylene core exhibit some additional phases characteristic to bent-core mesogens, that is, the B_4 phase (Lee et al. 2009) and some unique phases (Col_h and Cub) (Li et al. 2010, Kang et al. 2012) depending on the chemical structures of the arms (Figure 3.60). For R1 (Lee et al. 2007a), only the B_4 phase appears and shows characteristic features such as low birefringence and spontaneous chiral segregation. For R2 (Lee et al. 2009), $SmAP_A$ and B_4 emerge in addition to N and SmA for longer homologues. R3 (Li et al. 2010, Kang et al. 2012) induces more interesting phases, Cub and Col_h phases. The Cub phase appears for longer homologues and is attributed to $Pm3n$. For the Col_h phase, Li et al. (2010) proposed a very unique columnar structure formed by bent-core V-shaped molecules. The details are described in Section 3.5. The same group also introduced alkylthio terminal tails (R4 in Figure 3.60) in asymmetric bent-core molecules with an acute bend angle. These molecules with $n = 16$ and 18 exhibit a phase sequence SmA-Col_h-B_7 below Iso (Lee et al. 2015a). This phase sequence is

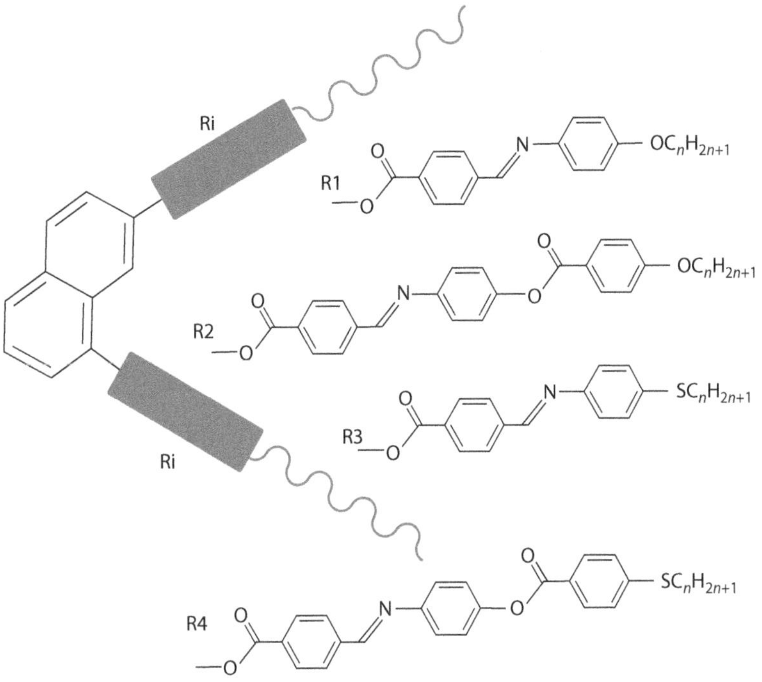

FIGURE 3.60 Chemical structures of V-shaped molecules with a 1,7-naphthylene core.

important, because Col_h might have a layer structure in between the phases with flat (SmA) and undulated (B_7) layers, although the Col_h phase usually lacks the layer order. The authors claim that this phase sequence supports the characteristic layering in the Col_h phase shown in Figure 3.48.

An interesting phase sequence, Iso–B_7–B_2, was reported in a molecule with a core shown in Figure 3.58a (Lee et al. 2014). An ortho-bistolane central core offers a 60° bend angle without conformational freedom. Observed ferroelectric (B_7) and antiferroelectric (B_2) switching behaviors ensure the molecular packing shown in Figure 3.59d.

Another type of unconventional molecular structure in bent-core molecules is those with highly asymmetric arms, sometimes called as hockey-stick-shaped bent-core molecules. The most important one is five-ring bent-core molecules with one arm terminally substituted with a highly polar cyano group (Sadashiva et al. 2002). As described in Section 3.2, this molecule shows the first SmA_dP_A phase. A similar molecule with a carbosilane terminal group is known to exhibit the SmA_dP_F phase (Guo et al. 2011a, Reddy et al. 2011). A molecule with a carbosilane terminal group and a CF_3 group at the end of the other arm instead of CN shows SmA_dP_{Fmod} in addition to SmA_dP_F (Zhu et al. 2012). The details of these SmAP phases are described in Section 3.2.

A different type of asymmetry arises in a four-ring system (Deb et al. 2010, Chakraborty et al. 2013). The phases emerged are strongly chemical structure dependent, the $B_{1RevTilt}$ or B_7 phases with spontaneous polar and chiral characteristics (Deb et al. 2010) and N with SmC-cybotactic clusters (Chakraborty et al. 2013).

Molecular asymmetry is also introduced by using both a four-ring core and halogen termination at one arm (Ghosh et al. 2014). The shape of these molecules is not hockey-stick type, because one arm with a longer core is terminated with a halogen atom and another arm with a shorter core has an alkyl chain. A wide nematic temperature range and ferroelectric-like switching were reported. However, same as the report on ferroelectricity in 1,2,4-oxadiazole derivative (Francescangeli et al. 2009), the contribution of smectic cybotactic clusters to the ferroelectric switching cannot be denied. Actually, the oxadiazole derivative was confirmed to be SHG inactive in the absence of an electric field (unpublished data from Takezoe Lab).

There are only a few examples of W-shaped molecules (Figure 3.61a). One of them synthesized by Rao et al. (2003) is extremely interesting. Miyake et al. (2005) carried out extensive measurements, including polarizing microscopy, circular dichroism, confocal fluorescent microscopy, and XRD using an x-ray microbeam. Upon cooling a cell, fine filaments grow (Figure 3.61b) and cover the entire viewing area, showing *Schlieren*-like texture with only two-brush defects (Figure 3.61c). By decrossing the polarizers, two opposite chiral domains were observed (Figure 3.61d), which were also confirmed by mirror images of the circular dichroism spectra in these two chiral domains. Polarized fluorescence signal periodically changed and the bright and dark regions interchanged under rotation of the input polarization by 90°. This suggests that the molecular orientation is symmetrical with respect to the stripe line and anticlinically tilted with the periodicity of half of the filament width. Other important structural information was obtained by the microbeam XRD; the layer is parallel to the filament axis and forms confocal cylinders. Detailed texture consideration including birefringence and optical activity together with a Muller matrix method lead to a model of molecular arrangement in the chiral filament, as shown in Figure 3.61e. This is a very unique structure, but further confirmation is necessary.

3.9 FIELD-INDUCED PHASE TRANSITION

In calamitic molecular systems, external-field-induced phase transitions rarely occur except for thermal effect and the antiferroelectric–ferroelectric phase transition. In bent-core molecular systems, however, there are many reports on the field-induced phase transitions. Some examples including chirality changes will be described in Section 5.3: electric-field-assisted control of four states in the B_2 phase (Jákli et al. 1998, 2003, Heppke et al. 1999, Zennyoji et al. 2000, Ortega et al. 2003, Pelzl et al. 2006), electric-field-induced DC–B_2 phase transition (Findeisen-Tandel et al. 2008, Nagaraj et al. 2014b), and electric-field-controllable different DC phases (see Section 5.3.2 for detail) (Deepa and Pratibha 2014).

One of the remarkable effects of an electric field on phase transitions is a field-induced Iso–SmCP transition (Weissflog et al. 2003). They observed field-induced nucleation of the SmCP phase above the clearing temperature. This behavior is very surprising, because the application of a high electric field usually depresses the clearing temperature as a result of the electric heating. The temperature, below which the nucleation is possible, increases almost linearly with increasing voltage applied and

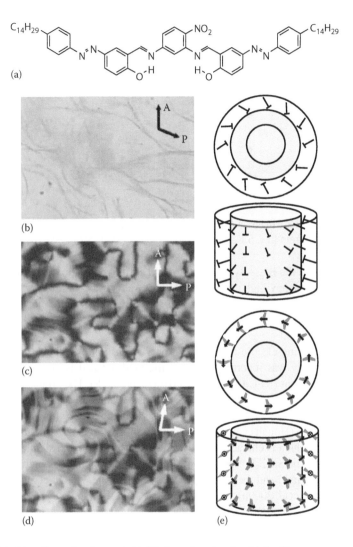

FIGURE 3.61 **(See color insert.)** (a) W-shaped molecule; (b) its filament texture growth; (c) characteristic texture under crossed polarizers; (d) under decrossed polarizers, two opposite chiral domains were observed; and (e) the phase structure of cylindrical architecture. (From Miyake, I., Takanishi, Y., Rao, N.V.S., Paul, M.K., Ishikawa, K., and Takezoe, H., Novel chiral filament in an achiral W-shaped liquid crystalline compound, *J. Mater. Chem.*, 15, 4688–4694, 2005. Reproduced by permission of The Royal Society of Chemistry.)

attains 9 K above the clearing temperature. Authors attributed this phenomenon to existing polar clusters in the Iso phase and succeeded in explaining the large temperature rise using a Helfrich model (Helfrich 1970) including polar interaction instead of the dielectric one.

There are a few reports on the field-induced B_1–B_2 phase transition (Ortega et al. 2004, Kirchhoff et al. 2007). Usually, the B_1 phase shows no electro-optic effect.

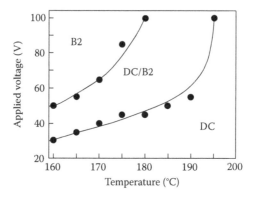

FIGURE 3.62 A phase diagram (*E*/temperature) showing field-induced changes between B_2 and DC. (Redrawn from Lee, S.K. et al., *J. Phys. Chem. B*, 111, 8698, 2007.)

Ortega et al. found the B_{1rev}–B_2 (homochiral and antiferroelectric) phase transition by applying low-frequency high enough (25 V/µm) fields in two bent-core compounds. Upon removal of the field, the B_1 structure is recovered. Later, Kirchhoff et al. (2007) reported that the B_1–B_2 phase transition can occur at much lower field strengths in different compounds.

Another interesting electric-field-induced phase transition is a DC–B_2 phase transition. The first study was reported as an electric-field-induced B_7–B_2 phase transition (Nakata et al. 2005) in (R)-Citronellyl-OPIMB (Figure 3.28). The ground state appears to be the B_7 phase (Coleman et al. 2003), which is concluded by characteristic texture and XRD patterns having multiple satellite peaks. Sometimes, very low birefringent DC textures can be seen (Hough et al. 2009b). Rapid cooling from Iso leads to the metastable $SmC_SP_F^*$ phase, which is characterized by a high birefringence and simple XRD peaks due to flat layers. The $SmC_SP_F^*$ phase is also induced by applying an electric field larger than the threshold field and thermally relaxes to the B_7 state. Hough et al. described in their paper (Hough et al. 2009b) that all three compounds used for DC phase studies were transformed to the chiral B_2 phase under the application of a sufficiently large electric field. The local chirality is maintained in this type of transition (Lee et al. 2007b). A typical temperature/field diagram is shown in Figure 3.62 (Lee et al. 2007b).

3.10 OPEN QUESTIONS AND PROSPECT

As described in this chapter, most of the phase structures, characteristic in bent-core molecular systems, have been clarified. However, there still exist open questions concerning the phase structures: (1) Is the B_7 phase actually a separate phase from the B_{1rev} phase? Both have periodic structures along the layer, and electron density map obtained by XRD in B_7 phase (the splay modulated Boulder model (Coleman et al. 2003)) can be replaced by the B_{1rev} phase structure. (2) Is DC phase actually always different from the B_4 phase? This question arises, since both phases are apparently very similar (low birefringence) and show distinct segregation into chiral domains with high optical activity. In both phases, the layers are associated with saddle-splay curvatures. The difference exists, however, in the fluidity; the B_4 phase is more or less crystalline, while the DC phase usually shows fluidity. The difference and simi-larity are well described in a recent minireview (Le et al. 2016). (3) The SmAP$_{AR}$ phase was reported only in a few compounds. Does this really exist, although the physical properties are distinctly different from the SmAP$_R$ phase? Can any other model be proposed? (4) Uncertainty still exists to identify phases showed by bent-core mesogens with acute angles. Further experiments are necessary.

REFERENCES

Achten, R., A. Koudijs, M. Giesbers, A. T. M. Marcelis, E. J. R. Sudholter, M. W. Schroedder, and W. Wiessflog. Liquid crystalline dimers with bent-core mesogenic units. *Liq. Cryst.* 34 (2007): 59–64.

Achten, R., A. Kousijs, M. Giesbers, R. A. Reddy, T. Verhulst, C. Tschierske, A. T. M. Marcelis, and E. J. R. Sudholter. Banana-shaped side chain liquid crystalline siloxanes. *Liq. Cryst.* 33 (2006): 681–688.

Alonso, I., J. Martinez-Perdiguero, J. Ortega, C. L. Folcia, J. Etxebarria, N. Gimeno, and M. B. Ros. The SmA phase of a bent-core V-shaped compound: Structure and electric-field response. *Liq. Cryst.* 37 (2010): 1465–1470.

Araoka, F., T. Otani, K. Ishikawa, and H. Takezoe. Spectral blue shift via intermolecular interactions in the B2 and B4 phases of a bent-core molecule. *Phys. Rev. E* 82 (2010): 041708-1–041708-5.

Araoka, F., G. Sugiyama, K. Ishikawa, and H. Takezoe. Highly ordered helical nanofilament assembly aligned by a nematic director field. *Adv. Funct. Mater.* 23 (2013): 2701–2707.

Araoka, F. and H. Takezoe. Columnar liquid crystal as a unique ferroelectric liquid crystal. *Jpn. J. Appl. Phys.* 53 (2014): 01AA01-1–01AA01-6.

Araoka, F., N. Y. Ha, Y. Kinoshita, B. Park, J. W. Wu, and H. Takezoe. Twist-grain-boundary structure in the B4 phase of a bent-core molecular system identified by second harmonic generaton circular dichroism measurement. *Phys. Rev. Lett.* 94 (2005): 137801-1–137801-4.

Attard, G. S. and A. G. Douglass. U-shaped dimeric liquid crystals derived from phthalic acid. *Liq. Cryst.* 22 (1997): 349–358.

Barbera, J., N. Gimeno, I. Pintre, M. B. Ros, and J. L. Serrano. Self-assembled bent-core side-chain liquid crystalline polymers. *Chem. Commun.* (2006): 1212–1214.

Barnes, P. J., A. G. Douglass, S. K. Heeks, and G. R. Luckhurst. An enhanced odd-even effect of liquid crystal dimers orientational order in the α,ω-bis(4′cyanobiphenyl-4-yl)alkanes. *Liq. Cryst.* 13 (1993): 603–613.

Barois, P., H. Gleeson, C. C. Huang, and R. Pindak. Application of X-ray resonant diffraction to structural studies of liquid crystals. *Eur. Phys. J. Special Topics* 208 (2012): 333–350.

Barois, P., J. Prost, and T. C. Lubensky. New critical points in frustrated smectics. *J. Phys. (France)* 46 (1985): 391–399.

Bialecka-Florjanczyk, E., I. Sleddzinska, E. Gorecka, and J. Przedmojski. Odd-even effect in biphenyl-based symmetrical dimers with methylene spacer-evidence of the B4 phase. *Liq. Cryst.* 35 (2008): 401–406.

Brand, H. R., P. E. Cladis, and H. Pleiner. Symmetry and defects in the C$_M$ phase of polymeric liquid crystals. *Macromolecules* 25 (1992): 7223–7226.

Bubnov, A., V. Novotna, D. Pociecha, M. Kaapar, V. Hamplova, G. Galli, and M. Glogarova. A liquid-crystalline co-polysiloxane with asymmetric bent side chains. *Macromol. Chem. Phys.* 212 (2011): 191–197.

Cady, A., R. Pindak, W. Caliebe, P. Barois, W. Weissflog, H. T. Nguyen, and C. C. Huang. Resonant x-ray scattering studies of the B2 phase formed by bent-core molecules. *Liq. Cryst.* 29 (2002): 1101–1104.

Cano, M., A. Sanchez-Ferrer, J. L. Serrano, N. Gimeno, and M. B. Ros. Supramolecular architectures from bent-core dendritic molecules. *Angew. Chem. Int. Ed.* 53 (2014): 13667–13671.

Chakraborty, L., N. Chakraborty, D. D. Sarkar, N. V. S. Rao, S. Aya, K. V. Le, F. Araoka et al. Unusual temperature dependence of smectic layer structure associated with the nematic-smectic C phase transition in a hockey-stick-shaped four-ring compound. *J. Mater. Chem. C* 1 (2013): 1562–1566.

Chattham, N., E. Korblova, R. Shao, D. M. Walba, J. E. Maclennan, and N. A. Clark. Triclinic fluid order. *Phys. Rev. Lett.* 104 (2010): 067801-1–067801-4.

Chattham, N., M.-G. Tamba, R. Stannarius, E. Westphal, H. Gallardo, M. Prehm, C. Tschierske, H. Takezoe, and A. Eremin. A leaning-type polar smectic-C phase in a freely suspended bent-core liquid crystal film. *Phys. Rev. E* 91 (2015): 030502(R)-1–030502(R)-5.

Chen, D., M.-S. Heberling, M. Nakata, L. E. Hough, J. E. Maclennan, M. A. Glaser, E. Korblova, D. M. Walba, J. Watanabe, and N. A. Clark. Structure of the B4 liquid crystal phase near a glass surface. *Chem. Phys. Chem.* 13 (2012): 155–159.

Chen, D., J. E. Maclennan, R. Shao, D. K. Yoon, H. Wang, E. Korblova, D. M. Walba, M. A. Glaser, and N. A. Clark. Chirality-preserving growth of helical filaments in the B4 phase of bent-core liquid crystals. *J. Am. Chem. Soc.* 133 (2011a): 12656–12663.

Chen, D., Y. Shen, J. Aguero, E. Korblova, D. M. Walba, N. Kapernaum, F. Giesselmann, J. Watanabe, J. E. Maclennan, M. A. Glaser, and N. A. Clark. Chiral isotropic sponge phase of hexatic smectic layers of achiral molecules. *ChemPhysChem* 15 (2014): 1502–1507.

Chen, D., Y. Shen, C. Zhu, L. E. Hough, N. Gimeno, M. A. Glaser, J. E. Maclennan, M. B. Ros, and N. A. Clark. Interface structure of the dark conglomerate liquid crystal phase. *Soft Matter* 7 (2011b): 1879–1883.

Chen, D., C. Zhu, R. K. Shoemaker, E. Korblova, D. M. Walba, M. A. Glaser, J. E. Maclennan, and N. A. Clark. Pretransitional orientational ordering of a calamitic liquid crystal by helical nanofilaments of a bent-core mesogen. *Langmuir* 26 (2010): 15541–15545.

Chen, W.-H., W.-T. Chuang, U.-S. Jeng, H.-S. Sheu, and H.-C. Lin. New SmCG phases in a hydrogen-bonded bent-core liquid crystal featuring a branched siloxane terminal group. *J. Am. Chem. Soc.* 133 (2011c): 15674–15685.

Chen, X., K. K. Tenneti, C. Y. Li, Y. Bai R. Shoou, X. Wan, X. Fan, and Q. F. Zhou. Design, synthesis, and characterization of bent-core mesogen-Jacketed liquid crystalline polymers. *Macromolecules* 39 (2006): 517–527.

Choi, E. J., J. C. Ahn, L. C. Chien, C. K. Lee, W. C. Zin, D. C. Kim, and S. T. Shen. Main chain polymers containing banana-shaped mesogens: Synthesis and mesomorphic properties. *Macromolecules* 37 (2004): 71–78.

Choi, E. J., X. Cui, C.-W. Ohk, W.-C. Zin, J.-H. Lee, T.-K. Lim, and W.-G. Jang. Smectic A phase in a new bent-core mesogen based on a 2,3-naphthalene central core with an acute-subtended angle. *J. Mater. Chem.* 20 (2010a): 3743–3749.

Choi, E. J., D. C. Kim, C. W. Ohk, W. C. Zin, J. H. Lee, and T. K. Lim. Synthesis and meso-morphic properties of main-chain polymers containing V-shaped bent-core mesogens with acute-subtended angle. *Macromolecules* 43 (2010b): 2865–2872.

Choi, S.-W., T. Izumi, Y. Hoshino, Y. Takanishi, K. Ishikawa, J. Watanabe, and H. Takezoe. Circular-polarization-induced enantiomeric excess in liquid crystals of an achiral, bent-core mesogen. *Angew. Chem. Int. Ed.* 45 (2006): 1382–1385.

Choi, S.-W., Y. Kinoshita, B. Park, H. Takezoe, T. Niori, and J. Watanabe. Second-harmonic generation in achiral bent-core liquid crystals. *Jpn. J. Appl. Phys.* 37 (1998): 3408–3411.

Coleman, D. A., J. Fernsler, N. Chattham, M. Nakata, Y. Takanishi, E. Korblova, D. R. Link et al. Polarization-modulated smectic liquid crystal phases. *Science* 301 (2003): 1204–1211.

Coleman, D. A., C. D. Jones, M. Nakata, and N. A. Clark. Polarization splay as the origin of modulation in the B1 and B7 smectic phases of bent-core molecules. *Phys. Rev. E* 77 (2008): 021703-1–021703-6.

Dantlgraber, G., U. Baumeister, S. Diele, H. Kresse, B. Luhmann, H. Lang, and C. Tschierske. Evidence for a new ferroelectric switching liquid crystalline phase formed by a carbosilane based dendrimer with banana-shaped mesogenic units. *J. Am. Chem. Soc.* 124 (2002a): 14852–14853.

Dantlgraber, G., S. Diele, and C. Tschierske. The first liquid crystalline dimers consisting of two banana-shaped mesogenic units: A new way for switching between ferroelectricity and antiferroelectricity with bent-core molecules. *Chem. Commun.* (2002b): 2768–2769.

Dantlgraber, G., A. Eremin, S. Diele, A. Hauser, H. Kresse, G. Pelzl, and C. Tschierske. Chirality and macroscopic polar order in a ferroelectric smectic liquid-crystalline phase formed by achiral polyphilic bent-core molecules. *Angew. Chem. Int. Ed.* 41 (2002c): 2408–2412.

Date, R. W., C. T. Imrie, G. R. Luckhurst, and J. M. Sedddon. Smectogenic dimeric liquid crystals. The preparation and properties of the α,ω-bis(4-n-alkylanilinebenzylidine-4′-oxy)alkanes. *Liq. Cryst.* 12 (1992): 203–238.

Deb, R., R. K. Nath, M. K. Paul, N. V. S. Rao, F. Tuluri, Y. Shen, R. Shao et al. Four-ring achiral unsymmetrical bent core molecules forming strongly fluorescent smectic liquid crystals with spontaneous polar and chiral ordered B7 and B1 phases. *J. Mater. Chem.* 20 (2010): 7332–7336.

Deepa, G. B. and R. Pratibha. Chiral symmetry breaking dictated by electric-field-driven shape transitions of nucleating conglomerate domains in a bent-core liquid crystal. *Phys. Rev. E* 89 (2014): 042504-1–042504-9.

Deepa, G. B., S. Radhika, B. K. Sadashiva, and R. Pratibha. Electric-field-induced switchable dark conglomerate phases in a bent-core liquid crystal exhibiting reverse columnar phases. *Phys. Rev. E* 87 (2013): 062508-1–062508-10.

de Gennes, P. G. and J. Prost. *Physics of Liquid Crystals*. Oxford, U.K.: Clarendon Press, 1993.

DiDonna, B. and R. Kamien. Smectic phases with cubic symmetry: The splay analog of the blue phase. *Phys. Rev. Lett.* 89 (2002): 215504-1–215504-4.

Dunemann, U., M. W. Schroder, R. A. Reddy, G. Pelzl, S. Diele, and W. Weissflog. The influence of lateral substituents on the mesophase behaviour of banana-shaped mesogens. Part II. *J. Mater. Chem.* 15 (2005): 4051–4061.

Earl, D. J., M. A. Osipov, H. Takezoe, Y. Takanishi, and M. R. Wilson. Induced and spontaneous deracemization in bent-core liquid crystal phases and in other phases doped with bent-core molecules. *Phys. Rev. E* 71 (2005): 021706-1–021706-11.

Emsley, J. W., G. R. Luckhurst, G. N. Shilstone, and I. Sage. The preparation and properties of the α,ω-bis(4,4′-cyanobiphenyloxy)alkanes: Nematogenic molecules with a flexible core. *Mol. Cryst. Liq. Cryst.*, 102 (1984): 223–233.

Eremin, A., S. Diele, G. Pelzl, H. Nadashi, W. Weissflog, J. Salfetnikova, and H. Kresse. Experimental evidence for an achiral orthogonal biaxial smectic phase without in-plane order exhibiting antiferroelectric switching behaviour. *Phys. Rev. E* 64 (2001): 051707-1–051707-6.

Eremin, A., S. Diele, G. Pelzl, and W. Weissflog. Field-induced switching between states of opposite chirality in a liquid-crystalline phase. *Phys. Rev. E* 67 (2003): 020702(R)-1–020702(R)-3.

Eremin, A., M. Floegel, U. Kornek, S. Stern, and R. Stannarius. Transition between paraelectric and ferroelectric phases of bent-core smectic liquid crystals in the bulk and in thin freely suspended films. *Phys. Rev. E* 86 (2012): 051701-1–051701-10.

Eremin, A. and A. Jákli. Polar bent-shape liquid crystals—From molecular bend to layer splay and chirality. *Soft Matter* 9 (2013): 615–637.

Eremin, A., H. Nadashi, G. Pelzl, S. Diele, H. Kresse, W. Weissflog, and S. Grande. Paraelecctric-antiferroelectric transitions in the bent-core liquid-crystalline materials. *Phys. Chem. Chem. Phys.* 6 (2004): 1290–1298.

Eremin, A., S. Stern, and R. Stannarius. Electrically induced tilt in achiral bent-core liquid crystals. *Phys. Rev. Lett.* 101 (2008): 247802-1–247802-4.

Fernandes, P., P. Barois, S. T. Wang, Z. Q. Liu, B. K. McCoy, C. C. Huang, R. Pindak, W. Caliebe, and H. T. Nguyen. Polarization studies of resonant forbidden reflections in liquid crystals. *Phys. Rev. Lett.* 99 (2007): 227801-1–227801-4.

Findeisen-Tandel, S., M. W. Schroder, G. Pelzl, U. Baumeister, W. Weissflog, S. Stern, A. Nemeş, R. Stannarius, and A. Eremin. Multistage polar switching in bent-core mesogens. *Eur. Phys. J. E* 25 (2008): 395–402.

Fitie, C. F. F., W. S. C. Roelofs, M. Kemerink, and R. P. Sijbesma. Remnant polarization in thin films from a columnar liquid crystal. *J. Am. Chem. Soc.* 132 (2010): 6892–6893.

Folcia, C. L., J. Ortega, J. Etxebarria, L.-D. Pan, S. Wang, C. C. Huang, V. Ponsinet, P. Barois, R. Pindak, and N. Gimeno. Polarization periodicity in the B1 columnar phase determined by resonant x-ray scattering. *Phys. Rev. E* 84 (2011): 010701(R)-1–010701(R)-5.

Francescangeli, O., V. Stanic, S. I. Torgova, A. Strigazzi, N. Scaramuzza, C. Ferrero, I. P. Dolbnya et al. Ferroelectric response and induced biaxiality in the nematic phase of a bent-core mesogen. *Adv. Funct. Mater.* 19 (2009): 2592–2600.

Garoff, S. and R. B. Meyer. Electroclinic effect at AC phase-change in a chiral smectic liquid-crystal. *Phys. Rev. Lett.* 38 (1977): 848–851.

Ghosh, S., N. Begum, S. Turlapati, S. K. Roy, A. K. Das, and N. V. S. Rao. Ferroelectric-like switching in the nematic phase of four-ring bent-core liquid crystals. *J. Mater. Chem.* 2 (2014): 425–431.

Gimeno, N., I. Pintre, M. Martinez-Abadia, J. L. Serráno, and M. B. Ros. Bent-core liquid crystal phases promoted by azo-containing molecules: From monomers to side-chain polymers. *RSC Adv.* 4 (2014): 19694–19702.

Gimeno, N., A. Sanchez-Ferrer, N. Sebastián, R. Mezzenga, and M. B. Ros. Bent-core based main-chain polymers showing the dark conglomerate liquid crystal phase. *Macromolecules* 44 (2011): 9586–9594.

Gomola, K., L. Guo, D. Pociecha, F. Araoka, K. Ishikawa, and H. Takezoe. An optically uniaxial antiferroelectric smectic phase in asymmetrical bent-core compounds containing a 3-aminophenol central unit. *J. Mater. Chem.* 20 (2010): 7944–7952.

Gorecka, E., D. Pociecha, F. Araoka, D. R. Link, M. Nakata, J. Thisayukta, Y. Takanishi, K. Ishikawa, J. Watanabe, and H. Takezoe. Ferroelectric phases in a chiral bent-core smectic liquid crystal: Dielectric and optical second-harmonic generation measurements. *Phys. Rev. E* 62 (2000): R4524–R4527.

Gorecka, E., D. Pociecha, J. Matraszek, J. Mieczkowski, Y. Shimbo, Y. Takanishi, and H. Takezoe. Polar order in columnar phase made of polycatenar bent-core molecules. *Phys. Rev. E* 73 (2006): 031704-1–031704-5.

Gorecka, E., D. Pociecha, J. Mieczkowski, J. Matraszek, D. Guillon, and B. Donnio. Axially polar columnar phase made of polycatenar bent-core molecules *J. Am. Chem. Soc.* 126 (2004): 15946–15947.

Gorecka, E., D. Pociecha, and N. Vaupotič. Chap. 15, Columnar liquid crystalline phases made of bent-core mesogens. In: J. W. Goodby, P. J. Collings, T. Kato, C. Tschierske, H. F. Gleeson, and P. Raynes, eds., *Handbook of Liquid Crystals*, Vol. 4. Weinheim, Germany: Wiley-VCH Verlag & Co. KGaA, 2014, pp. 743–768.

Gorecka, E., N. Vaupotič, A. Zep, and D. Pochiecha. From sponges to nanotubes: A change of nanocrystal morphology for acute-angle bent-core molecules. *Angew. Chem. Int. Ed.* 55 (2016): 12238–12242.

Gorecka, E., D. Pociecha, N. Vaupotič, M. Čepič, K. Gomola, and J. Mieczkowski. Modulated general tilt structures in bent-core liquid crystals. *J. Mater. Chem.* 18 (2008): 3044–3049.

Griffin, A. C. and T. R. Britt. Effect of molecular structure on mesomorphism. 12. Flexible-center siamese-twist liquid crystalline diesters—A prepolymer model. *J. Am. Chem. Soc.* 103 (1981): 4957–4959.

Guo, L., K. Gomola, E. Gorecka, D. Pociecha, S. Dhara, F. Araoka, K. Ishikawa, and H. Takezoe. Transition between two orthogonal polar phases in symmetric bent-core liquid crystals. *Soft Matter* 7 (2011b): 2895–2899.

Guo, L., E. Gorecka, D. Pociecha, N. Vaupotič, M. Čepič, R. A. Reddy, K. Gornik et al. Ferroelectric behavior of orthogonal smectic phase made of bent-core molecules. *Phys. Rev. E* 84 (2011a): 031706-1–031706-8.

Gupta, M., S. Datta, S. Radhika, B. K. Sadashiva, and A. Roy. Randomly polarized smectic A phase exhibited by bent-core molecules: Experimental and theoretical studies. *Soft Matter* 7 (2011): 4735–4741.

Hahn, H., C. Keith, H. Lang, R. A. Reddy, and C. Tschierske. First example of a third-generation liquid-crystalline carbosilane dendrimer with peripheral bent-core mesogenic units: Understanding of dark conglomerate phases. *Adv. Mater.* 18 (2007): 2629–2633.

Hegmann, T., J. Kain, S. Diele, G. Pelzl, and C. Tschierske. Evidence for the existence of the McMillan phase in a binary system of a metallomesogens and 2,4,7-trinitrofluorenone. *Angew. Chem. Int. Ed.* 40 (2001): 887–890.

Helfrich, W. Effect of electric fields on the temperature of phase transitions of liquid crystals. *Phys. Rev. Lett.* 24 (1970): 201–203.

Heppke, G., A. Jákli, S. Rauch, and H. Sawade. Electric-field-induced chiral separation in liquid crystals. *Phys. Rev. E* 60 (1999): 5575–5579.

Hough, L. E. Layer curvature and optical activity in bent core liquid crystalline phases, PhD thesis, Havard University, MA, 2002.

Hough, L. E., H. T. Jung, D. Kruerke, M. S. Heberling, M. Nakata, C. D. Jones, D. Chen et al. Helical nanofilament phases. *Science* 325 (2009a): 456–460.

Hough, L. E., M. Spannuth, M. Nakata, D. A. Coleman, C. D. Jones, G. Dantlgraber, C. Tschierske et al. Chiral isotropic liquids from achiral molecules. *Science* 325 (2009b): 452–456.

Imrie, C. T. and P. A. Henderson. Liquid crystal dimers and higher oligomers: Between monomers and polymers. *Chem. Soc. Rev.* 36 (2007): 2096–2124.

Izumi, T., S. Kang, T. Niori, Y. Takanishi, H. Takezoe, and J. Watanabe. Smectic mesophase behavior of dimeric compounds showing antiferroelectricity, frustration and chirality. *Jpn. J. Appl. Phys.* 45 (2006a): 1506–1514.

Izumi, T., Y. Naitou, Y. Shimbo, Y. Takanishi, H. Takezoe, and J. Watanabe. Several types of bilayer smectic liquid crystals with ferroelectric and antiferroelectric properties in binary mixture of dimeric compounds. *J. Phys. Chem. B* 110 (2006b): 23911–23919.

Jákli, A., D. Kruerke, H. Sawade, and G. Heppke. Evidence for triclinic symmetry in smectic liquid crystals of bent-shape molecules. *Phys. Rev. Lett.* 86 (2001a): 5715–5718.

Jákli, A., Y.-M. Huang, K. Fodor-Csorba, A. Vajda, G. Galli, S. Diele, and G. Pelzl. Reversible switching between optically isotropic and birefringent states in a bent-core liquid crystal. *Adv. Mater.* 15 (2003): 1606–1610.

Jákli, A., S. Rauch, D. Lotzsch, and G. Heppke. Uniform textures of smectic liquid-crystal phase formed by bent-core molecules. *Phys. Rev. E* 57 (1998): 6737–6740.

Kang, S., M. Harada, X. Li, M. Tokita, and J. Watanabe. Notable formation of a cubic phase from small bent-angle molecules based on the 1,7-naphthalene central core and alkyl-thio tails. *Soft Matter* 8 (2012): 1916–1922.

Kardas, D., M. Prehm, U. Baumeister, D. Pociecha, R. A. Reddy, G. H. Mehl, and C. Tschierske. End functionalized liquid crystalline bent-core molecules and first DAB derived dendrimers with banana shaped mesogenic units. *J. Mater. Chem.* 15 (2005): 1722–1733.

Kasha, M. Molecular excitons in small aggregates. In: B. DiBartolo, ed., *Spectroscopy of the Excited State*. New York: Plenum Press, 1976, p. 337.

Keith, C., G. Dantlgraber, R. A. Reddy, U. Baumeister, M. Prehm, H. Hahn, H. Lang, and C. Tschierske. The influence of shape and size of silyl units on the properties of bent-core liquid crystals—From dimers via oligomers and dendrimers to polymers. *J. Mater. Chem.* 17 (2007a): 3796–3805.

Keith, C., R. A. Reddy, U. Baumeister, H. Hahn, H. Lang, and C. Tschierske. Continuous transition from antiferroelectric to ferroelectric switching liquid crystalline phases in two homologous series of bent-core mesogenic dimers based on carbosilane spacer units. *J. Mater. Chem.* 16 (2006): 3444–3447.

Keith, C., R. A. Reddy, M. Prehm, U. Baumeister, H. Kresse, J. L. Chao, H. Hahn, H. Lang, and C. Tschierske. Layer frustration, polar order and chirality in liquid crystalline phases of silyl-terminated achiral bent-core molecules. *Chem. Eur. J.* 13 (2007c): 2556–2577.

Keith, C., R. A. Reddy, and C. Tschierske. The first example of a liquid crystalline side-chain polymer with bent-core mesogenic units: Ferroelectric switching and spontaneous achiral symmetry breaking in an achiral polymer. *Chem. Commun.* (2005): 871–873.

Kim, H., Y. H. Kim, S. Lee, D. M. Walba, N. A. Clark, S. B. Lee, and D. K. Yoon. Orientation control over bent-core smectic liquid crystal phases. *Liq. Cryst.* 41 (2014a): 328–341.

Kim, H., S. Lee, T. J. Shin, E. Korblova, D. M. Walba, N. A. Clark, S. B. Lee, and D. K. Yoon. Multistep hierarchical self-assembly of chiral nanopore arrays. *Proc. Natl. Acad. Sci. USA* 111 (2014b): 14342–14347.

Kirchhoff, J., L. S. Hirst, K. M. Fergusson, and M. Hird. Low electric-field-induced switching in the B1 bent-core liquid crystal phase. *Appl. Phys. Lett.* 90 (2007): 161905-1–161905-3.

Kishikawa, K., S. Nakahara, Y. Nishikawa, S. Kohmoto, and M. Yamamoto. A ferroelectrically switchable columnar liquid crystal phase with achiral molecules: Superstructures and properties of liquid crystalline ureas. *J. Am. Chem. Soc.* 127 (2005): 2565–2571.

Kuboshita, M., Y. Matsunaga, and H. Matsuzaki. Mesomorphic behavior of 1,2-phenylene bis[4-(4-alkoxybenzylideneamino)benzoates]. *Mol. Cryst. Liq. Cryst.* 199 (1991): 319–326.

Kumazawa, K., M. Nakata, F. Araoka, Y. Takanishi, K. Ishikawa, J. Watanabe, and H. Takezoe. Important role played by interlayer steric interactions for the emergence of the ferroelectric phase in bent-core mesogens. *J. Mater. Chem.* 14 (2004): 157–164.

Le, K. V., M. Mathews, M. Chambers, J. Harden, Q. Li, H. Takezoe, and A. Jákli. Electro-optic technique to study biaxiality of liquid crystals with positive dielectric anisotropy: The case of a bent-core material. *Phys. Rev. E* 79 (2009): 03701(R)-1–03701(R)-4.

Le, K. V., H. Takezoe, and F. Araoka. Chiral superstructure mesophases of achiral bent-core molecules—Hierarchical chirality amplification and physical properties. *Adv. Mater.* (2016) published online: 14 Dec 2016, DOI: 10.1002/adma.201602737.

Lee, E.-W., M. Hattori, F. Uehara, M. Tokita, S. Kawauchi, J. Watanabe, and S. Kang. Smectic A-hexagonal columnar-B7 phase transition of acute-angle bent-core molecules. *J. Mater. Chem. C* 3 (2015a): 2266–2273.

Lee, E.-W., K. Takimoto, M. Tokita, S. Kawauchi, J. Watanabe, and S. Kang. Bent molecules with a 60° central core angle that form B7 and B2 phases. *Angew. Chem. Int. Ed.* 53 (2014): 8216–8220.

Lee, G., H.-C. Jeong, F. Araoka, K. Ishikawa, J. G. Lee, K.-T. Kang, M. Čepič, and H. Takezoe. Anchoring transition of bent-rod liquid crystal dimers on different surfaces. *Liq. Cryst.* 37 (2010): 883–892.

Lee, S., H. Kim, T. J. Shin, E. Tsai, J. M. Pechardson, E. Korblova, D. M. Walba, N. A. Clark, S. B. Lee, and D. K. Yoon. Physico-chemical confinement of helical nanofilaments. *Soft Matter* 11 (2015b): 3653–3659.

Lee, S. K., S. Heo, J. G. Lee, K.-T. Kang, K. Kumazawa, K. Nishida, Y. Shimbo et al. Odd-even behavior of ferroelectricity and antiferroelectricity in two homologous series of bent-core mesogens. *J. Am. Chem. Soc.* 127 (2005): 11085–11091.

Lee, S. K., X. Li, S. Kang, M. Tokita, and J. Watanabe. Formation of banana phases in bent-core molecules with unusual bent angles as low as 60°. *J. Mater. Chem.* 19 (2009): 4517–4522.

Lee, S. K., Y. Naito, L. Shi, M. Tokita, H. Takezoe, and J. Watanabe. Mesomorphic behavior in bent-core molecules with side wings at different positions of a central naphthalene core. *Liq. Cryst.* 34 (2007a): 935–943.

Lee, S. K., L. Shi, M. Tokita, H. Takezoe, and J. Watanabe. Chirality transfer between weakly birefringent and electric-field-induced highly birefringent B2 phases in a bent-core mesogen. *J. Phys. Chem. B* 111 (2007b): 8698–8701.

Leube, H. F. and H. Finkelmann. New liquid-crystalline side-chain polymers with large trans-versal polarizability. *Macromol. Chem. Phys.* 191 (1990): 2707–2715.

Leube, H. F. and H. Finkelmann. Optical investigations on a liquid-crystalline side-chain poly-mers with biaxial nematics and biaxial smectic-A phase. *Macromol. Chem. Phys.* 192 (1991): 1317–1328.

Levelut, A.-M. and B. Pansu. Tensorial x-ray structure factor in smectic liquid crystals. *Phys. Rev. E* 60 (1999): 6803–6815.

Li, X., S. Kang, S. K. Lee, M. Tokita, and J. Watanabe. Unusual formation of switchable hex-agonal columnar phase by bent-core molecules with low bent-angle naphthalene central core and alkylthio tail. *Jpn. J. Appl. Phys.* 49 (2010): 121701-1–121701-6.

Lin, S.-C., R.-M. Ho, C.-Y. Chang, and C.-S. Hsu. Hierarchical superstructures with helical sense in self-assembled achiral banana-shaped liquid crystalline molecules. *Adv. Funct. Mater.* 18 (2008): 3386–3394.

Lin, S.-C., T.-F. Lin, R.-M. Ho, C.-Y. Chang, and C.-S. Hsu. Hierarchical superstructures with control of helicity from the self-assembly of chiral bent-core molecules. *Chem. Eur. J.* 18 (2012): 9091–9098.

Line, M. E. and M. Glass. *Principle and Applications of Ferroelectric and Related Materials.* Oxford, U.K.: Oxford University Press, 2001.

Link, D. R., G. Natale, R. Shao, J. E. Maclennan, N. A. Clark, E. Korblova, and D. M. Walba. Spontaneous formation of macroscopic chiral domains in a fluid smectic phase of achi-ral molecules. *Science* 278 (1997): 1924–1927.

Martínez-Perdiguero, J., I. Alonso, C. L. Folcia, and J. Etxebarria. Some aspects about the structure of the optically isotropic phase in a bent-core liquid crystal: Chiral, polar, or steric origin. *Phys. Rev. E* 74 (2006): 031701-1–031701-6.

Matsuzaki, H. and Y. Matsunaga. New mesogenic compounds with unconventional molecu-lar structures 1,2-phenylene and 2,3-naphthylene bis[4-(4-alkoxyphenyliminomethyl) benzoates] and related compounds. *Liq. Cryst.* 14 (1993): 105–120.

McMillan, V. L. Simple molecular model for the smectic A phase of liquid crystals. *Phys. Rev. A* 4 (1971): 1238–1246.

McMillan, V. L. Simple molecular theory of the smectic C phase. *Phys. Rev. A* 8 (1973): 1921–1929.

Milton, H. E., M. Nagaraj, S. Kaur, J. C. Jones, P. B. Morgan, and H. F. Gleeson. Field-induced refractive index variation in the dark conglomerate phase for polarization-independent switchable liquid crystal lenses. *Appl. Opt.* 53 (2014): 7278–7284.

Miyajima, D., F. Araoka, H. Takezoe, J. Kim, K. Kato, M. Takata, and T. Aida. Electric-field-responsive handle for large-area orientation of discotic liquid-crystalline molecules in millimeter-thick films. *Angew. Chem. Int. Ed.* 50 (2011): 7865–7869.

Miyajima, D., F. Araoka, H. Takezoe, J. Kim, K. Kato, M. Takata, and T. Aida. Ferroelectric columnar liquid crystal featuring confined polar groups within core–shell architecture. *Science* 336 (2012): 209–213.

Miyake, I., Y. Takanishi, N. V. S. Rao, M. K. Paul, K. Ishikawa, and H. Takezoe. Novel chiral filament in an achiral W-shaped liquid crystalline compound. *J. Mater. Chem.* 15 (2005): 4688–4694.

Nadasi, H., W. Weissflog, A. Eremin, G. Pelzl, S. Diele, B. Das, and S. Grande. Ferroelectric and antiferroelectric banana phases of new fluorinated five-ring bent-core mesogens. *J. Mater. Chem.* 12 (2002): 1316–1324.

Nagaraj, M., V. Görtz, J. W. Goodby, and H. F. Gleeson. Electrically tunable refractive index in the dark conglomerate phase of a bent-core liquid crystal. *Appl. Phys. Lett.* 104 (2014a): 7278–7284.

Nagaraj, M., J. C. Jones, V. P. Panov, H. Liu, G. Portale, W. Bras, and H. F. Gleeson. Understanding the unusual reorganization of the nanostructure of a dark conglomerate phase. *Phys. Rev. E* 91 (2015): 042504-1–042504-13.

Nagaraj, M., K. Usami, Z. Zhang, V. Gortz, J. W. Goodby, and H. F. Gleeson. Unusual electric-field-induced transformations in the dark conglomerate phase of a bent-core liquid crystal. *Liq. Cryst.* 41 (2014b): 800–811.

Nakata, M., D. R. Link, F. Araoka, J. Thisayukta, Y. Takanishi, K. Ishikawa, J. Watanabe, and H. Takezoe. A racemic layer structure in a chiral bent-core ferroelectric liquid crystal. *Liq. Cryst.* 28 (2001): 1301–1308.

Nakata, M., D. R. Link, Y. Takanishi, Y. Takahashi, J. Thisayukta, H. Niwano, D. A. Coleman et al. Electric field-induced transition between the polarization modulated and ferroelectric smectic CSPF* liquid crystalline states studied using microbeam x-ray diffraction. *Phys. Rev. E* 71 (2005): 011705-1–011705-6.

Nakata, Y. and J. Watanabe. Frustrated smectic phase with unusual density modulation along layer observed in main chain type of polymers. *Polym. J.* 29 (1997): 193–197.

Niigawa, Y., K. Nishida, W. J. Kim, S. K. Lee, S. Heo, J. G. Lee, F. Araoka et al. Polar structures in the binary mixtures of bent-core liquid crystals showing ferroelectric and antiferroelectric B2 phases. *Phys. Rev. E* 76 (2007): 031702-1–031702-7.

Niori, T., S. Adachi, and J. Watanabe. Smectic mesophase properties of dimeric compounds. 1. Dimeric compounds based on the mesogenic azobenzene unit. *Liq. Cryst.* 19 (1995): 139–148.

Niori, T., T. Sekine, J. Watanabe, T. Furukawa, and H. Takezoe. Distinct ferroelectric smectic liquid crystals consisting of banana shaped achiral molecules. *J. Mater. Chem.* 6 (1996): 1231–1233.

Nishida, K., W. J. Kim, S. K. Lee, S. Heo, J. G. Lee, K.-T. Kang, Y. Takanishi, K. Ishikawa, J. Watanabe, and H. Takezoe. Longer-terminal-chain-sensitive phase structures in mixtures and nonsymmetric molecules of bent-core mesogens. *Jpn. J. Appl. Phys.* 45 (2006): L329–L331.

Niwano, H., M. Nakata, J. Thisayukta, D. R. Link, H. Takezoe, and J. Watanabe. Chiral memory on transition between the B2 and B4 phases in an achiral banana-shaped molecular system. *J. Phys. Chem. B* 108 (2004): 14889–14896.

Okada, Y., S. Matsumoto, Y. Takanishi, K. Ishikawa, S. Nakahara, K. Kishikawa, and H. Takezoe. Polarization switching in a columnar liquid crystalline urea as studied by optical second-harmonic generation interferometry. *Phys. Rev. E* 72 (2005): 020701(R)-1–020701(R)-4.

Ortega, J., M. R. de la Fuente, J. Etxebarria, C. L. Folcia, S. Diez, J. A. Callastegui, N. Gimeno, M. B. Ros, M. A. Perez-Jubindo. Electric-field-induced B1-B2 transition in bent-core mesogens. *Phys. Rev. E* 69 (2004): 011703-1–011703-7.

Ortega, J., C. L. Folcia, J. Etxebarria, N. Gimeno, and M. B. Ros. Interpretation of unusual textures in the B2 phase of a liquid crystal composed of bent-core molecules. *Phys. Rev. E* 68 (2003): 011707-1–011707-4.

Ortega, J., C. L. Folcia, J. Etxebarria, J. Martinez-Perdiguero, J. A. Gallastegui, P. Eerrer, N. Gimeno, and M. B. Ros. Electric-field-induced phase transitions in bent-core mesogens determined by x-ray diffraction. *Phys. Rev. E* 84 (2011): 021707-1–021707-7.

Panarin, Y. P., M. Nagaraj, S. Sreenilayan, J. K. Vij, A. Lehmann, and C. Tschierske. Sequence of four orthogonal smectic phases in an achiral bent-core liquid crystal: Evidence for the SmAPα phase. *Phys. Rev. Lett.* 107 (2011): 247801-1–247801-5.

Panarin, Y. P., M. Nagaraj, J. K. Vij, C. Keith, and C. Tschierske. Field-induced transformations in the biaxial order of non-tilted phases in a bent-core smectic liquid crystal. *Eur. Phys. Lett.* 92 (2010): 26002-1–26002-6.

Pelz, K., W. Weissflog, U. Baumeister, and S. Diele. Various columnar phases formed by bent-core mesogens. *Liq. Cryst.* 30 (2003): 1151–1158.

Pelzl, G., S. Diele, S. Grande, A. Jákli, Ch. Lischka, H. Kresse, H. Schmalfuss, I. Wirth, and W. Weissflog. Structural and electro-optical investigations of the smectic phase of chlorine-substituted banana-shaped compounds. *Liq. Cryst.* 26 (1999a): 401–413.

Pelzl, G., S. Diele, and W. Weissflog. Banana-shaped compounds—A new field to liquid crystals. *Adv. Mater.* 11 (1999b): 707–724.

Pelzl, G., M. W. Schroder, A. Eremin, S. Diele, B. Das, S. Grande, H. Kresse, and W. Weissflog. Field-induced phase transitions and reversible field-induced inversion of chirality in tilted smectic phases of bent-core mesogens. *Eur. Phys. J. E* 21 (2006): 293–303.

Pociecha, D., M. Čepič, E. Gorecka, and J. Mieczkowski. Ferroelectric mesophase with randomized interlayer structure. *Phys. Rev. Lett.* 91 (2003): 185501-1–185501-4.

Pociecha, D., E. Gorecka, M. Čepič, N. Vaupotič, K. Gomola, and J. Mieczkowski. Paraelectric-antiferroelectric phase transition in achiral liquid crystals. *Phys. Rev. E* 72 (2005): 060701(R)-1–060701(R)-4.

Prasad, V. Liquid crystalline compounds with V-shaped molecular structures synthesis and characterization of new azo compounds. *Liq. Cryst.* 28 (2001a): 145–150.

Prasad, V., D. S. Shankar Rao, and S. Rishna Prasad. Liquid crystalline dimeric compounds with an alkylene spacer. *Liq. Cryst.* 28 (2001b): 761–767.

Pratibha, R., N. V. Madhusudana, and B. K. Sadashiva. An orientational transition of bent-core molecules in an anisotropic matrix. *Science* 288 (2000): 2184–2187.

Rao, N. V. S., M. K. Paul, I. Miyake, Y. Takanishi, K. Ishikawa, and H. Takezoe. A novel smectic liquid crystalline phase exhibited by W-shaped molecules. *J. Mater. Chem.* 13 (2003): 2880–2884.

Reddy, R. A., U. Baumeister, C. Keith, and C. Tschierske. Influence of the core structure on the development of polar order and superstructural chirality in liquid crystalline phases formed by silylated bent-core molecules: Naphthalene derivatives. *J. Mater. Chem.* 17 (2007): 62–75.

Reddy R. A. and B. K. Sadashiva. Direct transition from a nematic phase to a polar biaxial smectic A phase in a homologous series of unsymmetrically substituted bent-core compounds. *J. Mater. Chem.* 14 (2004): 310–319.

Reddy, R. A., C. Zhu, R. Shao, E. Korblova, T. Gong, Y. Shen, E. Garcia et al. Spontaneous ferroelectric order in a bent-core smectic liquid crystal of fluid orthorhombic layers. *Science* 332 (2011): 72–77.

Sadashiva, B. K., R. A. Reddy, R. Pratibha, and N. V. Madhusudana. Biaxial smectic A liquid crystal in a pure compound. *Chem. Commun.* (2001): 2140–2141.

Sadashiva, B. K., R. A. Reddy, R. Pratibha, and N. V. Madusudana. Biaxial smectic A phase in homologous series of compounds composed of highly polar unsymmetrically substituted bent-core molecules. *J. Mater. Chem.* 12 (2002): 943–950.

Sasaki, Y., H. Nagayama, F. Araoka, H. Yao, H. Takezoe, and K. Ema. Distinctive thermal behavior and nanoscale phase separation in the heterogeneous liquid-crystal B4 matrix of bent-core molecules. *Phys. Rev. Lett.* 107 (2011): 237802-1–237802-4.

Sekine, T., T. Niori, M. Sone, J. Watanabe, S.-W. Choi, Y. Takanishi, and H. Takezoe. Origin of helix in achiral banana-shaped molecular systems. *Jpn. J. Appl. Phys.* 36 (1997a): 6455–6463.

Sekine, T., Y. Takanishi, T. Niori, J. Watanabe, and H. Takezoe. Ferroelectric properties in banana-shaped achiral liquid crystalline molecular systems. *Jpn. J. Appl. Phys.* 36 (1997b): L2101–L2103.

Selinger, J. V. Chiral and achiral order in bent-core liquid crystals. *Phys. Rev. Lett.* 90 (2003): 165501-1–165501-4.

Semmler, K. J. K., T. J. Dingemans, and T. Edward. Biaxial smectic phases in non-linear optical properties and phase behaviour of an oxdiazole liquid crystal. *Liq. Cryst.* 24 (1998): 799–803.

Sepelj, M., A. Lesac, U. Baumeister, S. Diele, D. W. Bruce, and Z. Hamersak. Dimeric salicylaldimine-based mesogens with flexible spacers: Parity-dependent mesomorphism. *Chem. Mater.* 18 (2006): 2050–2058.

Sepelj, M., A. Lesac, U. Baumeister, S. Diele, H. L. Nguyen, and D. W. Bruce. Intercalated liquid-crystalline phases formed by symmetric dimers with an α,ω-diiminoalkylene spacer. *J. Mater. Chem.* 17 (2007): 1154–1165.

Shanker, G., M. Prehm, M. Nagaraj, J. K. Vij, and C. Tschierske. Development of polar order in a bent-core liquid crystal with a new sequence of two orthogonal smectic and an adjacent nematic phase. *J. Mater. Chem.* 21 (2011): 18711–18714.

Shanker, G., M. Prehm, and C. Tschierske. Laterally connected bent-core dimers and bent-core-rod couples with nematic liquid crystalline phases. *J. Mater. Chem.* 22 (2012): 168–174.

Shen, D., S. Diele, G. Pelzl, I. Wirth, and C. Tschierske. Designing banana-shaped liquid crystals without Schiff's base units: m-terphenyls, 2,6-diphenylpyridines and V-shaped tolane derivatives. *J. Mater. Chem.* 9 (1999): 661–672.

Shimbo, Y., E. Gorecka, D. Pociecha, F. Araoka, M. Goto, Y. Takanishi, K. Ishikawa, J. Mieczkowski, K. Gomola, and H. Takezoe. Electric-field-induced polar biaxial order in a nontilted smectic phase of asymmetric bent-core liquid crystal. *Phys. Rev. Lett.* 97 (2006b): 113901-1–113901-4.

Shimbo, Y., Y. Takanishi, K. Ishikawa, E. Gorecka, D. Pociecha, J. Mieczkowski, K. Gomola, and H. Takezoe. Ideal liquid crystal display mode using achiral banana-shaped liquid crystals. *Jpn. J. Appl. Phys.* 45 (2006a): L282–L284.

Sicilia, A., J. J. Arenzon, I. Dierking, A. J. Bray, L. F. Cugliandolo, J. Martinez-Perdiguero, I. Alonso, and I. C. Intre. Experimental test of curvature-driven dynamics in the phase ordering of a two dimensional liquid crystal. *Phys. Rev. Lett.* 101 (2008): 197801-1–197801-4.

Srinivasan, M. V., P. Kannan, and A. Roy. Photo and electrically switchable behavior of azobenzene containing pendant bent-ccore liquid crystalline polymers. *J. Polym. Sci. Part A Polym. Chem.* 51 (2013): 936–946.

Stern, S., R. Stannarius, A. Eremin, and W. Weissflog. A model for a field-induced ferrielectric state in a bent-core mesogen. *Soft Matter*, 5 (2009): 4136–4140.

Szydlowska, J., J. Mieczkowski, J. Matraszek, D. W. Bruce, E. Gorecka, D. Pociecha, and D. Guillon. Bent-core liquid crystals forming two- and three-dimensional modulated structures. *Phys. Rev. E* 67 (2003): 031702-1–031702-5.

Takanishi, Y., T. Izumi, J. Watanabe, K. Ishikawa, H. Takezoe, and A. Iida. Field-induced molecular reorientation keeping a frustrated structure in an achiral bent-core liquid crystal. *J. Mater. Chem.* 9 (1999): 2771–2774.

Takanishi, Y., Y. Ohtsuka, Y. Takahashi, and A. Iida. Microbeam resonant x-ray scattering from bromine-substituted bent-core liquid crystals. *Phys. Rev. E* 81 (2010): 011701-1–011701-4.

Takanishi, Y., H. Takezoe, and A. Fukuda. Visual observation of dispirations in liquid crystals. *Phys. Rev. B* 45 (1992): 7684–7689.

Takanishi, Y., H. Takezoe, J. Watanabe, Y. Takahashi, and A. Iida. Intralayer molecular orientation in the B1 phase of a prototype bent-core molecule P-6-O-PIMB studied by X-ray microbeam diffraction. *J. Mater. Chem.* 16 (2006a): 816–818.

Takanishi, Y., M. Toshimitsu, M. Nakata, N. Takada, T. Izumi, K. Ishikawa, H. Takezoe, J. Watanabe, Y. Takahashi, and A. Iida. Frustrated smectic layer structures in bent-core dimer liquid crystals studied by x-ray microbeam diffraction. *Phys. Rev. E* 74 (2006b): 051703-1–051703-10.

Takezoe, H., E. Gorecka, and M. Čepič. Antiferroelectric liquid crystals—Interplay of simplicity and complexity. *Rev. Mod. Phys.* 82 (2010): 897–937.

Takezoe, H., K. Kishikawa, and E. Gorecka. Switchable columnar phases. *J. Mater. Chem.* 16 (2006): 2412–2416.

Takezoe, H. and Y. Takanishi. Bent-core liquid crystals: Their mysterious and attractive world. *Jpn. J. Appl. Phys.* 45 (2006): 597–625.

Tamba, M. G., U. Baumeister, G. Pelzl, and W. Weissflog. Banana-calamitic dimers: Unexpected mesophase behavior by variation of the direction of ester linking groups in the bent-core unit. *Liq. Cryst.* 37 (2010): 853–874.

Tamba, M. G., A. Bobrovsky, V. Shibaev, G. Pelzl, U. Baumeister, and W. Weissflog. Photochromic azobenzene functionalized banana-calamitic dimers and trimers: Mesophase behavior and photo-orientational phenomena. *Liq. Cryst.* 38 (2011): 1531–1550.

Tamba, M. G., B. Kosata, K. Pelz, S. Diele, G. Pelzl, Z. Vakhovskaya, H. Kresse, and W. Weissflog. Mesogenic dimers composed of a calamitic and a bent-core mesogenic unit. *Soft Matter* 2 (2006): 60–65.

Tamba, M.-G., W. Weissflog, A. Eremin, J. Heuer, and R. Stannarius. Electro-optic characterization of a nematic phase formed by bent core mesogens. *Eur. Phys. J. E* 22 (2007): 85–95.

Tenneti, K. K., X. Chen, C. Y. Li, Z. Shen, X. Wan, X. Fan, Q. F. Zhou, L. Rong, and B. S. Hsiao. Influence of LC content on the phase structures of side-chain liquid crystalline block copolymers with bent-core mesogens. *Macromolecules* 42 (2009): 3510–3517.

Thisayukta, J., H. Takezoe, and J. Watanabe. Study on helical structure of the B4 phase formed from achiral banana-shaped molecule. *Jpn. J. Appl. Phys.* 40 (2001): 3277–3287.

Vaupotič, N., D. Pociecha, M. Čepič, K. Gomola, J. Mieczkowski, and E. Gorecka. Evidence for general tilt columnar liquid crystalline phase. *Soft Matter* 5 (2009a): 2281–2285.

Vaupotič, N., D. Pociecha, and E. Gorecka. Polar and apolar columnar phases made of bent-core mesogens. *Top. Curr. Chem.* 318 (2012): 281–302.

Vaupotič, N., J. Szydlowska, M. Salamonczyk, A. Kovarova, J. Svoboda, M. Osipov, D. Pociecha, and E. Gorecka. Structure studies of the nematic phase formed by bent-core molecules. *Phys. Rev. E* 80 (2009b): 030701(R)-1–030701(R)-4.

Vorländer, D. and A. Apel. Die Richtung der Kohlen-stoff-Valenzen in Benzolabkommlingen (II). *Ber. Dtsch. Chem. Ges.* 65 (1932): 1101–1109.

Vugmeister, B. E. and H. Rabitz. Coexistence of the critical slowing down and glassy freezing in relaxor ferroelectrics. *Phys. Rev. B* 61 (2000): 14448–14453.

Walba, D. M., E. Korblova, R. Shao, J. E. Maclennan, D. R. Link, M. A. Glaser, and N. A. Clark. A ferroelectric liquid crystal conglomerate composed of racemic molecules. *Science* 288 (2000): 2181–2184.

Wang, L.-Y., H.-Y. Tsai, and H.-C. Lin. Novel supramolecular side-chain banana-shaped liquid crystalline polymers containing covalent- and hydrogen-bonded bent cores. *Macromolecules* 43 (2010): 1277–1288.

Watanabe, J., H. Komura, and T. Niori. Thermotropic liquid crystals of polyesters having a mesogenic 4,4-bibenzoate unit: Smectic mesophase properties and structures in dimeric model compounds. *Liq. Cryst.* 13 (1993): 455–465.

Watanabe, J., T. Niori, S.-W. Choi, Y. Takanishi, and H. Takezoe. Antiferroelectric smectic liquid crystal formed by achiral twin dimer with two mesogenic groups linked by alkylene spacer. *Jpn. J. Appl. Phys.* 37 (1998a): L401–L403.

Watanabe, J., T. Niori, T. Sekine, and H. Takezoe. Frustrated structure induced on ferroelectric smectic phases in banana-shaped molecular system. *Jpn. J. Appl. Phys.* 37 (1998b): L139–L142.

Weissflog, W., M. W. Schroder, S. Diele, and G. Pelzl. Field-induced formation of the polar SmCP phase above the SmCP-isotropic transition. *Adv. Mater.* 15 (2003): 630–633.

Xu, Y., Q. Yang, Z. Shen, X. Chen, X. Fan, and Q. F. Zhou. Effects of mesogenic shape and flexibility on the phase structures of mesogen-jacketed liquid crystalline polymers with bent side groups containing 1,3,4-oxadiazole. *Macromolecules* 42 (2009): 2542–2550.

Yang, P.-J., L.-Y. Wang, C.-Y. Tang, and H.-C. Lin. Polymeric dopant effects of bent-core covalent-bonded and hydrogen-bonded structures on banana-shaped liquid crystalline complexes. *J. Polym. Chem. A Polym. Chem.* 484 (2010): 764.

Yelamaggad, C. V., S. Anitha Nagamani, U. S. Hiremath, D. S. Shankar Rao, and S. Krishna Prasad. Salicylaldimine-based symmetric dimers: Synthesis and thermal behavior. *Liq. Cryst.* 29 (2002): 1401–1408.

Yelamaggad, C. V., S. K. Prasad, G. G. Nair, I. S. Shashikala, D. S. Shankar Rao, C. V. Lobo, and S. Chandrasekhar. A low-molar-mass, monodispersive, bent-rod dimer exhibiting biaxial nematic and smectic A phases. *Angew. Chem. Int. Ed.*, 43 (2004a): 3429–3432.

Yelamaggad, C. V., I. Shashikala, D. S. Shankar Rao, and S. Krishna Prasad. Bent-core V-shaped mesogens consisting of salicylaldimine mesogenic segments: Synthesis and characterization of mesomorphic behavior. *Liq. Cryst.* 31 (2004b): 1027–1036.

Yoshizawa, A. and A. Yamaguchi. Kinetically induced intermolecular association: Unusual enthalpy changes in the nematic phase of a novel dimeric liquid-crystalline molecule. *Chem. Commun.* (2002): 2060–2061.

Yoon, D. K., Y. Yi, Y. Shen, E. D. Korblova, D. M. Walba, I. I. Smalyukh, and N. A. Clark. Orientation of a helical nanofilament (B4) liquid-crystal phase: Topographic control of confinement, shear flow, and temperature gradients. *Adv. Mater.* 23 (2011): 1962–1967.

Zennyoji, M., Y. Takanishi, K. Ishikawa, J. Thisayukta, J. Watanabe, and H. Takezoe. Partial mixing of opposite chirality in a bent-core liquid crystal molecular system. *J. Mater. Chem.* 9 (1999): 2775–2778.

Zennyoji, M., Y. Takanishi, K. Ishikawa, J. Thisayukta, J. Watanabe, and H. Takezoe. Electrooptic and dielectric properties in bent-core liquid crystals. *Jpn. J. Appl. Phys.* 39 (2000): 3536–3541.

Zhang, Y., U. Baumeister, C. Tschierske, M. J. O'Callaghan, and C. Walker. Achiral bent-core molecules with a series of linear or branched carbosilane termini: Dark conglomerate phases, supramolecular chirality and macroscopic polar order. *Chem. Mater.* 22 (2010): 2869–2884.

Zhang, C., N. Diorio, O. D. Lavrentovich, and A. Jákli. Helical nanofilaments of bent-core liquid crystals with a second twist. *Nat. Commun.* 5 (2014): 3302–3308.

Zhang, C., N. Diorio, S. Radhika, B. K. Sadashiva, S. N. Sprunt, and A. Jákli. Two distinct modulated layer structures of an asymmetric bent-shape smectic liquid crystal. *Liq. Cryst.* 39 (2012): 1149–1157.

Zhang, Y., M. J. O'Callaghan, U. Baumeister, and C. Tschierske. Bent-core mesogens with branched carbosilane termini: Flipping suprastructural chirality without reversing polarity. *Angew. Chem. Int. Ed.* 47 (2008): 6892–6896.

Zhu, C., R. Shao, R. A. Reddy, D. Chen, Y. Shen, T. Gong, M. A. Glaser et al. Topological ferroelectric bistability in a polarization-modulated orthogonal smectic liquid crystal. *J. Am. Chem. Soc.* 134 (2012): 9681–9687.

4 Peculiarities of the N Phases

The nematic (N) phase formed by bent-shaped mesogens seems to be the same as the conventional N phase consisting of rod-shaped molecules. Well-known physical phenomena such as flexoelectric effect and electroconvection have been observed in the "banana" N phase too. Orientational elasticity and formation of the blue phases occur in the bent-core nematics. However, these phenomena are quantitatively and sometimes even qualitatively quite unique in bent-shaped molecules. One example is the occurrence of a new nematic modification: the twist-bend nematic phase (N_{TB}). In this chapter, we describe such peculiarities observed in the N phases in bent-shaped molecular systems.

4.1 FLEXOELECTRIC EFFECT

The discovery of flexoelectric effect in nematic liquid crystals dates back to 1969 (Meyer 1969). Due to its remote similarity to the piezoelectric effect in crystals, flexoelectric effect has been thought to have potential applications for electromechanical devices such as touch sensors, strain gauges, and actuators. Flexoelectric effect is associated with curvature distortions of the director field, as shown in Figure 4.1. In the uniform bulk nematic state there is no preferred orientation of the molecular dipoles and the electric polarization is averaged out (Figure 4.1a and b). Such a structure complies with the $D_{\infty h}$ symmetry, although the molecules may have a lower symmetry and be polar. A distortion of the director field locally breaks the symmetry. For instance, the splay and bend distortions can be locally described by the splay $\mathbf{S} = \mathbf{n}(\nabla \cdot \mathbf{n})$ and the bend $\mathbf{B} = \mathbf{n} \times (\nabla \times \mathbf{n})$ vectors, respectively. Such a polar symmetry breaking would allow for the electric polarization that is proportional to $\mathbf{S}$ and $\mathbf{B}$:

$$\mathbf{P}_f = e_1 \mathbf{S} + e_3 \mathbf{B} \qquad (4.1)$$

or

$$\mathbf{P}_f = e_1 \mathbf{n}(\nabla \cdot \mathbf{n}) + e_3 \mathbf{n} \times (\nabla \times \mathbf{n}) \qquad (4.2)$$

This is called the flexoelectric polarization, and the coefficients e_1 and e_3 are the flexoelectric coefficients. The order of magnitude of the flexoelectric coefficients is given by the ratio of the elementary charge $e = 1.6 \cdot 10^{-19}$ C to a typical molecular length of 10 nm, yielding 10^{-11} C/m. There is no restriction on the signs of e_1 and e_3. On a microscopic level, the flexoelectric effect can be pictured as a coupling between the curvature distortions of the director field with the shape polarity of the consisting molecules. It is particularly well seen for the cone-shaped and the bent-shaped mesogenic forms. In the first case, the longitudinal polarization couples with the splay deformation of the director (Figure 4.1c). The polarization appears along the director. In the latter case, the transverse dipoles preferentially align under bend deformation

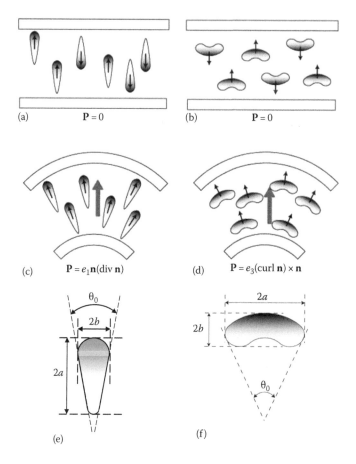

FIGURE 4.1 Development of the flexoelectric polarization in nematics in (a) a homeotropic cell containing pear-shaped molecules, and in (b) a planar cell with bent-core molecules. Polarizations are created by (c) a splay deformation, and (d) a bend deformation. Geometrical definition of (e) conical and (f) bent-shaped molecules for the Helfrich model. (From Helfrich, W., Z. *Naturforsch. A*, 26, 833, 1971; Reprinted from Le, K.V. et al., *Liq. Cryst.*, 36, 1119, 2009. With permission from Taylor & Francis.)

of the director resulting in a transverse polarization (Figure 4.1d). As mentioned in Section 1.1, bent-shaped molecules are quite interesting in terms of possible large flexoelectric coefficients.

First attempts to estimate the values of the flexoelectric coefficients using molecular statistical theory were given in the works by Helfrich (1971) and Derzhanski (1971). For a bent-core mesogen with the transversal electric dipole $\mu_\perp$ and the kink angle θ_0, e_3 reads

$$e_3 = \frac{\mu_\perp K_{33}}{2k_b T} \theta_0 \left(\frac{b}{a}\right)^{2/3} N^{1/3} \tag{4.3}$$

where
   K_{33} is the bend elastic constant
   a and b are the length and width of the molecule, respectively (Figure 4.1f)
   T is the absolute temperature
   N is the number density of the molecules

Depending on the angle θ_0, this formula gives the values of e_3 in the range of 0–10 pC/m when $\theta_0 < 1°$ and it is in the order of 1 nC/m for bent-core mesogens with $\theta_0 \approx 60°$.

Another shape-independent mechanism of developing the flexoelectric polarization was proposed by Prost and Marcerou (1977). In their model, the macroscopic polarization appears along the direction of the gradient of the average density of the molecular quadrupole moments. This mechanism does not require the existence of the dipole moments of single molecules to give rise to the flexoelectric effect. In recent years, both macroscopic and microscopic theories of the flexoelectric effect have been developed by various authors (Osipov 1983, 1984, Singh and Singh 1989, Ferrarini 2001, Zakharov and Dong 2001, Cheung et al. 2004, Billeter and Pelcovits 2010). Those theories take into account both steric repulsion and attraction, influence of dipole–dipole interactions, and molecular flexibility. Explicit expressions for the flexoelectric coefficients have been obtained. In the frame of the continuum theory, the upper limit of the flexoelectric coefficients is given by the energy conservation. Electrical energy produced by flexing cannot exceed the mechanical one, suggesting that the flexoelectric coefficients cannot exceed the values of few tenths of pC/m (Castles et al. 2011).

The flexoelectric polarization can be measured directly by flexing the sample and measuring the induced charges. The drawback of this technique is that flexoelectric charges can be screened by the ionic impurities. That is why this technique employs oscillating deformations. The flexoelectric current is measured with a lock-in amplifier. This method was used in lyotropic liquid crystals (Petrov and Sokolov 1996) and bent-core liquid crystals (Harden et al. 2006) too. In liquid crystals, however, most of the studies have been made using indirect methods. These methods include the converse flexoelectric effect, where an in-plane DC electric field E is applied across the homeotropically aligned nematic and the optical path difference is measured as a function of the field E (Figure 4.2a) (Schmidt et al. 1972). An electric field couples to the flexoelectric polarization, inducing a bend deformation of the director. Birefringence is induced in the homeotropic cell. The path difference is given by (Schmidt et al. 1972)

$$\Delta l = \frac{e_3^2}{K_3^2} E^2 n_o \left(1 - \frac{n_o^2}{n_e^2}\right) \frac{d^3}{24} \qquad (4.4)$$

where
   n_o and n_e are ordinary and extraordinary refractive indices
   d is the cell thickness

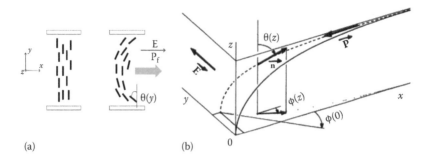

(a)   (b)

FIGURE 4.2 (a) Molecular alignment of a homeotropic cell for the measurement of the bend flexoelectric coefficient and (b) schematic diagram of hybrid-aligned cell. The continuous line shows the director in the field-free state between the upper and lower glass plates. The director has a splay-bend curvature distortion in the xz-plane. A DC field applied along the y-axis twists the director **n**. The twist $\varphi(z)$ is shown in the xy-projection by a dotted line. The polarization of the light propagating along the z-axis and polarized along the x-axis at the upper substrate rotates the polarization by the maximal deflection angle $\varphi(0)$ at the lower substrate. (a: Reprinted from Le, K.V. et al., *Liq. Cryst.*, 36, 1119, 2009. With permission from Taylor & Francis.); b: Reprinted from Dozov, I. et al., *J. Phys. Lett.*, 43, 365, 1982. With permission from EDP Science.)

This works, however, only when the anchoring force at the glass substrate is small and the dielectric torque is much smaller than the flexoelectric coupling (low fields, small dielectric anisotropy). The first estimation of e_3 was reported using N-(p-methoxybenzylidene)-4-butylaniline) (MBBA) by Schmidt et al. (1972). They experimentally verified the dependence $\Delta l \propto d^3$ and estimated $e_3 \approx 10^{12}$ C/m at $T = 22°$C.

An elegant method of measuring flexoelectric coefficient e^* in the hybrid-aligned nematic (HAN) cell was proposed by Dozov et al. (1982). In a HAN cell, the director is planarly aligned on one substrate and homeotropically aligned at the other substrate (Figure 4.2b). The director profile exhibits a splay-bend deformation $\theta(z)$ along the cell normal, leading to the flexoelectric polarization $\mathbf{P}_f(z)$. The distortion profile can be controlled by an external in-plane electric field, which couples to $\mathbf{P}_f(z)$. The field applied perpendicular to the rubbing direction exerts a torque on the director and rotates $\mathbf{P}_f(z)$. In small electric fields, the dielectric torque can be neglected and the maximum twist angle φ_m at the homeotropically aligned plate is given by

$$\varphi_m = -\frac{e^* E d}{\pi K} \tag{4.5}$$

where
 d is the sample thickness
 K is the Frank elastic constant

Experimentally, the twist angle can be determined by the measurement of the rotation of the optical polarization when the input light is plane-polarized along the rubbing direction of the planar substrate. This technique has been used to measure

$e*$ for a large number of nematic liquid crystals (Dozov et al. 1982, Takahashi et al. 1989, Blinov et al. 2001, Link et al. 2001, Mazzulla et al. 2001).

Flexoelectric instability method allows estimating the difference of the flexo-electric coefficients $e* = e_1 - e_3$ (Blinov 1979, Schiller et al. 1990). In planar-aligned nematic sample with low dielectric anisotropy $\Delta\varepsilon$ and strong anchoring at the substrate, the flexoelectric effect gives rise to a periodic pattern of stripes. In contrast to the electrohydrodynamic (EHD) instability, the stripes develop paral-lel to the initial director orientation (given by the rubbing direction of the cell). The director distortion, which was obtained by Bobylev and Pikin (1977), and an example showing the periodic instability pattern are shown in Figures 4.3 and 4.4, respectively (Le et al. 2009a).

The director distortion involves both the zenithal (xz plane) and azimuthal (xy plane) components described by the expression

$$\theta(y,z) = \theta_0 \cos(qy)\cos\left(\frac{\pi z}{d}\right) \tag{4.6}$$

$$\varphi(y,z) = \varphi_0 \sin(qy)\cos\left(\frac{\pi z}{d}\right) \tag{4.7}$$

where
  θ_0 and φ_0 are the distortion amplitude
  d is the cell thickness

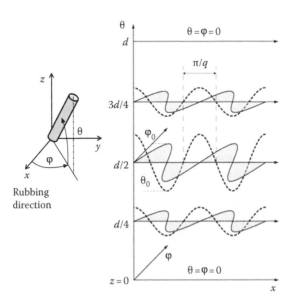

FIGURE 4.3 Two-dimensional flexoelectric distortion. (Reprinted from Le, K.V. et al., *Liq. Cryst.*, 36, 1119, 2009. With permission from Taylor & Francis.)

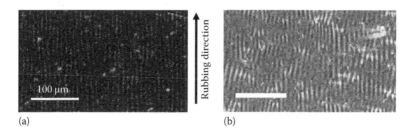

(a)   (b)

FIGURE 4.4   (a) Flexoelectric domain observed in a 5 μm thick cell of a compound shown in Figure 4.5a under a 5 Hz, 4.5 V_{rms} sinusoidal field and (b) texture observed by Wiant et al. (2005). Both textures are similar to each other. The length scale is 100 μm. The rubbing directions are in the vertical direction in both photomicrographs. (Reprinted from Le, K.V. et al., *Liq. Cryst.*, 36, 1119, 2009. With permission from Taylor & Francis.)

The width of the stripe $w = 2\pi/q$ is determined by e^*, $\Delta\varepsilon$, and the elastic constant K. In the one-constant approximation, by introducing $\mu = \varepsilon_0 \Delta\varepsilon K / e^{*2}$, the stripe width can be expressed as

$$w = \frac{d}{\pi}\sqrt{\frac{1+\mu}{1-\mu}}. \tag{4.8}$$

The instability occurs above a critical voltage given by

$$U_{th} = \frac{2\pi K}{|e^*|(1+\mu)}. \tag{4.9}$$

In materials with positive dielectric anisotropy, the flexoelectric threshold is lower than the Frederiksz one, since the nematic free-energy density is quadratic in the director distortion but the flexoelectric energy is linear. The theoretical description beyond the one-constant approximation becomes quite complicated (Krekhov et al. 2011). Bent-core LCs have usually negative dielectric anisotropy; this requires that the frequency of the driving field should be high enough to prevent the electrohydrodynamic instability (Wiant et al. 2005, Tanaka et al. 2008).

Most of the indirect measurements of the flexoelectric coefficients made in bent-core compounds show the values of the flexoelectric coefficients of the order of pC/m. The direct flexing method used by Harden et al. (2006, 2008) yielded unusually high values of the flexoelectric polarization in a series of bent-core chlorine-substituted mesogens (Figure 4.5). The flexoelectric coefficient e_3 was measured in flexible plastic sheets with indium–tin-oxide coated electrodes. Bending deformation of the cell was driven by a speaker. Flexoelectric polarization was determined by the current measured using a lock-in amplifier (Figure 4.6a).

The flexoelectric coefficient was found to be three orders of magnitude higher (nC/m) than in calamitic LCs, which gave the name giant flexoelectric effect (Harden et al. 2006). For instance, a value as high as $e_3 \approx 40$ nC/m was measured in 4-chloro-1,3-phenylene bis 4-[4'-(9-decenyloxy)-benzoyloxy] benzoate (Figure 4.5a) (Buka and Eber 2012).

FIGURE 1.3 A cigar box, in which many compounds synthesized by the Vorländer's group are stored.

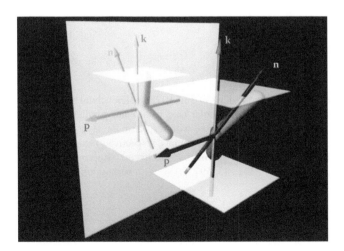

FIGURE 3.15 Mirror image of a bent-core molecule in the B₂ phase. A chiral structure formed by polar packed tilted bent-core molecules. The handedness is determined by three vectors **p, n, k**. The state of opposite handedness can be constructed by inversion of the tilt or polarization.

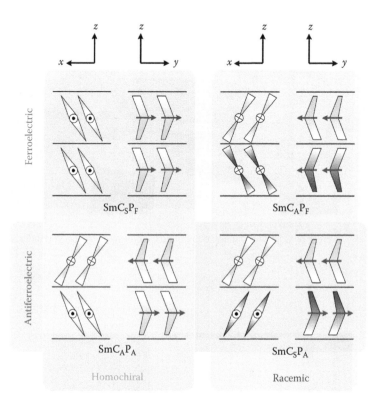

FIGURE 3.16 Two directional views of SmCP phase of nonchiral banana-shaped molecules, that is, views in the planes perpendicular (x–z plane) and parallel (y–z plane) to the layer polarization. The molecular plane is tilted with respect to the layer normal. Ferroelectric and antiferroelectric states (upper and lower columns, respectively) exist in homochiral and racemic layer chirality (left and right columns, respectively). (Reprinted from Takezoe, H. and Takanishi, Y., *Jpn. J. Appl. Phys.*, 45, 597, 2006. With permission from the Japan Society of Applied Physics.)

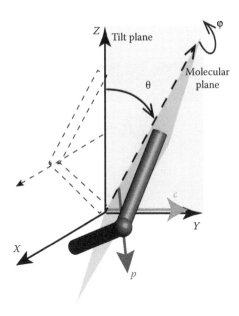

FIGURE 3.22 A schematic of a bent-core mesogen in the general polar SmC$_G$ phase. (Reprinted with permission from Chattham, N., Tamba, M.-G., Stannarius, R., Westphal, E., Gallardo, H., Prehm, M., Tschierske, C., Takezoe, H., and Eremin, A., *Phys. Rev. E*, 91, 030502(R)-1. Copyright 2015 by the American Physical Society.)

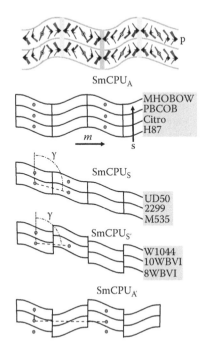

FIGURE 3.26 Layer structures in various modifications of the B$_7$ phase. (Partly reprinted from Kim, H. et al., *Liq. Cryst.*, 41, 2014a, 328. With permission from Taylor and Francis; Redrawn from Coleman, D.A. et al., *Phys. Rev. E*, 77, 021703-1, 2008.)

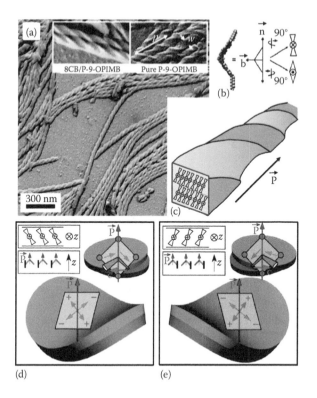

FIGURE 3.30 B$_4$ helical nanofilaments and their corresponding molecular organization. (a) FFTEM images of a mixture (8CB/P-9-OPIMB = 75/25) quenched at 37°C reveal large homochiral regions with either left- or right-handed helices. In an inset, an FFTEM image of pure P-9-OPIMB is also shown. Different morphologies in the pure and mixture samples can be noticed; Layer edges are visible in the pure P-9-OPIMB, whereas they are not visible because of the coverage by 8CB in the mixture. (b) The convention for the molecular director **n** and the molecular bow direction **b** of bent-core molecules, with the polarization **P** along **b**. The handedness of the helices is determined by the corresponding layer chirality. (c) The structure of a helical nanofilament. (d) and (e) The formation of helical nanofilaments. Each molecular arm can be viewed as an elastically isotropic slab that dilates parallel to and compresses perpendicular to the molecular tilt direction as a result of the hexagonal in-plane ordering. Because the tilt directions of the top and bottom molecular arms are orthogonal, the two elastic slabs adopt a saddle-splay curvature to relieve the intralayer mismatch. The local layer chirality, which is determined by the polarization and molecular tilt, results in distinct regions with orthogonal saddle-splay and opposite signs of filament twist. (a: Reprinted with permission from Chen, D., Zhu, C., Shoemaker, R.K., Korblova, E., Walba, D.M., Glaser, M.A., Maclennan, J.E., and Clark, N.A., Pretransitional orientational ordering of a calamitic liquid crystal by helical nanofilaments of a bent-core mesogen, *Langmuir*, 26, 15541. Copyright 2010 American Chemical Society; b, d, and e: Reprinted with permission from Chen, D., Maclennan, J.E., Shao, R., Yoon, D.K., Wang, H., Korblova, E., Walba, D.M., Glaser, M.A., and Clark, N.A., Chirality-preserving growth of helical filaments in the B4 phase of bent-core liquid crystals, *J. Am. Chem. Soc.*, 133, 12656. Copyright 2011a American Chemical Society.)

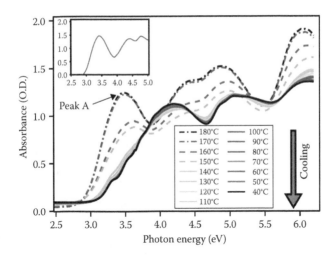

FIGURE 3.33 Temperature dependence of absorption spectra of P-12-O-PIMB on cooling. The broken, dashed, and solid lines represent spectra in the isotropic, B_2, and B_4 phases, respectively. An inserted spectrum is from P-12-O-PIMB dissolved in chloroform. (Reprinted with permission from Araoka, F., Otani, T., Ishikawa, K., and Takezoe, H., *Phys. Rev. E*, 82, 041708-1. Copyright 2010 by the American Physical Society.)

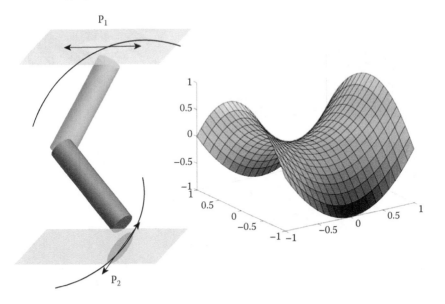

FIGURE 3.40 Formation of the saddle-splay structure in the DC phase: frustration between molecular fragments of a bent-core mesogen can be relieved by saddle-splay curvature of the layers. The projections P_1 and P_2 of the molecular wings on the smectic layer are nearly orthogonal in a tilted SmCP phase. In case of tilted phase of rod-shaped mesogens, the in-plane correlation function shows an elongation in the direction of the molecular tilt. This leads to frustration in case of tilted bent-core mesogens. This frustration can be relieved by layer curvature along the mutually orthogonal principal directions.

FIGURE 3.41 Textures of two different DC phases formed by the molecule shown at the top. Left (a, c, e) DC-B$_{1rev}$ process and right (b, d, f) DC-B$_{1revtiltM}$ process. (Reprinted with permission from Deepa, G.B., Radhika, S., Sadashiva, B.K., and Pratibha, R., *Phys. Rev. E*, 87, 062508-1. Copyright 2013 by the American Physical Society.)

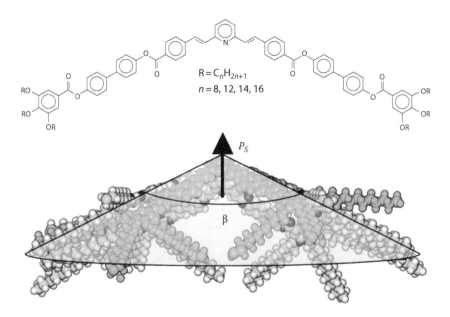

FIGURE 3.46 Polycatenar molecule and its organization to an umbrella assembly. (Reprinted with permission from Gorecka, E., Pociecha, D., Matraszek, J., Mieczkowski, J., Shimbo, Y., Takanishi, Y., and Takezoe, H., *Phys. Rev. E*, 73, 031704-1. Copyright 2006 by the American Physical Society.)

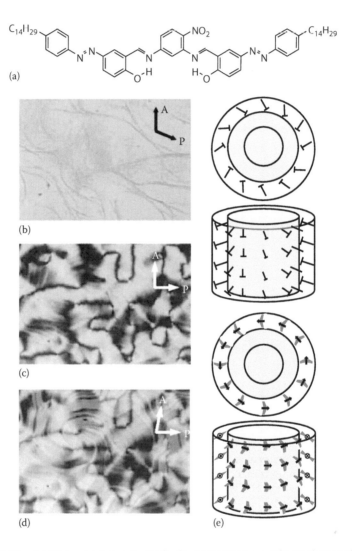

(a)

(b)

(c)

(d)

(e)

FIGURE 3.61 (a) W-shaped molecule; (b) its filament texture growth; (c) characteristic texture under crossed polarizers; (d) under decrossed polarizers, two opposite chiral domains were observed; and (e) the phase structure of cylindrical architecture. (From Miyake, I., Takanishi, Y., Rao, N.V.S., Paul, M.K., Ishikawa, K., and Takezoe, H., Novel chiral filament in an achiral W-shaped liquid crystalline compound, *J. Mater. Chem.*, 15, 4688–4694, 2005. Reproduced by permission of The Royal Society of Chemistry.)

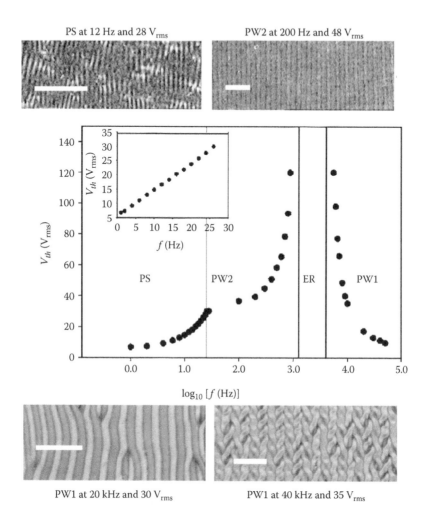

PS at 12 Hz and 28 V_{rms} | PW2 at 200 Hz and 48 V_{rms}

PW1 at 20 kHz and 30 V_{rms} | PW1 at 40 kHz and 35 V_{rms}

FIGURE 4.8 Threshold showing three nonstandard electroconvection regimes as a function of frequency. Textures at four frequencies are also shown. Two bottom textures are results by quickly changing the voltage from 0 to 30 V_{rms} (left) and 35 V_{rms} (right). The length scale is 100 μm. (Reprinted with permission from Wiant, D., Gleeson, J.T., Eber, N., Fodor-Csorba, K., Jákli, A., and Toth-Katona, T., *Phys. Rev. E*, 72, 041712-1. Copyright 2005 by the American Physical Society.)

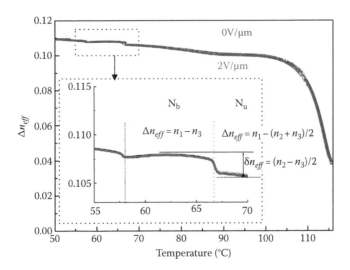

FIGURE 4.14 Effective birefringence as a function of temperature. (Reprinted with permission from Jang, Y., Panov, V.P., Kocot, A., Vij, J.K., Lehmann, A., and Tschierske, C., Optical confirmation of biaxial nematic (N_b) phase in a bent-core mesogen, *Appl. Phys. Lett.*, 95, 183304-1–183304-3. Copyright 2009 by the American Institute of Physics.)

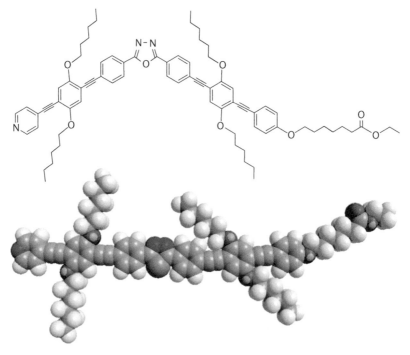

FIGURE 4.15 A molecule synthesized by Kim et al. to promote biaxiality. (From Kim, Y.-K., Majumdar, M., Senyuk, B.I., Tortora, L., Seltmann, J., Lehmann, M., Jákli, A., Gleeson, J.T., Lavrentovich, O.D., and Sprunt, S., Search for biaxiality in a shape-persistent bent-core nematic liquid crystal, *Soft Matter*, 8, 8880–8890, 2012. Reproduced by permission of The Royal Society of Chemistry.)

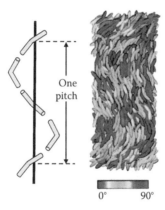

FIGURE 4.26 Heliconical structure with twist and bend deformations. Image at the right is a simulation by Memmer. (Reprinted from Memmer, R., *Liq. Cryst.*, 29, 483, 2002. With permission from Taylor & Francis.)

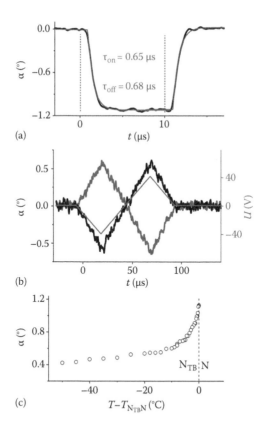

FIGURE 4.27 (a) Dynamic response of the induced tilt angle by applying a square pulsed field in a single domain; (b) response by applying a triangular wave voltage to two opposite chiral domains; and (c) temperature dependence of the induced tilt angle by applying a square wave voltage pulse of 40 V/1.6 μm. (Reprinted with permission from Meyer, C., Luckhurst, G.R., and Dozov, I., *Phys. Rev. Lett.*, 111, 067801-1. Copyright 2013 by the American Physical Society.)

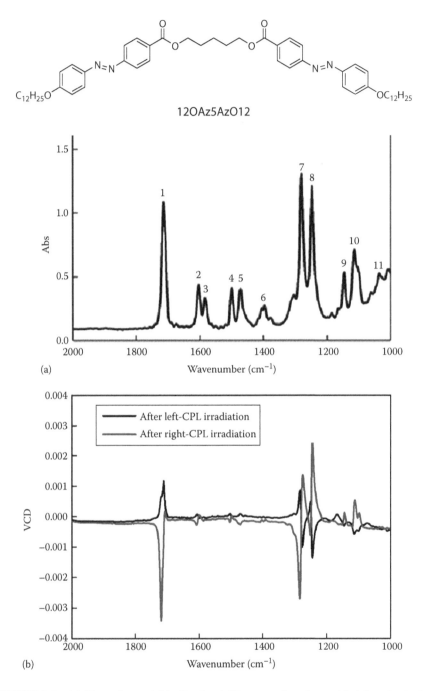

FIGURE 5.5 (a) Absorption and (b) vibrational CD spectra from homochiral domains in the B$_4$ phase of a twin dimer (12OAz5AzO12) obtained by left- and right-circular-polarized-light (CPL) light irradiations. For numbers in (a), please refer to the text. (Reprinted from Choi, S.-W., Kawauchi, S., Tanaka, S., Watanabe, J., and Takezoe, H., Vibrational circular dichroism spectroscopic study on circularly polarized light-induced chiral domains in the B$_4$ phase of a bent mesogen, *Chem. Lett.*, 36, 1018–1019, 2007. With permission from the Chemical Society of Japan.)

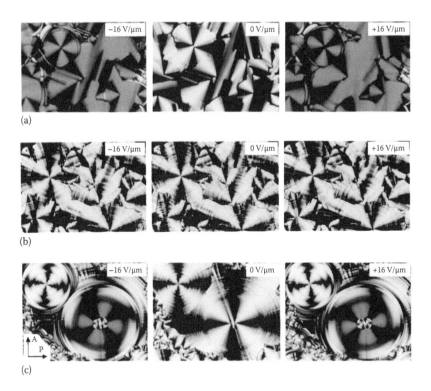

(a)

(b)

(c)

FIGURE 5.11 Photomicrographs of (a) homochiral (b) racemic and (c) partially mixed domains in the absence (middle) and presence of positive (right) and negative (left) electric fields. (Reprinted from Zennyoji, M., Takanishi, Y., Ishikawa, K., Thisayukta, J., Watanabe, J., and Takezoe, H., Electrooptic and dielectric properties in bent-shaped liquid crystals, *Jpn. J. Appl. Phys.*, 39, 3536–3541, 2000. With permission from the Japan Society of Applied Physics.)

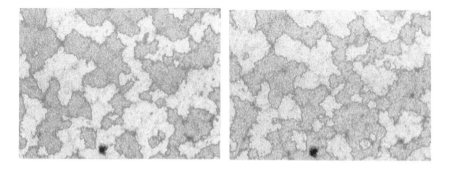

FIGURE 5.12 Photomicrographs of the B_4 phase under oppositely decrossed polarizers. (Reprinted with permission from Thisayukta, J., Nakayama, Y., Kawauchi, S., Takezoe, H., and Watanabe, J., Distinct formation of a chiral smectic phase in achiral banana-shaped molecules with a central core based on a 2,7-dihydroxynaphthalene unit, *J. Am. Chem. Soc.*, 122, 7441. Copyright 2000 by the American Chemical Society.)

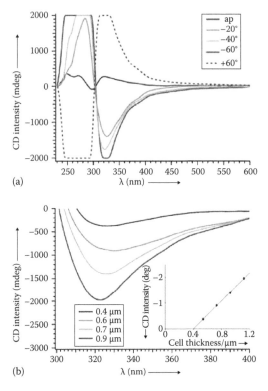

(a)

(b)

FIGURE 5.16 CD spectra in the B$_4$ phase cooled from twisted nematic structures. (a) Cells with different twist angles including an opposite twist and (b) cells with different thicknesses. CD peak intensity against cell thickness is shown in an inset. (From Ueda, T., Masuko, S., Araoka, F., Ishikawa, K., and Takezoe, H.: A general method for the enantioselective formation of helical nanofilaments. *Angew. Chem. Int. Ed.* 2013. 125. 7001–7004. Copyright Wiley-VCH Verlag GmbH & Co. KGaA. Reproduced with permission.)

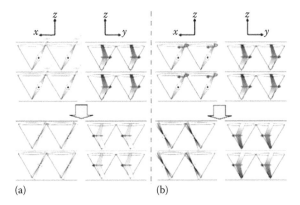

(a)   (b)

FIGURE 5.18 Schematic illustration of two switching processes: (a) rotation about the molecular axis associated with chirality switching and (b) rotation around a cone, the same as the switching in the SmC* phase. The layer chirality is preserved. (Reprinted from Takezoe, H. and Takanishi, Y., Bent-core liquid crystals: Their mysterious and attractive world, *Jpn. J. Appl. Phys.*, 45, 597–625, 2006. With permission from the Japan Society of Applied Physics.)

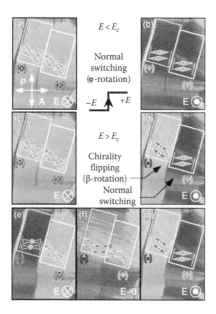

FIGURE 5.19 Photomicrographs showing different switching modes. (a, b) Under reversed field E below E_c, normal switching around a cone is observed. (c, d) Under reversed field E slightly larger than E_c, chirality switching in part occurs. (e–g) Under a triangular field, chirality switching never occurs. (Reprinted with permission from Nakata, M., Shao, R.-F., Maclennan, J.E., Weissflog, W., and Clark, N.A., *Phys. Rev. Lett.*, 96, 067802-1. Copyright 2006 by the American Physical Society.)

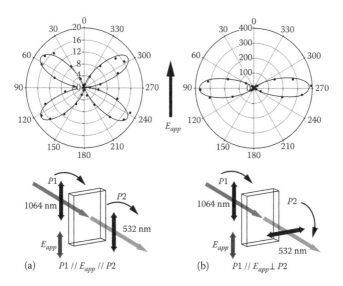

FIGURE 6.4 SHG in-plane anisotropy measured using well-aligned sample P8-O-PIMB6* ($SmC_A P_F$) under an applied electric field (shown by arrow). (a) Under parallel polarizers and (b) perpendicular polarizers. The polar angles denote the angle between the polarization direction of the normally incident fundamental beam and the field direction. (Reprinted with permission from Araoka, F., Thisayukta, J., Ishikawa, K., Watanabe, J., and Takezoe, H., *Phys. Rev. E*, 66, 021705-1. Copyright 2002 by the American Physical Society.)

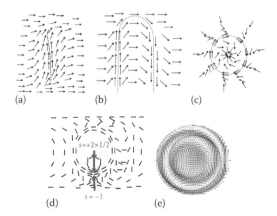

(a)   (b)   (c)

$s=+2\times1/2$

(d) $s=-1$ (e)

FIGURE 7.8 Different morphologies observed in free-suspending films: (a) single string; (b) pin; (c) spiral with an $s = +1$ vortex in the center; (d) boojum; and (e) target pattern. The arrows show the **c**-director. (Reprinted with permission from Eremin, A., Nemeş, A., Stannarius, R., and Weissflog, W., *Phys. Rev. E*, 78, 061705-1. Copyright 2008 by the American Physical Society.)

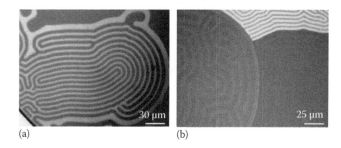

(a)   (b)

FIGURE 7.10 Morphology of labyrinthine patterns: (a) and (b) Examples of labyrinthine patterns of edge dislocation in freely suspended films. (Reprinted with permission from Eremin, A., Kornek, U., Stannarius, R., Weissflog, W., Nadasi, H., Araoka, F., and Takezoe, H., *Phys. Rev. E*, 88, 062512-1. Copyright 2013 by the American Physical Society.)

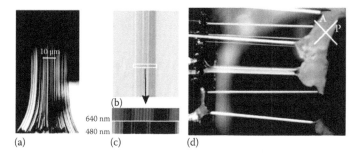

(a)   (b)   (c)   (d)

FIGURE 7.12 Images of freely suspended filaments in the PM-SmCP phase. (a) and (d) Images of filaments under a polarizing microscope. The image in (a) clearly demonstrates the striped optical pattern along the filament axis. (b) Single filament observed in transmission without polarizers. (c) Details of (b) observed under monochromatic illumination. (Reprinted with permission from Eremin, A., Nemeş, A., Stannarius, R., Schulz, M., Nadasi, H., and Weissflog, W., *Phys. Rev. E*, 71, 031705-1. Copyright 2005 by the American Physical Society.)

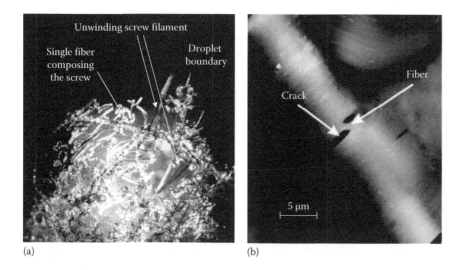

FIGURE 7.18 (a) Optical image of a screwlike filament growing in a large isotropic droplet under crossed polarizers. As soon as it touches the boundary, a longitudinal stress along the screw leads to unwinding of the small fibers and (b) "defect" cracks occur in the screw filament. These cracks show that the filament is composed of a tightly coiled fiber. The scale bar is 2 μm. (Reprinted from Eremin, A. et al., *Liq. Cryst.*, 33, 789, 2006. With permission from Taylor & Francis.)

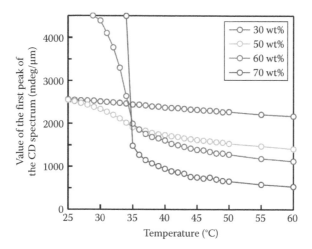

FIGURE 8.9 CD intensity as a function of temperature in four mixtures of P8-O-PIMB/5CB. The temperature dependence behavior changes at the B_4–B_x transition particularly for two mixtures with higher 5CB contents. (Reprinted with permission from Otani, T., Araoka, F., Ishikawa, K., and Takezoe, H., Enhanced optical activity by achiral rod-like molecules nanose-gregated in the B_4 structure of achiral bent-core molecules, *J. Am. Chem. Soc.*, 131, 12368. Copyright 2009 by the American Chemical Society.)

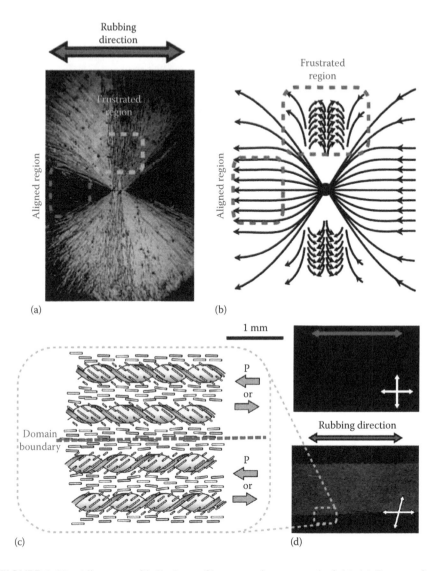

FIGURE 8.10 Alignment of helical nanofilaments using a nematic field. (a) Texture of a defect of strength 1; (b) orientation map of the defect region; (c) cartoon showing the oriented two domains with different chiralities; and (d) microphotographs of the oriented domains under crossed and decrossed polarizers. (From Araoka, F., Sugiyama, G., Ishikawa, K., and Takezoe, H.: Highly ordered helical nanofilament assembly aligned by a nematic director field. *Adv. Funct. Mater.* 2013. 23. 2701–2707. Copyright Wiley-VCH Verlag GmbH & Co. KGaA. Reproduced with permission.)

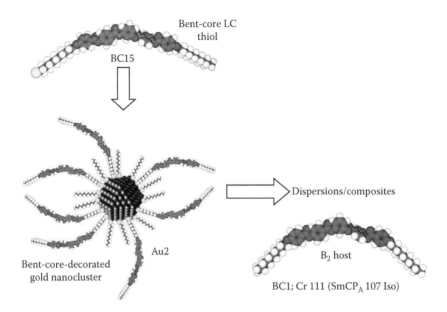

FIGURE 8.13 Gold nanocluster decorated with bent-core molecules. (From Marx, V.M., Girgis, H., Heiney, P.A., and Hegmann, T., Bent-core liquid crystal (LC) decorated gold nanoclusters: Synthesis, self-assembly, and effects in mixtures with bent-core LC hosts, *J. Mater. Chem.*, 18, 2983–2994, 2008. Reproduced by permission of The Royal Society of Chemistry.)

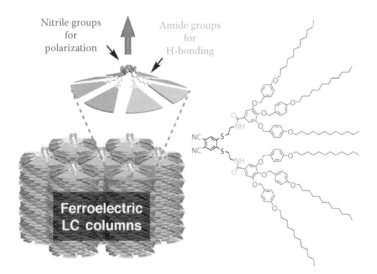

FIGURE 9.9 Molecular structure (right) and columnar structure (left) in the ferroelectric columnar phase formed by bent-core molecules. (From Miyajima, D., Araoka, F., Takezoe, H., Kim, J., Kato, K., Takata, M., and Aida, T., Ferroelectric columnar liquid crystal featuring confined polar groups within core-shell architecture, *Science*, 336, 209–213, 2012. Reprinted with permission of AAAS.)

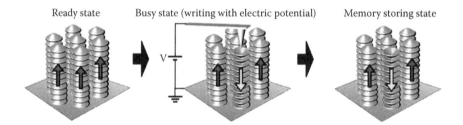

FIGURE 9.10 Cartoon of a high-density memory device using the ferroelectric columnar LC phase. If the polarization switching and retaining in each column are possible, a high-density memory device is realized. (Reproduced from Takezoe, H., *Ferroelectrics*, 468, 1, 2014. With permission from Taylor & Francis.)

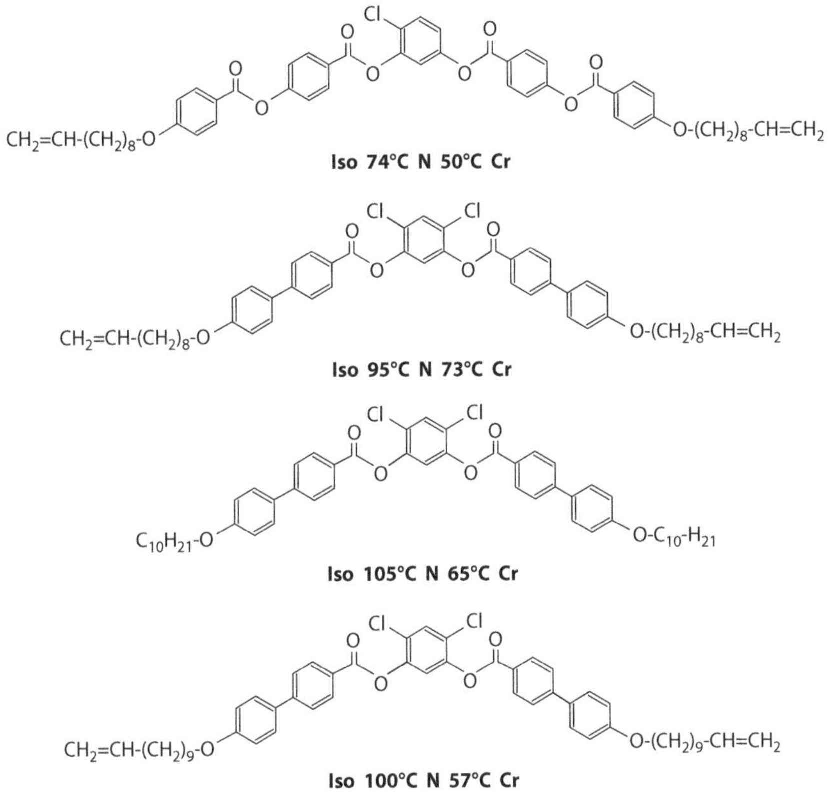

FIGURE 4.5 Compounds investigated using flexible cell technique. (From Buka, A. and Eber, N., eds., *Flexoelectricity in Liquid Crystals: Theory, Experiments and Application*, World Scientific, 2012.)

The same technique employed for 5CB showed only $e_3 \approx 40$ pC/m. A high flexoelectric response in bent-shaped liquid crystals (BLCs) was also confirmed by the measurements using the inverse flexoelectric effect, where the displacement of the flexible Mylar substrates in response to an applied electric field was measured using a Mirau interferometer. Both the direct and the inverse flexoelectric methods showed comparable values of e_3 in the investigated bent-core mesogens (Figure 4.6b).

Enhancement of the flexoelectric response has been demonstrated in mixtures of the calamitic and bent-core mesogens by direct flexing experiments (Buka et al. 2012). The flexoelectric coefficient e_3 increases by an order of magnitude as the weight fraction of the bent-core mesogens increases from 30 to 50 wt%, as shown in Figure 4.7 (Buka and Eber 2012).

However, indirect methods failed to confirm such a strong flexoelectric response in bent-core liquid crystals. For instance, measurements on the same 4-chloro-1,3-phenylene bis 4-[4'-(9-decenyloxy)-benzyloxy] benzoate (Figure 4.5a) performed by Le et al. using measurements of the field-induced director bend deformation and the measurement of the flexoelectric instabilities yielded the values

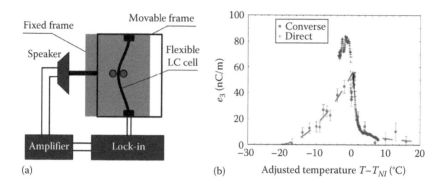

(a)   (b)   Adjusted temperature $T-T_{NI}$ (°C)

FIGURE 4.6   (a) A scheme of an experimental setup for the direct measurements of the flexo-electric polarization using a flexible cell employed by Harden et al. (2006) and (b) comparison of the flexoelectric coefficient determined using the direct and converse flexoelectric techniques in flexible cells. (Reprinted with permission from Harden, J., Teeling, R., Gleeson, J.T., Sprunt, S., and Jákli, A., *Phys. Rev. E*, 78, 031702-1. Copyright 2008 by the American Physical Society.)

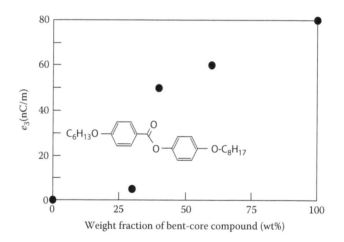

Weight fraction of bent-core compound (wt%)

FIGURE 4.7   Flexoelectric coefficient in mixtures of a calamitic compound shown in the inset and a bent-core compound shown in Figure 4.5a. The flexoelectric coefficient e_3 for the pure calamitic compound is 70 pC/m. (Redrawn from Buka, A. and Eber, N., eds., *Flexoelectricity in Liquid Crystals: Theory, Experiments and Application*, World Scientific, 2012.)

of only $e_3 = 15.8$ pC/m and $e^* = |e_1 - e_3| \approx 6$ pC/m, which is close to the values typically found in calamitics (Le et al. 2009a). Similar results were obtained by Kumar et al. (2009). In mixtures of calamitic and bent-core mesogens, optical techniques also showed the values in the range of pC/m (Wild et al. 2005, Aziz et al. 2008, Kundu et al. 2009b). However, the flexoelectric response of the host containing a small amount of bent-core molecules was enhanced by an order of magnitude as measured using HAN cells (Kundu et al. 2009b).

The large discrepancy between the values obtained by the direct and the indirect optical methods raised many questions related to the mechanisms of the flexoelectric polarization in bent-core systems. It is especially interesting, since the theoretical models based on dipolar and quadrupolar interactions cannot explain the giant flexoelectricity (values in the order of nC/m) (Helfrich 1974, Castles et al. 2011). It is tempting to assume that the giant flexoelectric effect may be attributed to the collective behavior of the bent-core mesogens such as the formation of polar clusters. Smectic clusters (or cybotactic groups) are often encountered in the nematic phases formed by bent-core mesogens (Balachandran et al. 2013, Shanker et al. 2014). Their presence can be detected by x-ray and light-scattering techniques. Different kinds of ferro- and antiferroelectric smectic clusters are also responsible for a strong polar and nonlinear-optic response in the nematic phases. Ferroelectric clusters are expected to strongly enhance the dipolar contribution to the flexoelectric effect, if a sterical asymmetry of the smectic clusters is assumed. One explanation of the discrepancy between the values obtained by the direct and HAN-cell methods is that flexoelectric polarization creates an internal electric field inducing an additional deformation of the director. All these factors may contribute to suppress the response of the cell. Another striking feature of the flexoelectric response in the converse flexoelectric experiment using flexible substrate is the nonlinear dependence of the flexing on an electric field. The presence of a critical threshold voltage in the flexing experiments may explain the failure to measure the flexoelectric coefficients in the Helfrich geometry, which is made at much smaller fields. Despite all that, the interpretation of the giant flexoelectric effect is still questionable. Further experimental and theoretical studies are required to understand the mechanisms underlying the response in those different geometries and establish the role of the smectic clusters.

4.2 ELECTROCONVECTION EFFECT

Since liquid crystals are neither perfect dielectric nor nonconducting materials, the application of an electric field induces ionic flows. This phenomenon is called electroconvection and has been extensively studied in the conventional nematic phase (Kramer and Pesch 1996, Buka et al. 2006). Some well-known examples are a Williams domain (Williams 1963) and a Helfrich deformation (Helfrich 1969). The electroconvection instability and the form of the patterns depend on dielectric ε_a and conductivity σ_a anisotropies. Compounds with negative dielectric and positive conductivity anisotropies have been most studied.

Electroconvection studies in bent-core molecules have been made by several groups (Stannarius et al. 2007, Heuer et al. 2008, Tanaka et al. 2008, 2009, Xiang et al. 2009, 2014, 2015, Tadapatri et al. 2010b, Kaur et al. 2011, Tadapatri and Krishnamurthy 2012, Krishnamurthy et al. 2014) after Wiant et al. (2005) showed nonstandard electroconvection, which cannot be explained by the Carr–Helfrich model (Helfrich 1969). Wiant et al. (2005) observed three nonstandard electroconvection regimes in a bent-core compound with a negative dielectric anisotropy, parallel stripes, prewavy 2 (PW2), and prewavy 1 (PW1) regimes with increasing frequency (Figure 4.8). Between two PW1 and 2, there exists a region where no instability can be observed. Similar or slightly different diagrams were also observed

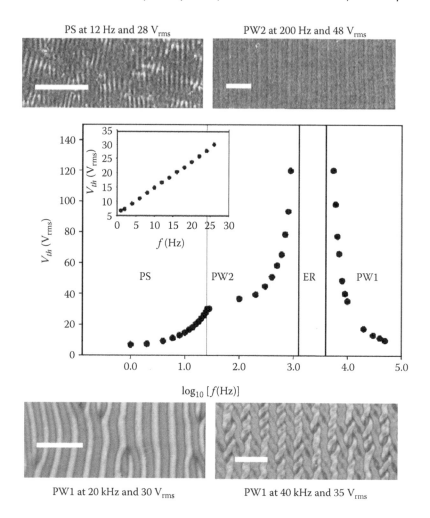

FIGURE 4.8 **(See color insert.)** Threshold showing three nonstandard electroconvection regimes as a function of frequency. Textures at four frequencies are also shown. Two bottom textures are results by quickly changing the voltage from 0 to 30 V_{rms} (left) and 35 V_{rms} (right). The length scale is 100 μm. (Reprinted with permission from Wiant, D., Gleeson, J.T., Eber, N., Fodor-Csorba, K., Jákli, A., and Toth-Katona, T., *Phys. Rev. E*, 72, 041712-1. Copyright 2005 by the American Physical Society.)

by other authors (Tanaka et al. 2008, 2009, Tadapatri et al. 2010b, Tadapatri and Krishnamurthy 2012). There are different types of stripes depending on frequency applied, some of which are displayed above (low frequency) and below (high frequency) the V_{th} vs frequency curve. In the micrographs, the rubbing direction is horizontal, and the scale bar is equal to 100 μm.

One interesting aspect is chirality. Tanaka et al. (2008) found that stripes in PW1 and PW2 regimes show no extinction position under crossed polarizers. This fact indicates that the director deformation does not occur within a single plane but the director orientation out of the plane is induced by the electroconvection. By measuring transmission spectra in the stripes with alternating colors and calculating the spectra, the structure was attributed to the similar structure of the twisted (splayed) director configurations observed in surface-stabilized ferroelectric liquid crystal cells. The adjacent domains with different colors are obtained for opposite twist senses.

A different type of electroconvection was also observed in a bent-core compound with the positive dielectric anisotropy and the negative conductivity anisotropy in homeotropically aligned cells. Xiang et al. (2009) reported two electroconvection instabilities depending on the uniaxial and biaxial nematics: PW2 both in the uniaxial and biaxial phases, whereas PW1 only in the uniaxial phase. Contrary to the previous observation (Wiant et al. 2005), the stripes in PW1 are parallel to the rubbing direction (longitudinal roll). The authors attributed such a difference to the insignificant change in the conductivity anisotropy (Xiang et al. 2009). It is noted, however, there is still a serious debate about the existence of the biaxial phase in thermotropic liquid crystals (see the next section). The stripes parallel to the rubbing direction were also observed in a bent-core compound with the negative dielectric and conductivity anisotropies in the high frequency range by Xiang et al. (2014). On the other hand, an oblique pattern was observed in the low-frequency range. In between, transformations from oblique to longitudinal and then to normal patterns were observed with increasing ac voltage. The pattern formation is strongly affected by the temperature and exhibits a complex behavior. Unusual effects such as a transition between longitudinal and normal rolls and a voltage-dependent reorientation of the convection rolls have also been reported (Heuer et al. 2008). A variety of electroconvection patterns were also observed, depending on the frequency, voltage, and temperature, by Tadapatri et al. (2010b). Another type of nonstandard electroconvection is the occurrence of three distinct bifurcation modes in a narrow frequency region (10–17 Hz); longitudinal, oblique, and normal rolls appear in the order of increasing threshold (Tadapatri and Krishnamurthy 2012). Similar complexity at low frequencies was also observed by Xiang et al. (2015). An example of the transformation of patterns is shown in Figure 4.9, in which the measurement was made at 12 Hz and 9.2 V, and the white arrow indicates the rubbing direction.

In this way, electroconvection effect in bent-core molecules is influenced by many physical parameters, so it is very complex. Most of them cannot be explained by Carr and Helfrich theory (Helfrich 1969). To clarify the difference from the phenomenon observed in calamitic molecules, the investigation of mixtures of bent- and rod-core molecules is also important, as was made by Tanaka et al. (2009). At low (<30 wt%) concentration of calamitic molecules, the peculiar features of bent-core molecules qualitatively persist. At higher concentration, morphological changes occur at high frequencies and the PW1 mode finally disappears at higher concentration (>70 wt%). This indicates that the nonstandard PW1 mode with a negative slope of threshold voltage against frequency is caused by the features of bent-core molecules.

In conclusion, it is evident that bent-core molecules exhibit nonstandard electroconvection, which cannot be explained using the existing theories. The electroconvection patterns are quite complex depending on a variety of physical parameters such as frequency, voltage, temperature, cell thickness, and dopants. This is related to unusual physical properties, such as high flexoelectric coefficients, unusual ratios

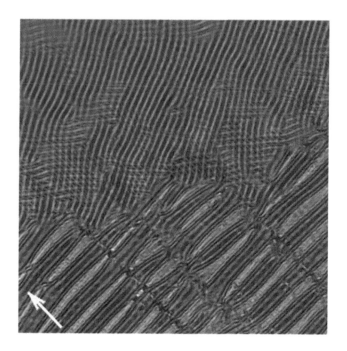

FIGURE 4.9   An example texture showing a transformation of patterns observed at 12 Hz and 9.2 V. The white arrow indicates the rubbing direction. (Reprinted with permission from Xiang, Y., Zhou, M.-J., Xu, M.-Y., Salamon, P., Eber, N., and Buka, A., *Phys. Rev. E*, 91, 042501-1. Copyright 2015 by the American Physical Society.)

of the elastic constants, chiral segregation ability, and tendency to form cybotactic smectic structures. Full understanding is a future problem.

4.3   BIAXIAL NEMATIC PHASE

In the conventional nematic phase composed of rod-shaped molecules, the long axes of the molecules tend to align along their average direction known as the director. However, the short axes are not ordered, resulting in the optically uniaxial nematic N_u phase with the phase symmetry $D_{\infty h}$. When the constituent molecules have lower symmetries than that of a cylindrical shape, the short molecular axes may have a long-range orientational order with the secondary director **m** in addition to the primary director **n** (Figure 4.10), reducing the symmetry of the phase to D_{2h}. Actually, such a biaxial nematic N_b phase is known to exist in lyotropic (Yu and Saupe 1980) and polymeric (Severing and Saalwachter 2004) systems. As candidates for thermotropic biaxial nematics, different molecular forms have been considered such as mixtures of rod-shaped and disc-shaped molecules (Stroobants and Lekkerkerker 1984), nematics of plate-shaped mesogens (Figure 4.10) (Madsen et al. 2004). Bent-shaped molecules are also an attractive candidate for the emergence of the biaxial nematic phase. Hence, two papers reporting the first thermotropic biaxial nematic phase using bent-core molecules that appeared in the same issue of

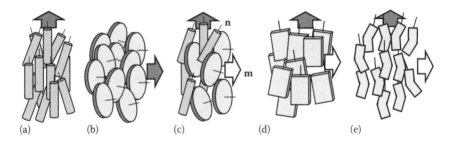

(a) (b) (c) (d) (e)

FIGURE 4.10 Uniaxial (a and b) an biaxial (c–e) orientation in (a) rod-shaped; (b) disc-shaped; (c) mixture of rod- and disc-shaped; (d) plate-shaped; and (e) bent-shaped molecules. (Redrawn from Madsen, L.A. et al., *Phys. Rev. Lett.*, 92, 145505-1, 2004.)

Phys. Rev. Lett. (Acharya et al. 2004, Madsen et al. 2004) have created considerable excitement. Later, however, a number of reports raised questions about the existence of the biaxial nematic phase. We will explain the background and the current status in the following text.

Let us describe the first two reports. Acharya et al. (2004) and Madsen et al. (2004) claimed the existence of the biaxial nematic phase using the same molecules with an oxadiazole core, one of which is shown in Figure 4.11a, and different techniques; polarizing microscopy, conoscopy, ^{2}H NMR spectroscopy (Madsen et al. 2004) and x-ray diffraction (XRD) (Acharya et al. 2004). In the uniaxial N phase, 2- and 4-brush defects can be observed, where 4-brush defects with a higher strength

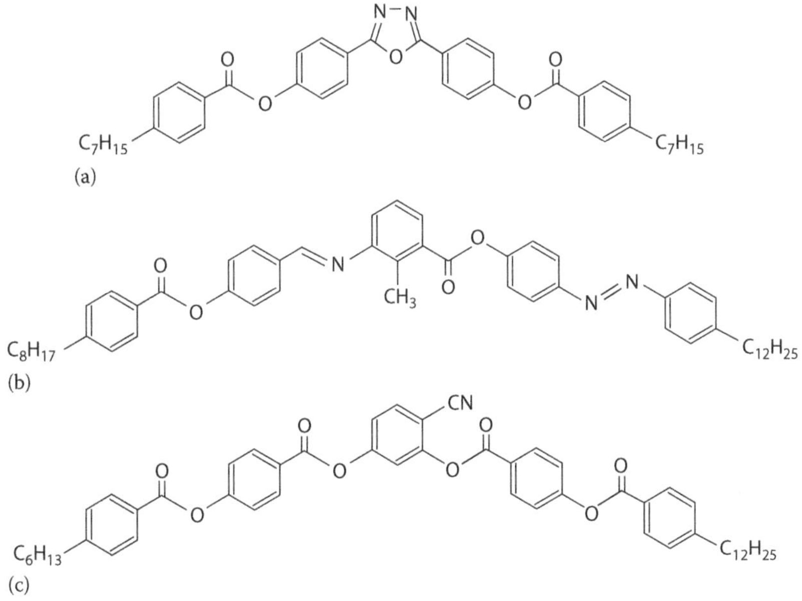

(a)

(b)

(c)

FIGURE 4.11 Typical molecules studied for the biaxial nematic phase.

($|k| = 1$) are known to be able to exist because of the escape from a line defect to a point defect in the director field. In the biaxial N phase, however, such escape is not possible because of a high cost in energy, and the line defects of $|k| = 1$ split into pairs of line defects of $|k| = 1/2$ and disappear by annihilation processes of these defects (Chiccoli et al. 2002). Hence, polarizing microscopy observation is one of the easy tests for the biaxiality. Another easy test also relies on optical microscopy in a conoscopy mode. An extinction cross (isogyre) observed in the uniaxial N phase opens into an image of a split cross in the biaxial N phase.

However, such methods are not conclusive because the image can be affected by other effects. For instance, anchoring transitions often occur among homeotropic, tilted, and planar orientations in the N phase. Actually, Senyuk et al. (2010) reported an anchoring transition from homeotropic to tilted orientation in a compound, in which both uniaxial and biaxial phases were reported to exist. In such a case, a small tilt results in the splitting of the isogyres, which could give misleading information on the existence of the biaxiality. The tilted state originates from the surface-induced birefringence associated with smectic layering near the surface. Some other optical observations such as topological point defects, which are pertinent only to the uniaxial order, also denied the biaxial N phase. It was also confirmed that biaxial nature can be easily lifted by applying a modest electric field, as Le et al. (2009b) also reported using the same molecule shown in Figure 4.11b. Thus, the apparent biaxiality could be caused by subtle surface effect rather than by the bulk biaxial phase.

An overview of research in the field of biaxial nematic liquid crystalline materials was given by Tschierske and Photinos (2010), in which theoretical concepts, material design, and identification methods were summarized. XRD and NMR measurements were believed to give more decisive techniques. However, there are still considerable debates supporting and denying the existence of the biaxial N phase. XRD patterns detected under a magnetic field showed two pairs of diffuse peaks at small angle unlike a uniaxial N phase, in which only one pair of diffuse peaks was observed (Acharya et al. 2004) (see Figure 4.12 later for similar XRD data). The authors ruled out the contribution from cybotactic SmC clusters existing in the N phase mainly on the basis of the fact that these peaks exist over the whole N phase region, contradicting the common knowledge that such clusters are due to the pretransitional effect and exist only in a very narrow temperature range near the N-Sm phase transition. Later, many researchers confirmed the smectic cybotactic clusters over the whole temperature range of the N phase (Prasad et al. 2005, Acharya et al. 2009, Vaupotič et al. 2009, Francescangeli and Samulski 2010, Hong et al. 2010, Keith et al. 2010, Tschierske and Photinos 2010, Francescangeli et al. 2011, Shanker et al. 2011, 2012, Picken et al. 2012, Nishiya et al. 2014) and even in the Iso phase (Hong et al. 2010). Except for a few reports (Prasad et al. 2005, Acharya et al. 2009), most of the reports support the formation of smectic cybotactic clusters in the uniaxial N phase, and deny the existence of the biaxial N phase.

Some examples of XRD patterns are shown in Figure 4.12 (Tschierske and Photinos 2010). Interestingly, most of the compounds, which show smectic cybotactic clusters, have the SmC, B_2, or Cr phases below N with a few exceptions showing the N-SmA (Vaupotič et al. 2009, Tschierske and Photinos 2010) and the N-SmAP$_R$ (Shanker et al. 2011) phase transitions. Hence, most of the small angle x-ray signals

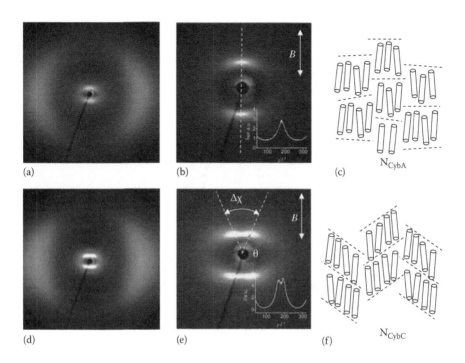

FIGURE 4.12 Some examples of wide-angle (a, d) and small-angle (b, e) XRD patterns showing uniaxial- (a–c) and biaxial-like (d–f) images. Possible cluster images responsible for uniaxial and biaxial images are also shown in (c) and (f), respectively. (From Tschierske, C. and Photinos, D.J., Biaxial nematic phases, *J. Mater. Chem.*, 20, 4263–4294, 2010. Reproduced by permission of The Royal Society of Chemistry.)

in the N phase show splitting, indicating SmC clusters (Figure 4.12d through f). The splitting in N and SmC is the same, though not always the case (Chakraborty et al. 2013). However, no splitting was observed in compounds with the N-SmA (or SmAP$_R$) transition (Vaupotič et al. 2009, Tschierske and Photinos 2010, Shanker et al. 2011) (Figure 4.12a through c). It is important to note an XRD experiment using a mixture of compounds showing the N-SmA and N-SmC phase transitions (Vaupotič et al. 2009). The addition of a small amount of the SmA compound results in the N-SmA–SmC transition, and the mixture shows no splitting in the small angle XRD signal in the N phase. It is hard to imagine that uniaxial or biaxial nature drastically changes by such moderate doping in an intrinsic biaxial N phase. The comparison between experimental and theoretical results was also successfully made based on the uniaxial N phase with smectic cybotactic clusters (Vaupotič et al. 2009, Hong et al. 2010). The N phases with cybotactic SmA or SmC clusters are sometimes called N$_{CybA}$ or N$_{CybC}$, respectively. In addition, Keith et al. (2010) reported an intermediate phase between N$_{CybC}$ and SmC, which composed of elongated but not yet fused cybotactic clusters, CybC phase.

Another important experimental result, which is provided as an evidence of the biaxial N phase, is NMR. Madsen et al. (2004) measured ^{2}H NMR spectra under the rotation of samples about an axis perpendicular to a magnetic field and fitted to a theoretical expression using a biaxiality parameter $\eta = (q_{xx} - q_{yy})/q_{zz}$, where q_{zz} is a nuclear quadrupole tensor component parallel to the major director $\mathbf{n}$ and q_{xx} and q_{yy} are perpendicular components. As shown in Figure 4.13, $\eta = 0.1$ gave a best fit, certifying the existence of biaxiality. However, there are some critiques. The N phase composed of smectic cybotactic clusters may account for the NMR biaxiality

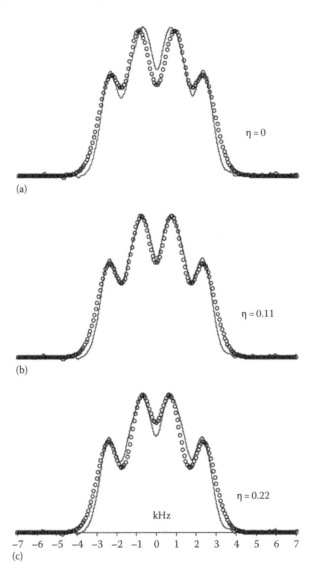

FIGURE 4.13 ^{2}H NMR spectra under rotation of samples about an axis perpendicular to a magnetic field. Fitting was made using different biaxial parameters η: (a) 0, (b) 0.11, and (c) 0.22. (Reprinted with permission from Madsen, L.A., Dingemans, T.J., Nakata, M., and Samulski, E.T., *Phys. Rev. Lett.*, 92, 145505-1. Copyright 2004 by the American Physical Society.)

observed (Francescangeli and Samulski 2010). Galerne (2006) casted doubt on the conclusion made by NMR measurements performed on spinning cells. Because of the viscosity problem, the overall orientation in the cell should be intermediate between the 2D and 3D isotropic orientations, producing biaxial-type NMR spectra even with uniaxial nematics. ^{13}C NMR spectroscopy was also used to support the uniaxial–biaxial nematic transition (Dong and Marini 2009), which will be discussed later.

Electron spin resonance (ESR) was also used to examine the biaxiality (Vaupotič et al. 2009). ESR spectrum was typical for uniaxial phase with the director aligned along a magnetic field. However, this method always requires the use of probe molecules, if the mesogenic molecules themselves are not paramagnetic. Hence, even if a uniaxial spectrum is recorded, there is the possibility that the probe molecules could have inhibited the biaxial order (Tschierske and Photinos 2010) or do not reflect behavior of the matrix.

Many researchers reported the uniaxial–biaxial phase transition (Prasad et al. 2005, Lee et al. 2007, Southern et al. 2008, Dong and Marini 2009, Francescangeli et al. 2009, Jang et al. 2009, Xiang et al. 2009, Yoon et al. 2010, Aluculesei et al. 2012, Picken et al. 2012). The materials used are classified into two: a five-membered ring at a core (Lee et al. 2007, Southern et al. 2008, Francescangeli et al. 2009, Xiang et al. 2009, Picken et al. 2012), which is the same as the original compound used for the two papers, which were published in Phys. Rev. Lett. (Acharya et al. 2004, Madsen et al. 2004) and is already shown in Figure 4.11a, and a phenyl ring with a substituent at the core (Prasad et al. 2005, Dong and Marini 2009, Jang et al. 2009, Yoon et al. 2010, Aluculesei et al. 2012), a typical example of which is shown in Figure 4.11b. Besides the techniques of XRD and NMR, optical measurements such as an electro-optic switching about the short and long molecular axis and birefringence (Lee et al. 2007, Jang et al. 2009) were conducted. However, the results of Lee were strongly criticized by Stannarius (2008) for their incorrect interpretation. Moreover, as mentioned earlier, biaxiality evidenced by XRD (Prasad et al. 2005) and NMR (Dong and Marini 2009) observed in the compound shown in Figure 4.11b (called A131) was disproved by optical measurements under a moderate electric field (Le et al. 2009b, Senyuk et al. 2010). However, there exist some results convincing the biaxiality. Figure 4.14, for example, shows one of the results obtained by Jang et al. (2009). They observed a jump in the effective birefringence at the N_u–N_b phase transition in both homeotropic (not shown) and planar (Figure 4.14) cells of a molecule shown in Figure 4.11c. However, the minor birefringence, n_2–n_3, obtained in these cells showed quite a large difference, 0.0085 in a homeotropic cell and 0.0023 in a planar cell. This fact still remains uncertain in convincing the biaxiality. More recent work by Aluculesei et al. (2012) also claim that the study of the proton NMR spin-lattice relaxation time clearly shows a transition within the nematic range. In this way, the re-evaluation of the identification of phase biaxiality is necessary from both viewpoints of methods and compounds.

Electro-optic measurements are also important, if we have real biaxial nematic materials from an application viewpoint. Berardi et al. (2008) conducted virtual molecular dynamics computer experiments, simulated the electro-optic switching of primary and secondary directors, and found a fast response of the secondary director. Electro-optic measurements have been made by many researchers (Berardi et al. 2007, Lee et al. 2007, Francescangeli et al. 2009, Le et al. 2009b, Yoon et al.

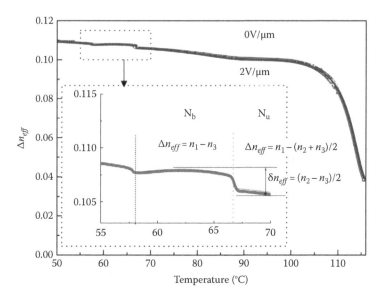

FIGURE 4.14 **(See color insert.)** Effective birefringence as a function of temperature. (Reprinted with permission from Jang, Y., Panov, V.P., Kocot, A., Vij, J.K., Lehmann, A., and Tschierske, C., Optical confirmation of biaxial nematic (N$_b$) phase in a bent-core mesogen, *Appl. Phys. Lett.*, 95, 183304-1–183304-3. Copyright 2009 by the American Institute of Physics.)

2010, Shanker et al. 2011). Lee et al. (2007) measured rising and relaxation times of both primary and secondary directors and found faster response time at smaller field strengths in the secondary director. Francescangeli et al. (2009) claimed ferro-electric switching and argued in connection with field-induced biaxiality. However, ferroelectricity in this nematic material is questionable, since this material is second-harmonic generation (SHG) inactive (unpublished data).

Despite such uncertainty of the existence of the thermotropic N$_b$ phase, molecular designs realizing real N$_b$ mesogens have been actively conducted. In their review article, Tschierske and Photinos (2010) suggested many molecular architectures designed to produce the N$_b$ phases. Recently, Kim et al. (2012) synthesized a compound, nonsymmetrically substituted thiadiazole shown in Figure 4.15, to promote the development of the secondary director. After many kinds of measurements such as conoscopy, their conclusion is quite cautious; significant biaxiality fluctuations and a remarkable flow-induced, transient realignment of the uniaxial director originate from the temperature-dependent formation of nanoscale, biaxial complexes of molecules. As they suggested, the design strategy based on a shape-persistent core structure is worth pursuing toward obtaining true biaxial nematic mesogens.

4.4 ELASTIC PROPERTIES

Elastic constants are one of the most important material constants in nematic (N) liquid crystals (NLCs), because they are directly correlated to the electro-optic performance in display devices. At the same time, they are also important from a basic

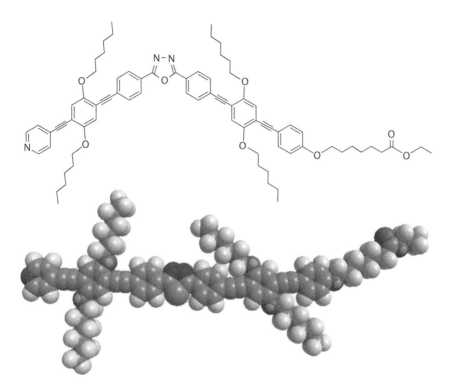

FIGURE 4.15 **(See color insert.)** A molecule synthesized by Kim et al. to promote biaxiality. (From Kim, Y.-K., Majumdar, M., Senyuk, B.I., Tortora, L., Seltmann, J., Lehmann, M., Jákli, A., Gleeson, J.T., Lavrentovich, O.D., and Sprunt, S., Search for biaxiality in a shape-persistent bent-core nematic liquid crystal, *Soft Matter*, 8, 8880–8890, 2012. Reproduced by permission of The Royal Society of Chemistry.)

science point of view. For instance, elastic constants diverge with decreasing temperature in the N phase when approaching the smectic (Sm) phase. This pretransitional phenomenon was observed quite a long time ago in many compounds having the N-Sm phase transition (Cheung and Meyer 1973, Cheung et al. 1973, Gruler 1973). Stabilization of the blue phase is known to be related to the elastic constant (Hur et al. 2011, Fukuda 2012).

Since NLCs are incompressible anisotropic fluids, the elastic response arises against the orientational distortion; the distortion of a uniform director orientation costs energy. Hence, we can naturally imagine that the elastic constants and their ratios sensitively depend on molecular shapes. There are three possible distortions, splay, twist, and bend, and the elastic constants against such distortions are denoted as K_{11}, K_{22}, and K_{33}, respectively.

In conventional calamitic NLCs, the magnitudes of elastic constants are in the order of $K_{33} > K_{11} > K_{22}$ (Schad and Osman 1981, Schadt and Gerber 1982) with some exceptions (de Jeu and Claassen 1977). However, bent-core LC molecules have reversed elastic anisotropy, that is, $K_{11} > K_{33}$ (Kundu et al. 2007, Sathyanarayana et al. 2010a, Tadapatri et al. 2010a, Hur et al. 2011, Majumdar et al. 2011), and T-shaped molecules has $K_{11} \approx K_{33}$ (Sathyanarayana et al. 2010b). Some data is given in the following text.

Dodge et al. (2000, 2003) reported that adding about 3 mol% of bent-core molecules can reduce K_{33} of a NLC made of rodlike molecules by a factor of 2. More systematic measurements were made in 2007 using mixtures of a bent-core molecule and a rodlike molecule shown in Figure 4.16a (Kundu et al. 2007). The temperature dependences of K_{11} and K_{33} in pure 8OCB and three mixtures are shown in Figure 4.16b and c, respectively. In both cases, the elastic constants decrease with increasing the contents of the bent-core molecule. But the decreasing rate is much more remarkable in K_{33} than in K_{11}. Let us look closely at K_{11} first. K_{11} increases with decreasing temperature as usual. But at a certain temperature, K_{11} shows a sharp decrease. At the same temperature, dielectric anisotropy also suddenly decreases. Hence, authors ascribed the reason to the formation of smectic-like short-range order. In this case, the bent-core molecules locate in the smectic structures as suggested in their previous work (Pratibha et al. 2000) (Figure 4.17b) and facilitate splay distortion of the director, resulting in the drop of K_{11}.

K_{33} becomes smaller than K_{11} at the smallest dopant ratio (11 mol%), although K_{33} is always larger than K_{11} in the whole temperature range of the N phase in pure 8OCB (Figure 4.16). K_{33} in a higher mixture (17 mol%) shows a unique temperature dependence; with decreasing temperature, K_{33} starts to increase, but turns to decrease, and again increases as the N-SmA transition temperature is approached. The very small K_{33} is attributed to the molecular orientation shown in Figure 4.17c, where the bent-shaped molecules promote bend deformation of rodlike molecular field.

The bent-core molecule used by Kundu et al. (2007) does not show the N phase, so they studied the elastic constants in mixtures. The first elastic constant measurements in a pure bent-core NLC (the same as in Figure 4.11b) were made by Sathyanarayana et al. (2010a). As shown in Figure 4.18, the value of K_{33} at a given temperature is always lower than K_{11} except for the temperature very close to the N-SmC transition, as observed in the mixtures. Later K_{22} was measured in addition to K_{11} and K_{33} (Majumdar et al. 2011), although the compound used was different from that used by Sathyanarayana et al. (2010a). They found that the reduction rate compared with calamitic molecules is larger in K_{22} than K_{33}, as predicted (Tamba et al. 2007); $K_{11} = 3.1$ pN, $K_{22} = 0.31$ pN, and $K_{33} = 0.88$ pN at 2° below the Iso-N transition. They also measured viscosity coefficients. The results show $\eta_{splay} \sim \eta_{bend} \sim 3\eta_{twist}$ and the anisotropy is quite different from calamitic LC (Majumdar et al. 2011). For instance, in 5CB, $\eta_{splay} \sim \eta_{twist} \sim 5\eta_{bend}$ (Chen et al. 1989, Majumdar et al. 2011). As a result, ratios of these viscosity coefficients between bent-core and calamitic molecules are quite different. Bent-core NLC is more than 10 times viscous for splay deformation. In contrast, for twist and bend, three or four times and about six times, respectively.

Let us describe the results on other compounds with different shapes, that is, hockey-stick-shaped molecules, T-shaped molecules, and bent-core molecules with long substituents in the core, to see further how the elastic constant anisotropy depends on the molecular shape. As for a hockey-stick-shaped molecule, the temperature dependence behavior of K_{11} and K_{33} is almost the same as in bent-core molecules (Figure 4.11)

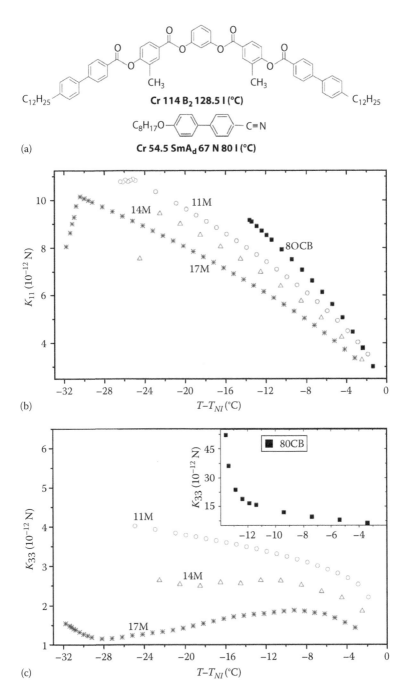

FIGURE 4.16 Temperature dependence of elastic constants (b) K_{11} and (c) K_{33} of mixtures of bent-core and rod-shaped molecules shown in (a). For instance, 11M stands for 11 mol% of bent-core molecules. (Reprinted with permission from Kundu, B., Pratibha, R., and Madhusudana, N.V., *Phys. Rev. Lett.*, 99, 247802-1. Copyright 2007 by the American Physical Society.)

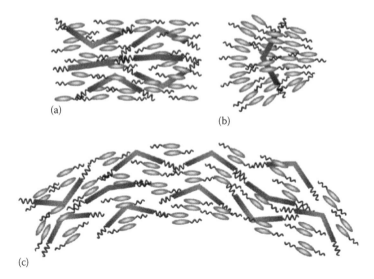

(a)

(b)

(c)

FIGURE 4.17 (a) Mutual alignment of rod-shaped and bent-core molecules in the nematic phase; (b) when smectic-like short-range order builds up, the bent-core molecules reorient and facilitate splay distortion of the director; and (c) in the nematic phase, bent-core molecules facilitate the band distortion of the director. (Reprinted with permission from Kundu, B., Pratibha, R., and Madhusudana, N.V., *Phys. Rev. Lett.*, 99, 247802-1. Copyright 2007 by the American Physical Society.)

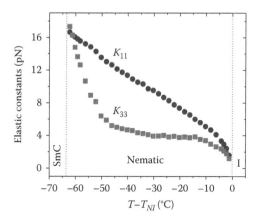

FIGURE 4.18 Temperature dependence of elastic constants K_{11} and K_{33}. (Reprinted with permission from Sathyanarayana, P., Mathew, M., Li, Q., Sastry, V.S.S., Kundu, B., Le, K.V., Takezoe, H., and Dhara, S., *Phys. Rev. E*, 81, 010702(R)-1. Copyright 2010 by the American Physical Society.)

aside from these absolute values (Sathyanarayana et al. 2012). In a T-shaped molecule, K_{11} is almost equal to K_{33}, and they are in the range of pN except for the temperature region near SmA, where both exhibit divergent increase (Sathyanarayana et al. 2010b, 2012). This suggests that the decreasing rate of K_{33} compared with that in calamitic molecules is not so remarkable in comparison with that in bent-core molecules.

The divergent increase of K_{11} (Sathyanarayana et al. 2010b, 2012) is also interesting to consider, since only K_{22} and K_{33} are known to show such behavior in calamitic molecules exhibiting the N-SmA phase transition since a long time ago (Cheung and Meyer 1973, Cheung et al. 1973, de Gennes 1973, Gruler 1973). However, it is also predicted that even K_{11} diverges as the N-SmC phase transition is approached (de Gennes 1973) and experimentally proved (Gruler 1973, de Jeu and Claassen 1977, Schadt and Gerber 1982). The T-shaped molecule used has the phase sequence of N-SmA. Because of the lack of the SmC phase in the T-shaped molecule used, the divergence should be attributed to the T-shape. Sathyanarayana et al. (2010b, 2012) pointed out the significant changes in the layer spacing when such molecule having a small molecular length/width ratio takes splay fluctuation near the N-SmA transition, leading to a large K_{11}.

The last example is a bent-core molecule with long substituents in the core (Figure 4.15), which is described as a material promoting biaxial N phase (Kim et al. 2012). This material is interesting; although the molecule has a bent-core. K_{33} is about double of K_{11}, and K_{22} is the smallest (Kaur et al. 2013). Namely, the elastic constant anisotropy is analogous to that commonly observed in calamitic NLCs. Kaur et al. ascribe this elastic behavior to the value of the bend angle in the molecular core. The bend angle increases from about 138° to about 164° on going from the oxadiazole to the thiadiazole system with the presence of the lateral chains in the latter.

As described in this section, elastic constants and their anisotropy critically depend on the shape of the constituent molecules forming the nematic phase. Cestari et al. (2011b) studied the elastic and flexoelectric properties of liquid crystal dimers (FFOnOCB, see at the top of Figure 4.19) as a case study using a molecular field theory with atomistic modeling. An unusually low bend elastic constant is predicted for bent dimers. Clear odd-even behavior is predicted as shown in Figure 4.19, where S_{CB} is the second rank order parameter for the para axis of the cyanobiphenyl group. It is interesting that even negative K_{33} values are predicted for odd ($n = 5$ and 7) dimers. The elastic constants particularly K_{33} is very important for the emergence of the blue phase and the twist-band N (N_{TB}) phase. These topics are described in Sections 4.5 and 4.6, respectively.

4.5 BLUE PHASE

The blue phase (BP) is one of the unique LC phases with three-dimensional superstructures stabilized by line defects and emerges below the Iso phase, generally in a very narrow temperature range (Wright and Mermin 1989, Crooker 2001). Hence, BP has never been considered for application. A breakthrough came from a pioneering work by Kikuchi et al. (2002); BP was stabilized over more than 60 K by partial polymerization. Extensive studies including theories and molecular designs have clarified that bent-shaped molecules are one of the ideal materials for stabilizing BP without polymerization. Coles and Pivnenko (2005) showed a wide-temperature BP using a bent-shaped molecule for the first time, and then lots of studies on BP using bent-core molecules followed. However, before these works, a pioneering work by Nakata et al. (2003) reported the usefulness of bent-core molecules for widening

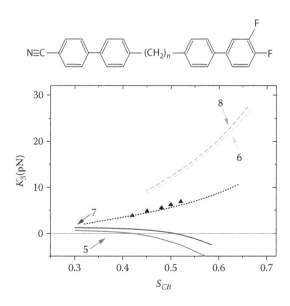

FIGURE 4.19   Calculation of K_{33} of FFOnOCB as a function of second-rank order parameter. Odd-even effect about the spacer length n is clearly shown. A dotted line and closed
triangle symbols are calculated and experimental results for a rod-shaped molecule 5OCB.
(From Cestari, M., Frezza, E., Ferrarini, A., and Luckhurst, G.R., Crucial role of molecular
curvature for the bend elastic and flexoelectric properties of liquid crystals: Mesogenic dimers
as a case study, *J. Mater. Chem.*, 21, 12303–12308, 2011. Reproduced by permission of the
Royal Society of Chemistry.)

BP; an addition of bent-shaped molecules to a conventional chiral host consisting of
calamitic NLC and chiral dopants stabilizes BP.

Let us first introduce the doping experiment. Figure 4.20 shows a phase behavior
of a commercial chiral host (nonchiral host and chiral dopant) as a function of content of a bent-core molecule (P-8-PIMB, see Figure 1.7) (Nakata et al. 2003). The
host chiral system without the bent-core molecule doping shows the Iso-N phase
transition. An addition of P-8-PIMB by only 6 wt% induces BP of about 1 K. The
widest BP of about 6 K was obtained at 15 wt%. Note that the dopant is a nonchiral
molecule. It is also true that doped rod-shaped molecules (TBBA) do not induce
BP, although another molecule (MHPOBC) showing the antiferroelectric phase does
induce BP to some extent. The latter effect was interpreted by the inherent bend
shape of MHPOBC. They pointed out two possible origins: chiral nature of bent-core
molecules (see Section 5.1) and low elastic constant (see Section 4.4).

In the wake of such pioneering works (Kikuchi et al. 2002, Nakata et al. 2003,
Coles and Pivnenko 2005) and possible application to display devices (Samsung
electronics, 2008 SID, Los Angeles, United States), many researchers started to be
involved in studying bent-shaped molecules as useful materials for widening BP. In
addition to the previous examples (Nakata et al. 2003, Coles and Pivnenko 2005), different types of bent-shaped molecules have been examined (Yelamaggad et al. 2006,
Lee et al. 2010, Taushanoff et al. 2010, Zheng et al. 2010, 2011, Hur et al. 2011,

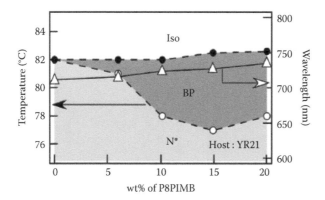

FIGURE 4.20 Phase diagram of mixtures of cholesteric liquid crystal YR21 (mixture of ZLI-2293 [Merck] and a chiral dopant MLC6248 [Merck] of 21%) and bent-core molecule P8-PIMB. Induced blue phase (BP) is clearly shown. Selective reflection wavelength is also shown. (Reprinted with permission from Nakata, M., Takanishi, Y., Watanabe, J., and Takezoe, H., *Phys. Rev. E*, 68, 041710-1. Copyright 2003 by the American Physical Society.)

Le et al. 2011, Mathews et al. 2011, Ocak et al. 2011, Wang et al. 2012a). Wang et al. (2012b) synthesized a series of oxadiazoles with different rigid cores (Figure 4.21) with C_5H_{11} for R and studied their BP ranges. C5H-OXD, C5T-OXD, and C5P-OXD show the N phase. C5P-OXD, which has a widest N range, doped with a chiral dopant showed the widest BP range of about 5 K. They found that the BP range varies with the core structures. Such oxadiazole derivatives have been widely used for the studies of the biaxial nematic (N_b) and BP; Zheng et al. (2010) also used CO4-OXP, CO8-OXP, C7P-OXD, and asymmetric CnP-OXD with C_7H_{15} and F at both ends. Lee et al. (2010) used a bent-core molecule (CO7P-OXP) shown in Figure 4.21 doped with a small amount of a chiral molecule with a high twisting power. They succeeded in obtaining BP at a doping rate of more than 5 wt% and extending BPI to more than 15 K. In contrast, BPIII was stably observed over or nearly 20 K by Taushanoff et al. (2010), Le et al. (2011), and Ocak et al. (2011) using different molecular systems. Particularly, the materials used by Le et al. showed BPIII even at room temperature.

Quite different molecular systems are also interesting for the emergence of BP. Yelamaggad et al. (2006) used novel dimers in which an achiral bent-core is covalently tethered to a bulky chiral rodlike mesogen through an alkylene spacer. Most of the compounds even with odd-numbered carbons in the alkylene spacer show a wide range of BPIII of about 20 K. In case of mesogens consisting of an achiral rod-shaped group and a cholesterol derivative connected by an alkylene spacer (Figure 4.22a), the influence of the number of carbons in the spacer on the phase structure is much more intriguing (Zep et al. 2013). (1) There is an odd-even effect in the phase sequence; Iso-N*-N_2 for odd carbons and Iso-N*-SmA for even carbons. (2) Multiple intermediate phases exist between N* and N_2. (3) N* is the cholesteric phase on heating from N_2, whereas N* is actually BP on cooling from Iso. N_2 is an N phase with twist-bend deformation (see Section 4.6). Here, we only focus on BP. Aya et al. (2014) studied the performance of electro-optic behavior in the BP of this

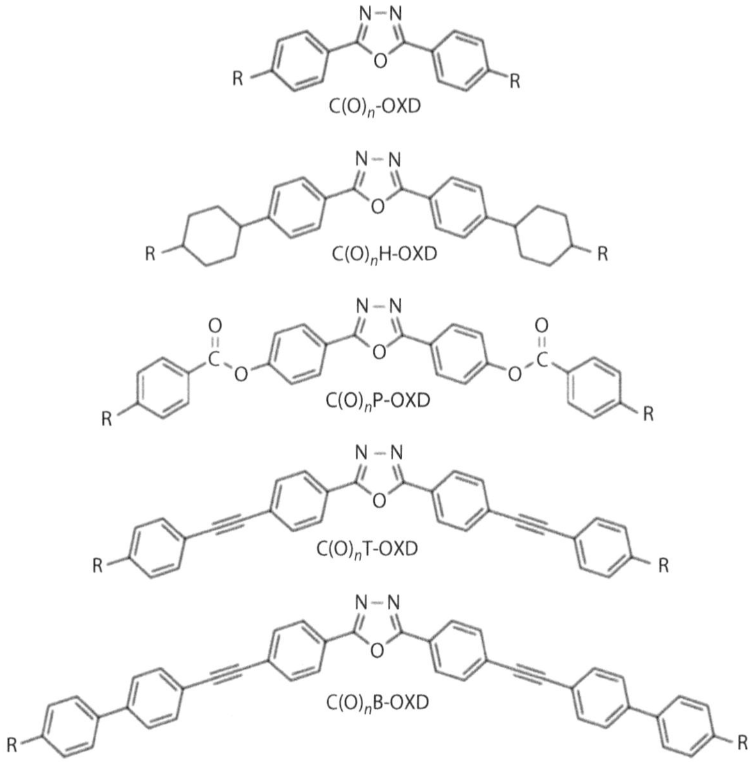

FIGURE 4.21   A series of oxadiazole molecules with different rigid cores synthesized and studied for their BP ranges. (Reprinted from Wang, L. et al., *Liq. Cryst.*, 39, 99, 2012. With permission from Taylor & Francis.)

compound with nine carbon atoms in the flexible spacer. The most important feature of this compound is the widest BP temperature range ~40 K among single component systems. A narrow temperature range (about 1 K) of BPIII exists, but the rest is BP_{cub} (BPI or BPII). There are some other compounds, which have comparable temperature range of 30–40 K, such as binaphthyl derivative (Yoshizawa et al. 2009) and eutectic LC mixtures composed of achiral or chiral hosts with chiral dopants (Coles and Pivnenko 2005, Kogawa and Yoshizawa 2011). Kerr constant was estimated by measuring induced birefringence Δn as a function of an electric field E, $\Delta n = \lambda KE^2$, where λ is the light wavelength used and K is the Kerr constant. The temperature dependence of K is shown in Figure 4.22b, where E^2 dependence of Δn at two temperatures is also shown in the inset (Aya et al. 2014). The attainable Kerr constant over 1.3×10^{-9} mV^{-2} is a relatively large value among those in other BP materials. Electro-optic performance is shown in Figure 4.22c and d. Much faster response such as μs was also reported in polymer-stabilized BP (Hisakado et al. 2005). But, still, very high applied voltages necessary for electro-optic switching are one of the serious problems for the practical application as a display.

As mentioned earlier, one of the possible reasons of the usefulness of bent-shaped molecules for stabilizing BP is small K_{33}. Very important results supporting this idea

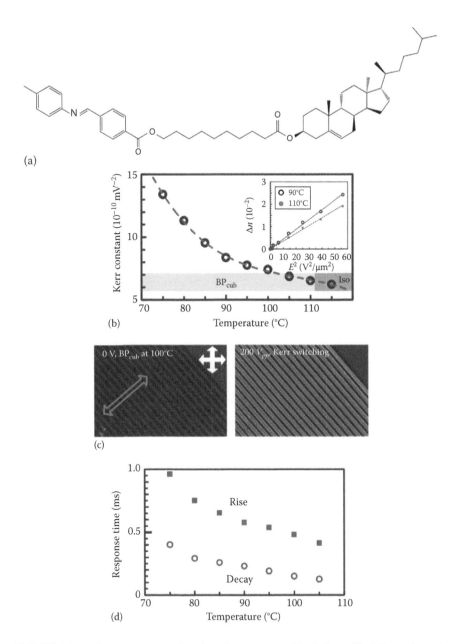

FIGURE 4.22 Kerr constant as a function of temperature (b) of a bent chiral dimer shown in (a). Quadratic field dependence of the induced birefringence is shown in the inset. (c) Textures showing an electro-optic response by applying an in-plane electric field. (d) Response time of electrooptic switching for rise and decay processes. (From Aya, S., Zep, A., Aihara, K., Ema, K., Pociecha, D., Gorecka, E., Araoka, F., Ishikawa, K., and Takezoe, H., Stable electro-optic response in wide-temperature blue phases realized in chiral asymmetric bent dimers, *Opt. Mater. Exp.*, 4, 662–671, 2014. With permission of Optical Society of America.)

have been reported by Hur et al. (2011). They examined the relationship between the thermal stability (temperature range) of BP and elastic anisotropy K_{33}/K_{11}. According to the elastic constant measurements, with increasing the content of the bent-core molecule, K_{33} sharply decreases, and K_{11} gradually increases, becoming equal to each other at a content of 24–25 wt%. Figure 4.23 is phase diagrams of binary mixtures

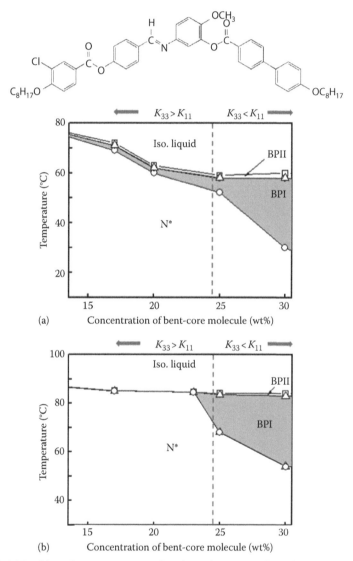

FIGURE 4.23 Blue phase ranges as a function of concentration of bent-core molecule shown at the top. Host is a commercial NLC mixture with two different chiral dopants: (a) S811 and (b) Iso-(6OBA)$_2$. (Reprinted from Hur, S.-T., Gim, M.-J., Yoo, H.-J., Choi, S.-W., and Takezoe, H., Investigation for correlation between elastic constant and thermal stability of liquid crystalline blue phase I, *Soft Matter*, 7, 8800–8803, 2011. Reproduced by permission of The Royal Society of Chemistry.)

of a commercially available NLC mixture and a bent-core LC shown at the top of Figure 4.23; chiral dopants used are (a) 28 wt% of S811 (Merck) and (b) 6 wt% of Iso-(6OBS)[2], which has a higher twisting power than that of S811 (Lee et al. 2010). It is very interesting to know that a BP temperature range (mostly BPI) sharply increases or BP is remarkably enhanced when K_{33} becomes smaller than K_{11}. This result is well predicted by numerical calculation (Alexander and Yeomans 2006, Fukuda 2012), which is described in the following text.

Yoshizawa (2013) reported material design for BP and summarized ideal molecular design for stabilizing BP in his review article. It appears that biaxial molecules such as bent-shaped, U-shaped, and T-shaped molecules may stabilize the BP phase. As we described earlier, indeed many bent-shaped molecules actually stabilize BP. Then, is molecular biaxiality an important factor for stabilizing BP? The review by Wright and Mermin (1989) discussed factors stabilizing BP, but does not give any answer for the importance of molecular biaxiality. It is true that BP is a strongly biaxial system, because, in the high-chirality limit, the free energy is given by any linear combination of biaxial helical order parameters and the orientational order around disclination cores, which stabilize BP, even if constituent molecules are uniaxial. However, the symmetry of the ordered phase does not need to directly reflect the symmetry of the individual constituent molecules (Wright and Mermin 1989). In this way, we need some more theoretical considerations why biaxial molecules enhance the stability of BP.

Then, what is important in bent-shaped molecules for stabilizing BP? The answer is elastic constants. We introduce two theoretical works on the relationship between stable BPs and elastic constants. Alexander and Yeomans (2006) simulated the phase diagram of N* LC within the framework of modified Landau-de Gennes theory, which incorporates all three elastic constants and allows for a temperature-dependent pitch. The simulation was made by minimizing the modified free energy,

$$F = \frac{1}{V} \int d^3r \left\{ \frac{\tau}{4} \mathrm{tr}\left(\mathbf{Q}^2\right) - \sqrt{6\mathrm{tr}}\left(\mathbf{Q}^3\right) + \left[\mathrm{tr}\left(\mathbf{Q}^2\right)\right]^2 + \frac{\kappa^2}{4}\left(\nabla \times \mathbf{Q} + \mathbf{Q}\right)^2 + \frac{L_{22}}{L_{21}}\left(\nabla \cdot \mathbf{Q}\right)^2 \right.$$
$$\left. + \frac{L_{31}}{L_{21}} \mathbf{Q}^2 \cdot \nabla \times \mathbf{Q} + \frac{L_{34}}{L_{21}} Q_{ab}\nabla_a Q_{cd}\nabla_b Q_{cd} + \frac{L_{38}}{L_{21}} Q_{ab}\nabla_c Q_{ad}\nabla_c Q_{bd} \right\},$$

$$(4.10)$$

where
   $\mathbf{Q}$ is a traceless symmetric second-rank tensor
   κ is the chirality parameter
   τ is the temperature parameter
   L_{ij}'s are parameters related to the Frank elastic constants K_{ii}

The latter term inside angled brackets comes from a gradient contribution, accounting for the energy cost associated with defect structures necessary for the stabilization of BP. By minimizing the modified free energy, phase diagrams, which include Iso, BPI, BPII, and N* phases, were obtained. The results show that the temperature ranges of BPI and BPII strongly depend on the elastic constants, being reduced when K_{33} is larger than K_{11} and when K_{22} is smaller than the others. This condition is normally the case in calamitic NLCs, being consistent with a very narrow BP range. The smaller K_{33} than K_{11} in bent-shaped NLCs enhances the stability of BP, as shown earlier (see Figure 4.23, Hur et al. 2011).

The same behavior was also theoretically concluded by Fukuda (2012). He modified Meiboom et al.'s theory (Meiboom et al. 1981), where one-constant approximation was used. The free energy of BP per unit length of a disclination line was written as

$$F = F_{core} + F_{interface} + F_2 + F_1, \qquad (4.11)$$

where each term stands for the excess free energy of the disclination core, the interfacial energy between the bulk and the disclination core, the free energy obtained by the integral over the surface surrounding the disclination line, and the Frank elastic energy in a region surrounding the $-1/2$ disclination, respectively. With and without considering the director profile of double-twist cylinders, BP range was respectively obtained as (Fukuda 2012)

$$\Delta T = \frac{\tilde{K}}{8aR_{max}^2} \exp\left(\frac{2K_{11} + 0.57K_{22} + 0.50K_{33}}{\tilde{K}} - 1 \right) \qquad (4.12)$$

$$\Delta T = \frac{\tilde{K}}{8aR_{max}^2} \exp\left(\frac{2(K_{11} + K_{22})}{\tilde{K}} - 1 \right), \qquad (4.13)$$

where R_{max} is a cutoff radius of the disclination core, and

$$\tilde{K} = \frac{1}{2}(K_{11} + K_{33})\left[1 - \frac{25}{72}\left(\frac{K_{11} - K_{33}}{K_{11} + K_{33}} \right)^2 \right]. \qquad (4.14)$$

Figure 4.24 shows the relative temperature range t with respect to that with one constant approximation as a function of (a) K_{11}/K (dotted curves) and K_{33}/K (solid curves), and (b) K_{22}/K, obtained by using Equation 4.12 (Fukuda 2012). A dramatic change was obtained in the effect of K_{33}; smaller K_{33} significantly enhances the stability of BP. As seen in Figure 4.24a, larger K_{22} amplifies the effect of K_{33}. Actually, larger K_{22} stabilize BP, as shown in Figure 4.24b. The results obtained using Equation 4.13 give similar behavior, although the dependence of K_{22} is much moderate. In conclusion, the stability of BP is enhanced by larger K_{11} and K_{22} and particularly smaller K_{33}. Hence, bent-shaped mesogens, which have a smaller K_{33}/K_{11}, are good choice for obtaining BP with a wide temperature range, as experimentally observed.

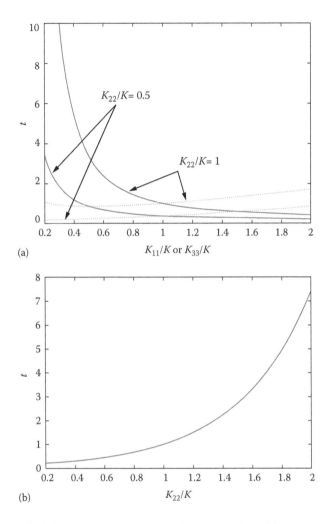

(a)

(b)

FIGURE 4.24 Relative temperature range t with respect to that with one constant approximation as a function of (a) K_{11}/K (dotted curves), K_{33}/K (solid curves) and (b) K_{22}/K, obtained by using Equation 4.12. Smaller K_{33} and larger K_{22} values expand the BP range. (Reprinted with permission from Fukuda, J., *Phys. Rev. E*, 85, 020701(R)-1. Copyright 2012 by the American Physical Society.)

4.6 TWIST-BEND NEMATIC PHASE

A new type of nematic phase was found and attracted considerable attention in bent-core and dimeric systems such as molecules called CB*n*CB with odd *n* (Figure 4.25), which has distinctly a bent shape.

This phase occurs below a conventional N phase and it is now designated as the twist-bend nematic N_{TB} phase. In this phase, the molecules form a heliconical structure with the twist and bend deformations, where the molecular axes tilt from the

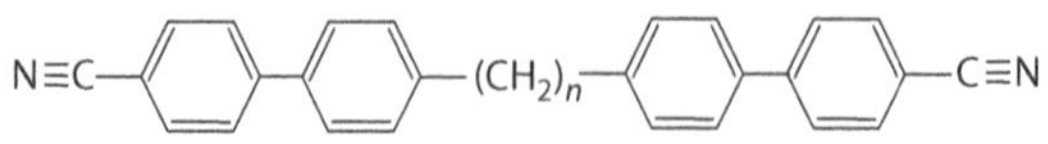

$$N{\equiv}C\!-\!\!\bigcirc\!\!\bigcirc\!\!-(CH_2)_n\!-\!\!\bigcirc\!\!\bigcirc\!\!-C{\equiv}N$$

FIGURE 4.25 Molecules, CBnCB with odd n, which show the N_{TB} phase.

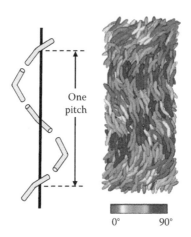

0°   90°

FIGURE 4.26 **(See color insert.)** Heliconical structure with twist and bend deformations. Image at the right is a simulation by Memmer. (Reprinted from Memmer, R., *Liq. Cryst.*, 29, 483, 2002. With permission from Taylor & Francis.)

helical axis by an angle θ_{TB} (Figure 4.26). Here, "heliconical" is used instead of "helical" to distinguish this unique arrangement from the helix in the cholesteric (Ch or N*) phase. The local average orientation of the long axis of one of the molecular cores $\mathbf{n}(z)$ is given by $\mathbf{z} \cos \theta_{TB} + \sin \theta_{TB} (\mathbf{x} \sin \varphi + \mathbf{y} \cos \varphi)$, where φ is given by $2\pi z/p_{TB}$ with a helical pitch p_{TB}. This structure is a result of a spontaneous reflection symmetry breaking, that is, two types of macroscopic chiral domains are formed. The existence of such a phase was originally proposed by Meyer as the result of the spontaneous appearance of the bend flexoelectric polarization (Meyer 1976). Later the phase was theoretically predicted by Dozov based on an elastic instability, where the bend elastic constant turns negative (Dozov 2001). Higher-ordered terms in the free-energy expansion are required to stabilize the phase. Depending on the ratio of the other two elastic constants in the high-temperature nonmodulated nematic phase, the low-temperature nematic phase can have a heliconical twist-bend structure of the local director, when $K_{11} > 2K_{22}$, or a splay-bend structure when $K_{11} < 2K_{22}$. An anomalously strong reduction of the bend elastic constant in the high-temperature N phase of the dimeric compounds on decreasing the temperature toward the N_{TB} phase (see the following text for details) sparkled undivided interest to this model. Memmer (2002) obtained a nematic-like phase with a heliconical structure using Monte Carlo simulation of the dynamics of rigid bent-core particles interacting via Gay–Berne potential (Figure 4.26, right).

It is impossible to fill space with a uniformly bent nematic director. However, a combination of twist and bend or splay and bend allows filling the space without voids. The structure resembles the SmC* structure, but the N_{TB} phase has no layer periodicity (it does not exhibit the density modulation along the helical axis) and the helical pitch is very short (less than 10 nm) compared with that of SmC* (Borshch et al. 2013, Chen et al. 2013, Meyer et al. 2013). In fact, the phase also resembles a cholesteric phase.

Although the existence of a N–N transition was reported by several authors (Ungar et al. 1992, Warrier et al. 1998, Sepelj et al. 2007, Kundu et al. 2009a, Tamba et al. 2010), the active investigations on the lower N phase under the recognition of the Dozov's N_{TB} phase started around 2010 (Panov et al. 2010, Cestari et al. 2011a). The nematic character of the twist-bend phase was confirmed by XRD by Panov et al. (2010). Cestari et al. reported XRD, dielectric, calorimetric, ESR, and ^{2}H NMR measurements and theoretical consideration. Most of the important observations, such as a ropelike texture (chirality), a focal conic texture (periodic structure), absence of the layer structure (nematic), mesogenic tilt and yet the optical uniaxiality (different from biaxial nematic), were made and all are consistent with the model phase structure shown in Figure 4.26. Those studies revealed an astonishingly small pitch of about 8–10 nm, which makes the N_{TB} phase also resemble the SmA phase. Smectic-like textures such as focal conics and the ability to form metastable freely suspended films up to 10° above the transition into the smectic phase are among important "smectic-like" features of the N_{TB} phase (Sebastián et al. 2016). Yet no electron-density modulation along the helical axis is present.

The in-plane polarization in the N_{TB} phase (perpendicular to the heliconical axis) is twisted too. Because of this short pitch, polarization cannot be measured directly. Indirect observation of the existence of polarization can be made by applying an electric field perpendicular to the heliconical axis (Panov and Vij 2011, Meyer et al. 2013). The electro-optic effect is of the flexoelectric origin and analogous to the electroclinic effect in the SmA* phase (Garoff and Meyer 1977) and the flexoelectric effect in the N* phase (Meyer 1969); a tilt of the optical axis linear with E is induced by an electric field. Figure 4.27 shows the tilt responses observed in CB7CB (Meyer et al. 2013): (a) the dynamic response of the induced tilt angle to an applied square-pulse field in a single domain, (b) the response to an applied triangular-wave voltage to two chiral domains of opposite handedness, (c) the temperature dependence of the induced tilt angle response to a square-wave voltage pulse of 40 V/1.6 µm. The response times for both rise and decay processes are very fast (~0.7 µs), as shown in Figure 4.27a. The response is linear in field and the tilt directions are opposite in domains of opposite chirality (Figure 4.27b). The induced tilt angle is very small (~1°) (Figure 4.27c). Very fast (less than µs) linear electro-optic response (electroclinic effect) indicates the existence of the in-plane polarization. Panov and Vij (2011) observed the linear electro-optic response in the homologous compound CB9CB and CB11CB with positive dielectric anisotropy ($\Delta\varepsilon > 0$), and also in other compounds with negative dielectric anisotropy ($\Delta\varepsilon < 0$).

Meyer et al. (2013) obtained theoretical expressions for the induced tilt $\alpha(E)$ and response time (decay) τ using the Dozov's theory with negative K_{33} and $K_{22} < K_{11}/2$

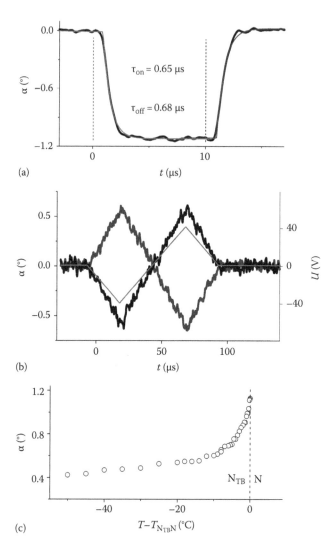

FIGURE 4.27   **(See color insert.)** (a) Dynamic response of the induced tilt angle by applying a square pulsed field in a single domain; (b) response by applying a triangular wave voltage to two opposite chiral domains; and (c) temperature dependence of the induced tilt angle by applying a square wave voltage pulse of 40 V/1.6 μm. (Reprinted with permission from Meyer, C., Luckhurst, G.R., and Dozov, I., *Phys. Rev. Lett.*, 111, 067801-1. Copyright 2013 by the American Physical Society.)

(Dozov 2001), generating spontaneous bend, $\mathbf{b}_s = q_s \sin\theta_s \cos\theta_s [\sin(q_s z), -\cos(q_s z), 0]$ and flexoelectric polarization $\mathbf{P}_f = -e_3 \mathbf{b}_s$:

$$\tan\alpha(E) = \frac{G_e}{K_N}E = \frac{(e_1 - e_3)E}{q_s(K_{11} + K_{22})}, \tag{4.15}$$

$$\tau_{off} = \frac{\gamma_1}{K_N} = \frac{2\gamma_1}{q_s^2\left(K_{11} + K_{22}\right)\sin^2\theta_s}, \tag{4.16}$$

where

e_1 and e_3 are the flexoelectric coefficients

q_s is the wave number of the helix

θ_s is the saturated tilt angle

γ_1 is the rotational viscosity coefficient

Using appropriate parameters from previous works and the experimental results obtained, the helical pitch was found to be 7 nm, and θ_s to be 11°. The value of the helical pitch is considerably smaller than that predicted by the elastic model (Dozov 2001) (approx. 300 nm) and comparable with the simulations by Memmer (2002) (a few molecular lengths).

Meyer et al. (2015) precisely measured the birefringence in both the N and N_{TB} phases in a quasi-planar state. They determined the tilt angle θ_{TB} of the heliconical twist bend structure. As shown in Figure 4.28, θ_{TB} is 9° at the N–N_{TB} transition temperature, indicating the weak first-order transition. The tilt tends to saturate at about 37°. The birefringence of the domain walls between the areas of opposite chirality is considerably higher than that in the bulk of the N_{TB} phase. Assuming the heliconical model, Meyer et al. suggested a local splay-bend director structure inside the walls to explain the anomaly of the birefringence.

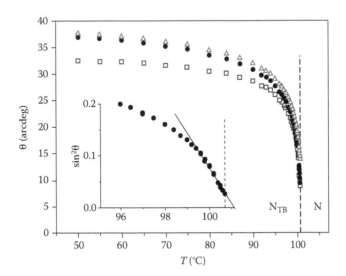

FIGURE 4.28 Temperature dependence of the tilt angle θ_{TB} of the heliconical twist bend structure. (From Meyer, C., Luckhurst, G.R., and Dozov, I., The temperature dependence of the heliconical tilt angle in the twist-bend nematic phase of the odd dimer CB7CB, *J. Mater. Chem. C*, 3, 318–328, 2015. Reproduced by permission of The Royal Society of Chemistry.)

Direct observation of the periodic structure was first made by Chen et al. (2013) using freeze-fracture transmission electron microscopy (FFTEM); the 2D fracture faces observed by FFTEM show quasiperiodic nanoscale stripes. It was confirmed that no stripes are seen, but only the larger-scale roughness was observed in the replicas cooled from the normal N phase, certifying that the periodic structure is characteristic of the N_{TB} phase. Careful analysis of the image estimates the periodicity of 8.3 nm. Only a very small temperature dependence of the periodicity was observed. The N_{TB} phase has no layer structure, but a continuous network of overlapping dimers, as shown in Figure 4.26. Thus, this structure cannot be detected by the nonresonant x-ray because it has no density modulation typical of the smectics. Better FFTEM images of CB7CB shown in Figure 4.29 were taken by Borshch et al. (2013). Figure 4.29a shows a typical FFTEM image, where the periodic structure is the same as that observed by Chen et al. (2013). The corresponding fast Fourier transform (FFT) pattern in the inset reveals a distinct uniform periodic structure with a pitch of 8.05 nm. Bouligand arches shown in Figure 4.29b are known to be indicative of an oblique helical structure, so providing evidence that the periodic structure is actually the heliconical structure of the N_{TB} phase. The observation of the Bouligand arches also indicates that the fractured plane is almost perpendicular to the helical axis of N_{TB}, as schematically shown in the inset. It is important to note that the same heliconical structure was reported in a crystalline structure of CB9CB in 2004 (Hori et al. 2004). The pitch (lattice constant) is 8.73 nm, being very close to that in the N_{TB} phase.

Gorecka et al. (2015b) recently applied atomic force microscopy (AFM) to observe periodic structures in CBnCB. They found surface crystallization occurs quite easily in samples after being frozen in liquid nitrogen and brought to room temperature within several minutes for $n = 9$ and 11 and less than an hour for $n = 7$. They warn that, although a periodic structure of about 8 nm can be found using AFM, this observation is made on crystalized surface, but not in the N_{TB} phase, and that it

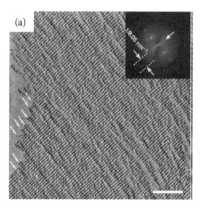 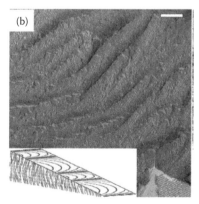

FIGURE 4.29 Freeze-fracture transmission electron microscopy (FFTEM) images of CB7CB. (a) A typical image and (b) an image showing Bouligand arches in a fractured plane. (Reprinted by permission from Macmillan Publishers Ltd. *Nat. Commun.*, Borshch, V., Kim, Y.-K., Xiang, J., Gao, M., Jákli, A., Panov, V.P., Vij, J.K. et al., Nematic twist-bend phase with nanoscale modulation of molecular orientation, 4, 2635-1–2635-8. Copyright 2013.)

is ambiguous whether or not the 8 nm stripes are indicative of heliconical structure. However, the compressibility measurements (see the following text) also indicate the existence of very short pitch, although indirectly.

A direct evidence of a bulk helical periodic modulation in the N_{TB} phase in CB7CB was given using the polarized resonant x-ray by Zhu et al. (2016). This technique has been shown to be an effective probe to elucidate the interlayer structure of synclinic and anticlinic smectics where the absence of the transversal electron-density modulations made these structures indistinguishable in nonresonant x-ray. Resonant x-ray studies using K-edge soft x-ray revealed that the helical pitch in the N_{TB} phase exhibits a weak temperature dependence increasing from 8 nm at a low temperature up to 9.5 nm at the transition to the N phase.

Some other features have been reported: novel textures such as focal conics and their electro-optic response, characteristic elastic constant anisotropy, and NMR particularly for chirality-related issues. Here we will describe them in the following.

Beguin et al. (2012) reported the use of NMR spectroscopy to determine experimentally whether the phase is chiral or not. They showed that all of the methylene groups in the heptane spacer have equivalent pairs of C–H groups in the N phase. However, this equivalence is lost for the six prochiral methylene groups with their enantiotopic protons on passing to the N_{TB} phase, whereas this equivalence is not lost for the central methylene group where the two protons are homotopic. Figure 4.30 shows the temperature dependence of the quadrupolar splitting for

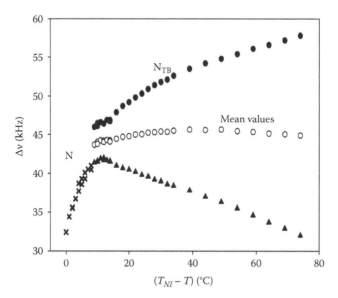

FIGURE 4.30 Temperature dependence of the quadrupolar splitting for partially deuterized 8CB-d_2 dissolved in CB7CB. (Reprinted with permission from Beguin, L., Emsley, J.W., Lelli, M., Lesage, A., Luckhurst, G.R., Timimi, B.A., and Zimmermann, H., The chirality of a twist-bend nematic phase identified by NMR spectroscopy, *J. Phys. Chem. B*, 116, 7940. Copyright 2012 by the American Chemical Society.)

partially deuterated 8CB-d_2 dissolved in CB7CB. The mean value shows a jump at the N–N$_{TB}$ phase transition but then changes only slightly and decreases at low temperatures in N$_{TB}$. The effect of the phase chirality on the quadrupolar splitting for 8CB-d_2 increases significantly with decreasing temperature. A similar observation was also made by Cestari et al. (2011a). NMR was also used to identify a single domain of uniform handedness by the addition of the dopant chiral solute of just a few % by weight (Emsley et al. 2013).

Characteristic textures in the N$_{TB}$ phase are striped texture and focal conics. Panov et al. (2010) reported a characteristic striped texture, which is parallel to the rubbing direction and has a period equal to double the cell gap, in the absence of a field. Other researchers (Cestari et al. 2011a, Sebastián et al. 2014, Tamba et al. 2015) also observed similar patterns. In these reports, a very characteristic rope texture, that is, tilted bands across the stripes, was observed. Panov et al. (2012) also demonstrated periodic pattern formation perpendicular to the rubbing direction by applying rather high ac electric field (10 V/μm) and frequency (>1.5 kHz) to bend dimers with negative Δε. The period depends on the field amplitude and the frequency; higher fields induce narrower and better-defined domain patterns but the visibility becomes worse at high frequency because of narrower periodicity such as 2 μm. The periodic pattern consists of two chiral domains, and the visibility is caused by small opposite tilts in two chiral domains.

Focal conic structures are known to be observed in the phases with long-range periodic structures such as smectic and cholesteric phases. Many researchers observed the focal conic structures in the N$_{TB}$ phase (Cestari et al. 2011a, Henderson and Imrie 2011, Mandle et al. 2014a, Gorecka et al. 2015b), indicating long-range periodic 1D ordering. Periodic structures could be the origin of finite compressibility. Normal N phase is noncompressible because of the absence of the translational order. In the smectic phases, it is known to exhibit the layer compressibility B of the order of 10^7 N/m. It is interesting that the compounds in the N$_{TB}$ phase showed the large compressibility of the same order of that in SmA (Gorecka et al. 2015a). Conventional N* phase shows small compressibility of the order of 10^4 N/m depending on the wavenumber of the helical structure q as $B \sim K_{22}q^2$, where K_{22} is a twist elastic constant. Although we have to include K_{33} as well, in addition to K_{22}, in the equation, short pitch of the order of 10 nm in N$_{TB}$ compared with that in N* (~1 μm) must be the reason of the large compressibility.

Elastic properties of the N$_{TB}$ phase remain still unknown. The model, based on the elastic instability accompanied by the inversion of the bend elastic constant, suggests negative K_{33} in the N$_{TB}$ phase. The stability of the phase, however, requires a bounded form of the elastic free energy. This was investigated in detail by Virga (2014) who developed a coarse-grained model of the N$_{TB}$ elasticity. In heliconical twist-bend phase, K_{33} loses its meaning of the elastic constant since the reference uniform director configuration is not the ground state for the N$_{TB}$. K_{33} is the local curvature of the free energy density $f(q)$ at the point of an infinite pitch $p = 2\pi/q$.

It is not easy to determine these elastic constants in the N$_{TB}$ phase or even no methodology has developed yet, but we can infer the negative K_{33} in N$_{TB}$ by measuring them in the N phase as a function of temperature. It is actually known that bend-shaped molecules have small K_{33} compared with K_{11} and K_{22} as described in Section 4.4.

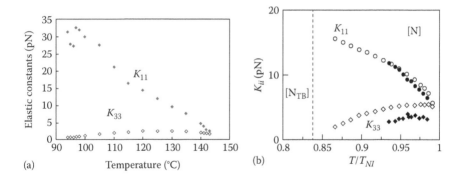

FIGURE 4.31 Temperature dependence of K_{11} and K_{33} determined by Fréedericksz transition method in two different materials (a) and (b). (Reprinted by permission from Macmillan Publishers Ltd. *Nat. Commun.*, Borshch, V., Kim, Y.-K., Xiang, J., Gao, M., Jákli, A., Panov, V.P., Vij, J.K. et al., Nematic twist-bend phase with nanoscale modulation of molecular orientation, 4, 2635-1–2635-8. Copyright 2013; From Sebastián, N., Lopez, D.O., Robles-Hernandez, B., de la Fuente, M.R., Salud, J., Perez-Jubindo, M.A., Dunmur, D.A., Luckhurst, G.R., and Jackson, D.J.B., Dielectric, calorimetric and mesophase properties of 1″-(2′,4-difluorobiphenyl-4′-yloxy)-9″-(4-cyano-biphenyl-4′-yloxy) nonane: An odd liquid crystal dimer with a monotropic mesophase having the characteristics of a twist-bend nematic phase, *Phys. Chem. Chem. Phys.*, 16, 21391–21406, 2014. Reproduced by permission of The Royal Society of Chemistry.)

Measurements in compounds with the N–N$_{TB}$ phase transition have been done by several groups (Atkinson et al. 2012, Adlem et al. 2013, Borshch et al. 2013, Sebastián et al. 2014, Sebastián et al. 2016). All groups reported decreasing K_{33} (abnormal) and increasing K_{11} (normal) with decreasing temperature in the N phase using different mixtures consisting of bent dimers (Figures 4.31 and 4.32), although in the close vicinity of the transition K_{33} increases again. Figure 4.31 shows the temperature dependence of K_{11} and K_{33} in different materials determined by Fréedericksz transition method by different groups (Borshch et al. 2013, Sebastián et al. 2014). Adlem et al. (2013) used dynamic light scattering in addition to the Fréedericksz method to determine all K_{ii} (i = 1, 2, and 3). Although the theoretical base is not established for the analysis of dynamic light scattering signal in the N$_{TB}$ phase, unexpected behavior in K_{33} was observed in the close vicinity of the N–N$_{TB}$ phase transition temperature (Figure 4.32). As seen in enlarged figure (Figure 4.32), a shallow minimum was observed at about 2 K above the N–N$_{TB}$ transition. Thus, still no distinct proof for negative K_{33} in N$_{TB}$ has been obtained. Similarly, K_{11} and K_{22} also increase with decreasing temperature (Figure 4.32). According to the Dozov's prediction, the condition of the twist bend phase is $K_{11} > 2K_{22}$, and it seems likely to be the case for CB7CB.

Rheological measurements have been done in the N$_{TB}$ phase (Salili et al. 2014). At shear stresses below 1 Pa, the apparent viscosity of N$_{TB}$ is 1000 times larger than in the N phase. But at the shear stress above 10 Pa, the N$_{TB}$ viscosity drops by two orders of magnitude, showing Newtonian fluid behavior. This is because the heliconical axis becomes normal to the shear plane via shear-induced alignment; the N$_{TB}$ phase is strongly shear thinning. Effective flexoelectric coefficient in N$_{TB}$ was found to be larger than that in the normal N (Atkinson et al. 2012, Balachandran

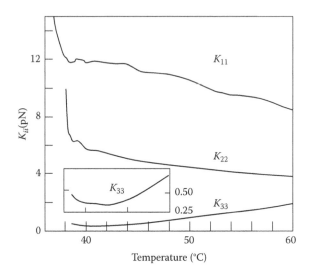

FIGURE 4.32   Temperature dependence of K_{11} and K_{33} determined by light scattering method in different materials. (Redrawn from Adlem, K. et al., *Phys. Rev. E*, 88, 022503-1, 2013.)

et al. 2014). Detailed dielectric measurements have also been made in bent dimers (Ribeiro de Almeida et al. 2014, Sebastián et al. 2014), but only a little information was obtained particularly for N_{TB}.

Extensive efforts have been made in synthesizing compounds that show the N_{TB} phase (Henderson and Imrie 2011, Ivsic et al. 2014, Mandle et al. 2014a) including unsymmetrical bimesogens (Chan et al. 2012, Mandle et al. 2014b) and a hydrogen-bonded system (Jansze et al. 2015). Ivsic et al. (2014) synthesized odd-membered imino-linked cyanobiphenyl, performed computational and liquid-state NMR, and showed that molecules have to have transient chiral conformation in order to exhibit the N_{TB} phase. Jansze et al. (2015) realized a bent shape using mesogens with an open hydrogen-bonded unit. Recently, the N_{TB} phase was also found in achiral rigid bent-core mesogens (stick shaped) (Chen et al. 2014). Similar to the N_{TB} phase observed in bent dimers, most of the characteristic features such as focal conic textures, no layer structure, short helical pitch (14 nm), and electroclinic effect, were observed. Actually, Pelzl et al. (2002) observed a chiral nematic structure in nonchiral bent-shaped molecule, and suggested the twist-bend structure, but it has not been proved yet.

Even more interesting is the chiral version of the N_{TB} phase. Asymmetric dimers with a cholesterol derivative and a rod-shaped mesogenic group (Chan et al. 2012, Zep et al. 2013) have been synthesized. No N_{TB} phase was found in the work by Chan et al. (2012). Zep et al. (2013) found an N_{TB}-like chiral phase (N_{TB}^*, where * designates a phase formed by *chiral* molecules) in every homologue they studied (Figure 4.22 top, $n = 3, 5, 7, 9$, and 15). In this phase, no layer reflection was observed, and a characteristic texture, which cannot be observed in the conventional N_{TB} phase, was found (Figure 4.33a and b). A striped pattern, perpendicular to the rubbing direction, can be seen using light polarized along the rubbing direction. The stripe was interpreted to be visible by the focusing of light due to a splay-bend modulated structure.

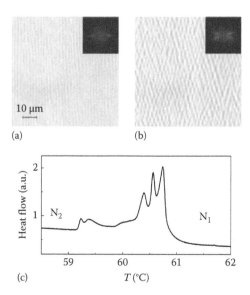

(a)   (b)

(c) T (°C)

FIGURE 4.33 (a, b) Textures in N_{TB}^* close to N* and (c) high definition DSC chart between two nematic phases. (From Zep, A., Aya, S., Aihara, K., Ema, K., Pociecha, D., Madrak, K., Bernatowicz, P., Takezoe, H., and Gorecka, E., Multiple nematic phases observed in chiral mesogenic dimers, *J. Mater. Chem. C*, 1, 46–49, 2013. Reproduced by permission of The Royal Society of Chemistry.)

With slightly decreasing temperature, the stripes become inclined and crossed, which could not be interpreted yet. The transition from N to N_{TB}^* phase is associated with a sequence of several nematic phases, as shown in Figure 4.33c (see also Section 4.5).

More recently, the chiral dimer was investigated from a viewpoint whether N_{TB}^* is the same as N_{TB} or not (Gorecka et al. 2015a). According to a miscibility test, N_{TB}^* is identical to N_{TB} of CB7CB like the relation between N* and N. However, there are many differences, that is, N_{TB}^* and N_{TB} have positive and negative optical anisotropy, respectively, and the periodicities observed by AFM are ~50 and 8 nm for N_{TB}^* and N_{TB}, respectively. The AFM images in (a) pure N_{TB}^*, (b) pure N_{TB}, and (c) a mixture consisting of 5% chiral dimer are shown in Figure 4.34. In (b) and (c) focal conic textures are clearly observed. In addition to defect lines of 8 nm periodicity, irregular defect lines are very characteristic for N_{TB}^* in Figure 4.34c.

Theoretical works have been made by several groups (Shamid et al. 2013, Greco et al. 2014, Vaupotič et al. 2014, Virga 2014). Shamid et al. (2013) developed Landau theory for bend flexoelectricity. The theory leads to a second-order transition from high-temperature uniform N phase to low-temperature nonuniform polar phase composed of twist-bend or splay-bend deformations. In addition to the normal elastic contributions due to splay, twist, and bend deformations (Oseen–Frank free energy), they introduced a polar interaction

$$F_{pp} = \frac{1}{2}\mu \mathbf{P}^2 \qquad (4.17)$$

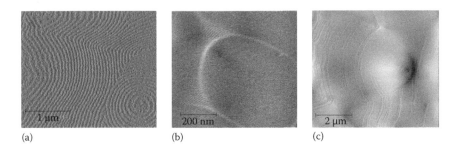

(a)   (b)   (c)

FIGURE 4.34 AFM images in (a) pure N_{TB}^{*} of chiral dimer; (b) pure N_{TB} of CB7CB; and (c) a mixture containing 5% chiral dimer. (From Gorecka, E., Vaupotič, N., Zep, A., Pociecha, D., Yoshioka, J., Yamamoto, J., and Takezoe, H.: A twist-bend nematic (N_{TB}) phase of chiral materials. *Angew. Chem. Int. Ed.* 2015. 54. 10155–10159. Copyright Wiley-VCH Verlag GmbH & Co. KGaA. Reproduced with permission.)

and a coupling energy between the bend flexoelectricity and the bend deformation

$$\mathbf{B} = \hat{\boldsymbol{n}} \times (\nabla \times \hat{\boldsymbol{n}}) \tag{4.18}$$

$$F_{np} = -\lambda \mathbf{B} \cdot \mathbf{P} \tag{4.19}$$

then wrote the free energy using effective renormalized bend elastic constant given by

$$K_{33}^{eff} = K_{33} - \frac{\lambda^2}{\mu} = K_{33} - \frac{\lambda^2}{\mu'(T - T_0)}. \tag{4.20}$$

However, they could not conclude that the low-temperature phase is the N_{TB} phase. Greco et al. described the N–N_{TB} phase transition using generalized Maier–Saupe molecular field theory (Greco et al. 2014). On approaching the transition, a gradual softening of the bend mode in the N phase was suggested and is the crucial role for the formation of the modulated nematic phase. Virga (2014) treated the N_{TB} phase as mixtures of two different N phases, each with a heliconical molecular arrangement of the same pitch but opposite helicities. A quadratic elastic theory was put forward for each helical nematic phase, which extends the classical theory of Frank for ordinary nematics and still features four positive elastic constants. The question whether the N–N_{TB} transition is a spontaneous chiral instability or it requires chiral configurations of the mesogens has been raised and discussed by several researchers. A cluster-type structure composed by twisted mesogens was proposed by Vaupotič et al. (2016) and also used by Ramou et al. (2016). At the same time, Greco and Ferrarini (2015) showed that the transition can be entropy driven in a system of achiral crescent-shaped particles, and demonstrated in molecular dynamic simulations that no chirality of the mesogens is necessary to develop the twist bend structure. In this case, the reflection symmetry breaking is spontaneous.

In summary, the N_{TB} phase is one of the most recently discovered phases consisting of bent-shaped molecules. The essence of the structural feature is more or less clear: doubly degenerate heliconically modulated nematic structure with possibly a local polar order. Still there is strong debate for the details. We still need further experimental and theoretical studies (Čopič 2013).

REFERENCES

Acharya, B. R., S.-W. Kang, V. Prasad, and S. Kumar. Role of molecular structure on x-ray diffraction in uniaxial and biaxial phases of thermotropic liquid crystals. *J. Phys. Chem. B* 113 (2009): 3845–3852.

Acharya, B. R., A. Primak, and S. Kumar. Biaxial nematic phase in bent-core thermotropic mesogens. *Phys. Rev. Lett.* 92 (2004): 145506-1–145506-4.

Adlem, K., M. Čopič, G. R. Luckhurst, A. Mertelj, O. Parri, R. M. Richardson, B. D. Snow, B. A. Timimi, R. P. Tuffin, and D. Wilkes. Chemically induced twist-bend nematic liquid crystals, liquid crystal dimers, and negative elastic constants. *Phys. Rev. E* 88 (2013): 022503-1–022503-8.

Alexander, G. P. and J. M. Yeomans. Stabilizing the blue phases. *Phys. Rev. E* 74 (2006): 061706-1–061706-9.

Aluculesei, A., F. Vaca Chavez, C. Cruz, P. J. Sebastiao, N. G. Nagaveni, V. Prasad, and R. Y. Dong. Proton NMR relaxation study on the nematic–nematic phase transition in A131 liquid crystal. *J. Phys. Chem. B* 116 (2012): 9556–9563.

Atkinson, K. L., S. M. Morris, F. Castles, M. M. Qasim, D. J. Gardiner, and H. J. Coles. Flexoelectric and elastic coefficients of odd and even homologous bimesogens. *Phys. Rev. E* 85 (2012): 012701-1–012701-4.

Aya, S., A. Zep, K. Aihara, K. Ema, D. Pociecha, E. Gorecka, F. Araoka, K. Ishikawa, and H. Takezoe. Stable electro-optic response in wide-temperature blue phases realized in chiral asymmetric bent dimers. *Opt. Mater. Exp.* 4 (2014): 662–671.

Aziz, N., S. M. Kelly, W. Duffy, and M. Goulding. Banana-shaped dopants for flexoelectric nematic mixtures. *Liq. Cryst.* 35 (2008): 1279–1292.

Balachandran, R., V. P. Panov, Y. P. Panarin, J. K. Vij, M. G. Tamba, G. H. Mehl, and J. K. Song. Flexoelectric behavior of bimesogenic liquid crystals in the nematic phase— Observation of a new self-assembly pattern at the twist-bend nematic and the nematic interface. *J. Mater. Chem. C* 2 (2014): 8179–8184.

Balachandran, R., V. P. Panov, J. K. Vij, A. Lehmann, and C. Tschierske. Effect of cybotactic clusters on the elastic and flexoelectric properties of bent-core liquid crystals belonging to the same homologous series. *Phys. Rev. E* 88 (2013): 032503-1–032503-11.

Beguin, L., J. W. Emsley, M. Lelli, A. Lesage, G. R. Luckhurst, B. A. Timimi, and H. Zimmermann. The chirality of a twist-bend nematic phase identified by NMR spectroscopy. *J. Phys. Chem. B* 116 (2012): 7940–7951.

Berardi, R., L. Muccioli, and C. Xannoni. Field response and switching times in biaxial nematics. *J. Chem. Phys.* 127 (2007a): 024905–024916.

Berardi, R., A. Costantini, L. Muccioli, S. Orlandi, and C. Zannoni. A computer simulation study of the formation of liquid crystal nanodroplets from a homogeneous solution. *J. Chem. Phys.* 126 (2007b): 044905–044909.

Berardi, R., L. Muccioli, and C. Zannoni. Field response and switching times in biaxial nematics. *J. Chem. Phys.* 128 (2008): 024905-1–024905-12.

Billeter, J. L. and R. A. Pelcovits. Molecular shape and flexoelectricity. *Liq. Cryst.* 27 (2010): 1151–1160.

Blinov, L. M. Domain instabilities in liquid crystals. *J. Phys. (France)* 40 (1979): C3-247–C3-258.

Blinov, L. M., M. I. Barnik, H. Ohoka, M. Ozaki, and K. Yoshino. Surface and flexoelectric polarization in nematic and smectic A phases of liquid crystal 4-octyloxy-4′-cyanobiphenyl. *Jpn. J. Appl. Phys.* 40 (2001): 5011–5018.

Bobylev, Yu. P. and S. A. Pikin. Threshold piezoelectric instability in a liquid crystal. *Sov. Phys. JETP* 45 (1977): 195–198.

Borshch, V., Y.-K. Kim, J. Xiang, M. Gao, A. Jákli, V. P. Panov, J. K. Vij et al. Nematic twist-bend phase with nanoscale modulation of molecular orientation. *Nat. Commun.* 4 (2013): 2635-1–2635-8.

Buka, A. and N. Eber, eds. *Flexoelectricity in Liquid Crystals: Theory, Experiments and Application.* World Scientific, 2012.

Buka, A., N. Eber, K. Fodor-Csorba, A. Jákli, and P. Salamon. Physical properties of a bent-core nematic liquid crystal and its mixtures with calamitic molecules. *Phase Trans.* 85 (2012): 872–887.

Buka, A., N. Eber, W. Pesch, and L. Kramer. Convective patterns in liquid crystals driven by electric field. In: A. A. Golovin and A. A. Nepomnyashchy, eds., *Self Assembly, Pattern Formation and Growth Phenomena in Nano-Systems.* Dordrecht, the Netherlands: Springer, 2006, pp. 55–82.

Castles, F., S. M. Morris, and H. J. Coles, The limits of flexoelectricity in liquid crystals. *AIP Adv.* 1 (2011): 032120.

Cestari, M., S. Diez-Berart, D. A. Dunmur, A. Ferrarini, M. R. de la Fuente, D. J. B. Jackson, D. O. Lopez et al. Phase behavior and properties of the liquid-crystal dimer 1″,7″-bis(4-cyanobiphenyl-4′-yl)heptane: A twist-bend nematic liquid crystal. *Phys. Rev. E* 84 (2011a): 031704-1–031704-20.

Cestari, M., E. Frezza, A. Ferrarini, and G. R. Luckhurst. Crucial role of molecular curvature for the bend elastic and flexoelectric properties of liquid crystals: Mesogenic dimers as a case study. *J. Mater. Chem.* 21 (2011b): 12303–12308.

Chakraborty, L., N. Chakraborty, D. D. Sarkar, N. V. S. Rao, S. Aya, K. V. Le, F. Araoka et al. Unusual temperature dependence of smectic layer structure associated with the nematic–smectic C phase transition in a hockey-stick-shaped four-ring compound. *J. Mater. Chem. C* 1 (2013): 1562–1566.

Chan, T.-N., Z. Lu, W.-S. Yam, G.-Y. Yeap, and C. T. Imrie. Non-symmetric liquid crystal dimers containing an isoflavone moiety. *Liq. Cryst.* 39 (2012) 393–402.

Chen, D., M. Nakata, R. Shao, M. R. Tuchband, M. Shuai, U. Baumeister, W. Weissflog et al. Twist-bend heliconical chiral nematic liquid crystal phase of an achiral rigid bent-core mesogen. *Phys. Rev. E* 89 (2014): 022506-1–022506-5.

Chen, D., J. H. Porada, J. B. Hooper, A. Klittnick, Y. Shen, M. R. Tuchband, E. Korblova et al. Chiral heliconical ground state of nanoscale pitch in a nematic liquid crystal of achiral molecular dimers. *Proc. Natl. Acad. Sci.* 1101 (2013): 15931–15936.

Chen, G. P., H. Takezoe, and A. Fukuda. Determination of K_i ($i = 1$–3) and μ_j ($j = 2$–6) in 5CB by observing the angular dependence of Rayleigh line spectral widths. *Liq. Cryst.* 5 (1989): 341–347.

Cheung, D., S. J. Clark, and M. R. Wilson, Calculation of flexoelectric coefficients for a nematic liquid crystal by atomistic simulation. *J. Chem. Phys.* 121 (2004): 9131.

Cheung, L. and R. B. Meyer. Pretransitional anomaly in bend elastic-constant for a nematic to smectic A transition. *Phys. Lett.* 43A (1973) 261–262.

Cheung, L., R. B. Meyer, and H. Gruler. Measurements of nematic elastic constants near a second order nematic–smectic-A phase change. *Phys. Rev. Lett.* 31 (1973): 349–352.

Chiccoli, C., I. Feruli, O. D. Lavrentovich, P. Pasini, S. V. Shiyanovskii, and C. Zannoni. Topological defects in schlieren textures of biaxial and uniaxial nematics. *Phys. Rev. E* 66 (2002): 030701(R)-1–030701(R)-4.

Coles, H. J. and M. N. Pivnenko. Liquid crystal 'blue phases' with a wide temperature range. *Nature* 436 (2005): 997–1000.

Čopič, M. Nematic phase of achiral dimers spontaneously bends and twists. *Proc. Natl. Acad. Sci.* 110 (2013): 15855–15856.

Crooker, P. P. Chapter 7. In: Blue Phases, H.-S. Kitzerow and C. Bahr, eds., *Chirality in Liquid Crystals.* New York: Springer, 2001.

de Gennes, P. G. Some remarks on the polymorphism of smectics. *Mol. Cryst. Liq. Cryst.* 21 (1973): 49–76.

de Jeu, W. H. and W. A. P. Claassen. The elastic constants of nematic liquid crystalline terminally substituted azoxybenzenes. *J. Chem. Phys.* 67 (1977): 3705–3712.

Derzhanski, A. and A. G. Petrov. A molecular-statistical approach to the piezoelectric properties of nematic liquid crystals, *Phys. Lett. A* 36 Oct. (1971): 6, 483–484.

Dodge, M. R., R. G. Petschek, C. Rosenblatt, M. E. Neubert, and M. E. Walsh. Light scattering investigation above the nematic–smectic-A phase transition in binary mixtures of calamitic and bent-core mesogens. *Phys. Rev. E* 68 (2003): 031703-1–031703-6.

Dodge, M. R., C. Rosenblatt, R. G. Petschek, M. E. Neubert, and M. E. Walsh. Bend elasticity of mixtures of V-shaped molecules in ordinary nematogens. *Phys. Rev. E* 62 (2000): 5056–5063.

Dong, R. Y. and A. Marini. Conformational study of a bent-core liquid crystal: ^{13}C NMR and DFT computation approach. *J. Phys. Chem. B* 13 (2009): 14062–14072.

Dozov, I. On the spontaneous symmetry breaking in the mesophases of achiral banana-shaped molecules. *Europhys. Lett.* 56 (2001): 247–253.

Dozov, I., P. Martinot-Lagarde, and G. Durand, Flexoelectrically controlled twist of texture in a nematic liquid crystal. *J. Phys. Lett.* 43 (1982): 365–369.

Emsley, J. W., P. Lesot, G. R. Luckhurst, A. Meddour, and D. Merlet. Chiral solutes can seed the formation of enantiomorphic domains in a twist-bend nematic liquid crystal. *Phys. Rev. E* 87 (2013): 040501(R)-1–040501(R)-4.

Ferrarini, A. Shape model for the molecular interpretation of the flexoelectric effect. *Phys. Rev. E* 64 (2001): 021710-1–021710-11.

Francescangeli, O. and E. Samulski. Insights into the cybotactic nematic phase of bent-core molecules. *Soft Matter* 6 (2010): 2413–2420.

Francescangeli, O., F. Vita, C. Ferrero, T. Dingemans, and E. Samulski. Cybotaxis dominates the nematic phase of bent-core mesogens: A small-angle diffuse x-ray diffraction study. *Soft Matter* 7 (2011): 895–901.

Francescangeli, O., V. Stanic, S. I. Torgova, A. Strigazzi, N. Scaramuzza, C. Ferrero, I. P. Dolbnya et al. Ferroelectric response and induced biaxiality in the nematic phase of a bent-core mesogen. *Adv. Func. Mater.* 19 (2009): 2592–2600.

Fukuda, J. Stabilization of blue phases by the variation of elastic constants. *Phys. Rev. E* 85 (2012): 020701(R)-1–020701(R)-4.

Galerne, E. Comment on Thermotropic biaxial nematic liquid crystals. *Phys. Rev. Lett.* 96 (2006): 219803-1.

Garoff, S. and R. B. Meyer. Electro-clinic effect at AC phase-change in a chiral smectic liquid-crystal. *Phys. Rev. Lett.* 38 (1977): 848–851.

Gorecka, E., M. Salamonczyk, A. Zep, D. Pociecha, C. Welch, Z. Ahmed, and G. H. Mehl. Do the short helices exist in the nematic TB phase? *Liq. Cryst.* 42 (2015b): 1–7.

Gorecka, E., N. Vaupotič, A. Zep, D. Pociecha, J. Yoshioka, J. Yamamoto, and H. Takezoe. A twist-bend nematic (N_{TB}) phase of chiral materials. *Angew. Chem. Int. Ed.* 54 (2015a): 10155–10159.

Greco, C. and A. Ferrarini. Entropy-driven chiral order in a system of achiral bent particles. *Phys. Rev. Lett.* 115 (2015): 147801–147805.

Greco, C., G. R. Luckhurst, and A. Ferrarini. Molecular geometry, twist-bend nematic phase and unconventional elasticity: A generalized Maier-Saupe theory. *Soft Matter* 10 (2014): 9318–9323.

Gruler, H. Elastic constants of nematic liquid crystals. *Z. Naturforsch.* 28a (1973): 474–483.

Harden, J., B. Mbanga, N. Eber, K. Fodor-Csorba, S. Sprunt, J. T. Gleeson, and A. Jákli. Giant flexoelectricity of bent-core nematic liquid crystals. *Phys. Rev. Lett.* 97 (2006): 157802-1–157802-4.

Harden, J., R. Teeling, J. T. Gleeson, S. Sprunt, and A. Jákli. Converse flexoelectric effect in a bent-core nematic liquid crystal. *Phys. Rev. E* 78 (2008): 031702-1–031702-5.

Helfrich, W. Conduction-induced alignment of nematic liquid crystals: Basic model and stability considerations. *J. Chem. Phys.* 51 (1969): 4092–4105.

Helfrich, W. The strength of piezoelectricity in liquid crystals. *Z. Naturforsch. A* 26 (1971): 833–835.

Helfrich, W. Inherent bounds to the elasticity and flexoelectricity of liquid crystals *Mol. Cryst. Liq. Cryst.* 26 (1974): 1–5.

Henderson, P. A. and C. T. Imrie. Methylene-linked liquid crystal dimers and the twist-bend nematic phase. *Liq. Cryst.* 38 (2011): 1407–1414.

Heuer, J., R. Stannarius, M.-G. Tamba, and W. Weissflog. Longitudinal and normal electroconvection rolls in a nematic liquid crystal with positive dieletric and negative conductivity anisotropy. *Phys. Rev. E* 77 (2008): 056206-1–056206-11.

Hisakado, Y., H. Kikuchi, T. Nagamura, and T. Kajiyama. Large electro-optic Kerr effect in polymer-stabilized liquid-crystalline blue phases. *Adv. Mater.* 17 (2005): 96–98.

Hong, S. H., R. Verduzco, J. C. Williams, R. J. Twieg, E. DiMasi, R. Pindak, A. Jákli, J. T. Gleeson, and S. Sprunt. Short-range smectic order in bent-core nematic liquid crystals. *Soft Matter* 6 (2010): 4819–4827.

Hori, K., M. Iimuro, A. Nakao, and H. Toriumi. Conformational diversity of symmetric dimer mesogens, α,ω-bis(4,4′-cyanobiphenyl)octane, -nonane, α,ω-bis(4-cyanobiphenyl-4′-yloxycarbonyl)propane, and -hexane in crystal structures. *J. Mol. Struc.* 699 (2004): 23–29.

Hur, S.-T., M.-J. Gim, H.-J. Yoo, S.-W. Choi, and H. Takezoe. Investigation for correlation between elastic constant and thermal stability of liquid crystalline blue phase I. *Soft Matter* 7 (2011): 8800–8803.

Ivsic, T., M. Vinkovic, U. Baumeister, A. Mikleusevic, and A. Lesac. Milestone in the N_{TB} phase investigation and beyond: Direct insight into molecular self-assembly. *Soft Matter* 10 (2014): 9334.

Jang, Y., V. P. Panov, A. Kocot, J. K. Vij, A. Lehmann, and C. Tschierske. Optical confirmation of biaxial nematic (N_b) phase in a bent-core mesogen. *Appl. Phys. Lett.* 95 (2009): 183304-1–183304-3.

Jansze, S. M., A. Martinez-Felipe, J. M. D. Storey, A. T. M. Marcelis, and C. T. Imrie. A twist-bend nematic phase derived by hydrogen bonding. *Angew. Chem. Int. Ed.* 54 (2015): 643–646.

Kaur, S., A. Belaissaour, J. W. Goodby, V. Gortz, and H. F. Gleeson. Nonstandard electroconvection in a bent-core oxadiazole material. *Phys. Rev. E* 83 (2011): 041704-1–041704-12.

Kaur, S., L. Tian, H. Liu, C. Greco, A. Ferrarini, J. Seltmann, M. Lehmann, and H. F. Gleeson. The elastic and optical properties of a bent-core thiadiazole nematic liquid crystal: The role of the bend angle. *J. Mater. Chem. C* 1 (2013): 2416–2425.

Keith, C., A. Lehmann, U. Baumeister, M. Prehm, and C. Tschierske. Nematic phases of bent-core mesogens. *Soft Matter* 6 (2010): 1704–1721.

Kikuchi, H., M. Yokota, Y. Hisakado, H. Yang, and H. Kajiyama. Polymer-stabilized liquid crystal blue phases. *Nat. Mater.* 1 (2002): 64–68.

Kim, Y.-K., M. Majumdar, B. I. Senyuk, L. Tortora, J. Seltmann, M. Lehmann, A. Jákli, J. T. Gleeson, O. D. Lavrentovich, and S. Sprunt. Search for biaxiality in a shape-persistent bent-core nematic liquid crystal. *Soft Matter* 8 (2012): 8880–8890.

Kogawa, Y. and A. Yoshizawa. Chiral effects of blue phase stabilization of a binaphthyl derivative. *Liq. Cryst.* 38 (2011): 303–307.

Kramer, L. and W. Pesch. Electrohydrodynamic instabilities in nematic liquid crystals. In: A. Buka and L. Kramer, eds., *Pattern Formation in Liquid Crystals*. New York: Springer-Verlag, 1996, pp. 221–256.

Krekhov, A., W. Pesch, and A. Buka. Flexoelectricity and pattern formation in nematic liquid crystals. *Phys. Rev. E* 83 (2011): 051706-1–051706-13.

Krishnamurthy, K. S., P. Tadapatri, and P. Viswanath. Dislocations and metastable chevrons in the electroconvective inplane normal roll state of a bent core nematic liquid crystal. *Soft Matter* 10 (2014): 7316–7327.

Kumar, P., Y. G. Marinov, H. P. Hinov, U. S. Hiremath, C. V. Yelamaggad, K. S. Krishnamurthy, and A. G. Petrov. Converse flexoelectric effect in bent-core nematic liquid crystals. *J. Phys. Chem. B* 113 (2009): 9168–9174.

Kundu, B., S. K. Pal, S. Kumar, R. Pratibha, and N. V. Madhusudana. Unusual odd-even effects depending on the monomer chain length in nematic liquid crystals made of rod-like dimers. *Europhys. Lett.* 85 (2009a): 36002-1–36002-6.

Kundu, B., R. Pratibha, and N. V. Madhusudana. Anomalous temperature dependence of elastic constants in the nematic phase of binary mixtures made of rodlike and bent-core molecules. *Phys. Rev. Lett.* 99 (2007): 247802-1–247802-4.

Kundu, B., A. Roy, R. Pratibha, and N. V. Madhusudana. Flexoelectric studies on mixtures of compounds made of rodlike and bent-core molecules. *Appl. Phys. Lett.* 95 (2009b): 081902-1–081902-3.

Le, K. V., F. Araoka, K. Fodor-Csorba, K. Ishikawa, and H. Takezoe. Flexoelectric effect in a bent-core mesogen. *Liq. Cryst.* 36 (2009a): 1119–1124.

Le, K. V., S. Aya, Y. Sasaki, H. Choi, F. Araoka, K. Ema, J. Mieczkowski, A. Jákli, K. Ishikawa, and H. Takezoe. Liquid crystalline amorphous blue phase and its large electrooptical Kerr effect. *J. Mater. Chem.* 21 (2011): 2855–2857.

Le, K. V., M. Mathews, M. Chambers, J. Harden, Q. Li, H. Takezoe, and A. Jákli. *Phys. Rev. E* 79 (2009b): 03701(R).

Lee, M., S.-T. Hur, H. Higuchi, K. Song, S.-W. Choi, and H. Kikuchi. Liquid crystalline blue phase I observed for a bent-core molecule and its electro-optical performance. *J. Mater. Chem.* 20 (2010): 5813–5816.

Lee, J.-H., T.-K. Lim, W.-T. Kim, and J.-I. Jin. Dynamics of electro-optical switching processes in surface stabilized biaxial nematic phase found in bent-core liquid crystal. *J. Appl. Phys.* 101 (2007): 034105-1–034105-9.

Link, D. R., M. Nakata, Y. Takanishi, K. Ishikawa, and H. Takezoe. Flexoelectric polarization in hybrid nematic film. *Phys. Rev. E* 65 (2001): 010701-1–010701-4.

Madsen, L. A., T. J. Dingemans, M. Nakata, and E. T. Samulski. Thermotropic biaxial nematic liquid crystals. *Phys. Rev. Lett.* 92 (2004): 145505-1–145505-4.

Majumdar, M., P. Salamon, A. Jákli, J. T. Gleeson, and S. Sprunt. Elastic constants and orientational viscosities of a bent-core nematic liquid crystal. *Phys. Rev. E* 83 (2011): 031701-1–031701-8.

Mandle, R. J., E. J. Davis, C. T. Archbold, S. J. Cowling, and J. W. Goodby. Microscopy studies of the nematic N_{TB} phase of 1,11-di-(1″-cyanobiphenyl-4-yl)undecane. *J. Mater. Chem. C* 2 (2014a): 556–566.

Mandle, R. J., E. J. Davis, S. A. Lobato, C.-C. A. Vol, S. J. Cowling, and J. W. Goodby. Synthesis and characterization of an unsymmetrical, ether-linked, fluorinated bimesogen exhibiting a new polymorphism containing the N_{TB} or 'twist-bend' phase. *Phys. Chem. Chem. Phys.* 16 (2014b): 6907–6915.

Mathews, M., R. S. Zola, D. Yang, and Q. Li. Thermally, photochemically and electrically switchable reflection colors from self-organized chiral bent-core liquid crystals. *J. Mater. Chem.* 21 (2011): 2098–2103.

Mazzulla, A., F. Ciuchi, and J. R. Sambles. Optical determination of flexoelectric coefficients and surface polarization in a hybrid aligned nematic cell. *Phys. Rev. E* 64 (2001): 021708-1–021708-6.

Meiboom, S., J. P. Sethna, P. W. Anderson, and W. F. Brinkman. Theory of the blue phase of cholesteric liquid-crystals. *Phys. Rev. Lett.* 46 (1981): 1216–1219.

Memmer, R. Liquid crystal phases of achiral banana-shaped molecules: A computer simulation study. *Liq. Cryst.* 29 (2002): 483–496.

Meyer, R. B. Piezoelectric effects in liquid crystals. *Phys. Rev. Lett.* 22 (1969): 918–921.

Meyer, R. B. Structural problems in liquid crystal physics. In: R. Balian and G. Weil, eds., *Les
Houches Summer School in Theoretical Physics, 1973.* Molecular Fluids. New York:
Gordon and Breach, 1976, pp. 273–373.

Meyer, C., G. R. Luckhurst, and I. Dozov. Flexoelectrically driven electroclinnic effect in the
twist-bend nematic phase of achiral molecules with bent shapes. *Phys. Rev. Lett.* 111
(2013): 067801-1–067801-5.

Meyer, C., G. R. Luckhurst, and I. Dozov. The temperature dependence of the heliconical tilt
angle in the twist-bend nematic phase of the odd dimer CB7CB. *J. Mater. Chem. C* 3
(2015): 318–328.

Nakata, M., Y. Takanishi, J. Watanabe, and H. Takezoe. Blue phases induced by doping chiral nematic liquid crystals with nonchiral molecules. *Phys. Rev. E* 68 (2003):
041710-1–041710-6.

Nishiya, W., Y. Takanishi, J. Yamamoto, and A. Yoshizawa. Molecular design for a cybotactic
nematic phase. *J. Mater. Chem. C* 2 (2014): 3677–3685.

Ocak, H., B. Eilgin-Eran, M. Prehm, S. Schymura, J. P. F. Lagerwall, and C. Tschierske.
Effects of chain branching and chirality on liquid crystalline phases of bent-core mol-
ecules: Blue phases, de Vries transitions and switching of diastereomeric states. *Soft
Matter* 7 (2011): 8266–8280.

Osipov, M. A. Molecular theory of flexoelectric effect in nematic liquid crystals. *JETP* 85
(1983): 2011–2018.

Osipov, M. A. The order parameter dependence of the flexoelectric coefficients in nematic
liquid crystals. *J. Phys. Lett.* 45 (1984): 823–826.

Panov, V. P., R. Balachandran, J. K. Vij, M. G. Tamba, A. Kohlmeier, and G. H. Mehl. Field-
induced periodic chiral pattern in the N_x phase of achiral bimesogens. *Appl. Phys. Lett.*
101 (2012): 234106-1–234106-4.

Panov, V. P., M. Nagaraj, J. K. Vij, Yu. P. Panarin, A. Kohlmeier, M. G. Tamba, R. A. Lewis,
and G. H. Mehl. Spontaneous periodic deformations in nonchiral planar-aligned bimesogens with a nematic–nematic transition and a negative elastic constant. *Phys. Rev.
Lett.* 105 (2010): 167801-1–167801-4.

Panov, V. P. and J. K. Vij. Microsecond linear optical response in the unusual nematic phase of
achiral bimesogens. *Appl. Phys. Lett.* 99 (2011): 261903-1–261903-3.

Pelzl, G., A. Eremin, S. Diele, H. Kresse, and W. Weissflog. Spontaneous chiral ordering in
the nematic phase of an achiral banana-shaped compound. *J. Mater. Chem.* 12 (2002):
2591–2593.

Petrov, A. G. and V. S. Sokolov. Curvature-electric effect in black lipid membranes. *Eur.
Biophys. J.* 13 (1996): 139–155.

Picken, S. J., T. J. Dingemans, L. A. Madsen, O. Francescangeli, and E. Samulski. Uniaxial to
biaxial nematic phase transition in a bent-core thermotropic liquid crystal by polarizing
microscopy. *Liq. Cryst.* 39 (2012): 19–23.

Prasad, V., S.-W. Kang, K. A. Suresh, L. Joshi, Q. Wang, and S. Kumar. Thermotropic uniaxial
and biaxial nematic and smectic phases in bent-core mesogens. *J. Am. Chem. Soc.* 127
(2005): 17224–17227.

Pratibha, R., N. V. Madhusudana, and B. K. Sadashiva. An orientational transition of bent-core
molecules in an anisotropic matrix. *Science* 288 (2000): 2184–2187.

Prost, J. and J. P. Marcerou. On the microscopic interpretation of flexoelectricity. *J. Phys.
(France)* 38 (1977): 315–324.

Ramou, E., Z. Ahmed, C. Welch, P. K. Karahaliou, and G. H. Mehl. The stabilisation of the Nx
phase in mixtures. *Soft Matter* 12 (2016): 888–899.

Ribeiro de Almeida, R. R., C. Zhang, O. Parri, S. N. Sprunt, and A. Jákli. Nanostructure and dielectric properties of a twist-bend nematic liquid crystal mixture. *Liq. Cryst.* 41 (2014): 1661–1667.

Salili, S. M., C. Kim, S. Sprunt, J. T. Gleeson, O. Parri, and A. Jákli. Flow properties of a twist-bend nematic liquid crystal. *RSC Adv.* 4 (2014): 57419–57423.

Sathyanarayana, P., M. Mathew, Q. Li, V. S. S. Sastry, B. Kundu, K. V. Le, H. Takezoe, and S. Dhara. Splay bend elasticity of a bent-core nematic liquid crystal. *Phys. Rev. E* 81 (2010a): 010702(R)-1–010702(R)-4.

Sathyanarayana, P., M. C. Varia, A. K. Prajapati, B. Kundu, V. S. S. Sastry, and S. Dhara. Splay-bend elasticity of a nematic liquid crystal with T-shaped molecules. *Phys. Rev. E* 82 (2010b): 050701(R)-1–050701(R)-4.

Sathyanarayana, P., S. Radhika, B. K. Sadashiva, and S. Dhara. Structure-property correlation of a hockey stick-shaped compound exhibiting N-SmA-SmCa phase transitions. *Soft Matter* 8 (2012): 2322–2327.

Schad, Hp. and M. A. Osman. Elastic constants and molecular association of cyano-substituted nematic liquid crystals. *J. Chem. Phys.* 75 (1981): 880–885.

Schadt, M. and P. R. Gerber. Class specific physical properties of liquid crystals and correlatiions with molecular structure and static electrooptical performance in twist cells. *Z. Naturforsch. A* 37 (1982): 165–178.

Schiller, P., G. Pelzl, and D. Demus. Analytical theory for flexo-electric domains in nematic layers. *Cryst. Res. Technol.* 25 (1990): 111–116.

Schmidt, D., M. Schadt, and W. Helfrich. Liquid-crystalline curvature electricity: The bending mode of MBBA. *Z. Naturforsch. A* 27 (1972): 277–280.

Sebastián, N., D. O. Lopez, B. Robles-Hernandez, M. R. de la Fuente, J. Salud, M. A. Perez-Jubindo, D. A. Dunmur, G. R. Luckhurst, and D. J. B. Jackson. Dielectric, calorimetric and mesophase properties of 1″-(2′,4-difluorobiphenyl-4′-yloxy)-9″-(4-cyano-biphenyl-4′-yloxy) nonane: An odd liquid crystal dimer with a monotropic mesophase having the characteristics of a twist-bend nematic phase. *Phys. Chem. Chem. Phys.* 16 (2014): 21391–21406.

Sebastián, N., M. G. Tamba, R. Stannarius, M. R. de la Fuente, M. Salamonczyk, G. Cukrov, J. Gleeson et al. Mesophase structure and behaviour in bulk and restricted geometry of a dimeric compound exhibiting a nematic-nematic transition. *Phys. Chem. Chem. Phys.* (2016).

Senyuk, B., H. Wonderly, M. Mathews, Q. Li, S. V. Shiyanovskii, and O. D. Lavrentovich. Surface alignment, anchoring transitions, optical properties, and topological defects in the nematic phase of thermotropic bent-core liquid crystal A131. *Phys. Rev. E* 82 (2010): 041711-1–041711-13.

Sepelj, M., A. Lesac, U. Baumeister, S. Diele, H. L. Nguyen, and D. W. Bruce. Intercalated liquid-crystalline phases formed by symmetric dimers with an α,ω-diiminoalkylene spacer. *J. Mater. Chem.* 17 (2007): 1154–1165.

Severing, K. and K. Saalwachter. Biaxial nematic phase in a thermotropic liquid-crystalline side-chain polymer. *Phys. Rev. Lett.* 92 (2004): 125501-1–125501-4.

Shamid, S. M., S. Dhakal, and J. V. Selinger. Statistical mechanics of bend flexoelectricity and the twist-bend phase in bent-core liquid crystals. *Phys. Rev. E* 87 (2013): 052503-1–052503-11.

Shanker, G., M. Nagaraj, A. Kocot, J. K. Vij, M. Prehm, and C. Tschierske. Nematic phases in 1,2,4-oxadiazole-based bent-core liquid crystals: Is there a ferroelectric switching? *Adv. Funct. Mater.* 22 (2012): 1671–1683.

Shanker, G., M. Prehm, M. Nagaraj, J. K. Vij, and C. Tschierske. Development of polar order in a bent-core liquid crystal with a new sequence of two orthogonal smectic and an adjacent nematic phase. *J. Mater. Chem.* 21 (2011): 18711–18714.

Shanker, G., M. Prehm, M. Nagaraj, J. K. Vij, M. Weyland, A. Eremin, and C. Tschierske. 1,2,4-Oxadiazole-based bent-core liquid crystals with cybotactic nematic phases. *ChemPhysChem* 15 (2014): 1323–1335.

Singh, Y. and U. P. Singh. Density-functional theory of the flexoelectric effect in nematic liquids. *Phys. Rev. A* 39 (1989): 4254–4262.

Southern, C. D., P. D. Brimicombe, S. D. Siemianowski, S. Jaradat, N. Roberts, V. Gortz, J. W. Goodby, and H. F. Gleeson. Thermotropic biaxial nematic order parameters and phase transitions deduced by Raman scattering. *Eur. Phys. Lett.* 82 (2008): 56001-1–56001-6.

Stannarius, R. Comment on Dynamics of electro-optical switching processes in surface stabilized biaxial nematic phase found in bent-core liquid crystal. *J. Appl. Phys.* 104 (2008): 036104-1–036104-3.

Stannarius, R., A. Eremin, M.-G. Tamba, G. Pelzl, and W. Weissflog. Field-induced texture transitions in a bent-core nematic liquid crystal. *Phys. Rev. E* 76 (2007): 061704-1–061704-7.

Stroobants, A. and H. N. W. Lekkerkerker. Liquid crystal phase transitions in a solution of rodlike and disklike particles. *J. Phys. Chem.* 88 (1984): 3669–3674.

Tadapatri, P., U. S. Hiremath, C. V. Yelamaggad, and K. S. Krishnamurthy. Permittivity, conductivity, elasticity, and viscosity measurements in the nematic phase of a bent-core liquid crystal. *J. Phys. Chem. B* 114 (2010a): 1745–1750.

Tadapatri, P. and K. S. Krishnamurthy. Competing instability modes in an electrically driven bent-core nematic liquid crystal. *J. Phys. Chem. B* 116 (2012): 782–793.

Tadapatri, P., K. S. Krishnamurthy, and W. Weissflog. Multiple electroconvection scenarios in a bent-core nematic liquid crystal. *Phys. Rev. E* 82 (2010b): 031706-1–031706-15.

Takahashi, T., S. Hashidate, H. Nishijou, M. Usui, M. Kimura, and T. Akahane. Novel measurement method for flexoelectric coefficients of nematic liquid crystals. *Jpn. J. Appl. Phys.* 37 (1989): 1865–1869.

Tamba, M. G., U. Baumeister, G. Pelzl, and W. Weissflog. Banana-calamitic dimers: Unexpected mesophase behavior by variation of the direction of ester linking groups in the bent-core unit. *Liq. Cryst.* 37 (2010): 853–874.

Tamba, M. G., S. M. Salili, C. Zhang, A. Jákli, G. H. Mehl, R. Stannarius, and A. Eremin. A fibre forming smectic twist-bent liquid crystalline phase. *RSC Adv.* 5 (2015): 11207–11211.

Tamba, M.-G., W. Weissflog, A. Eremin, J. Heuer, and R. Stannarius. Electro-optic characterization of a nematic phase formed by bent core mesogens. *Eur. Phys. J. E* 22 (2007): 85–95.

Tanaka, S., S. Dhara, B. K. Sadashiva, Y. Shimbo, Y. Takanishi, F. Araoka, K. Ishikawa, and H. Takezoe. Alternating twist structures formed by electroconvection in the nematic phase of an achiral bent-core molecule. *Phys. Rev. E* 77 (2008): 041708-1–041708-5.

Tanaka, S., H. Takezoe, N. Eber, K. Fodor-Csorba, A. Vajda, and A. Buka. Electroconvection in nematic mixtures of bent-core and calamitic molecules. *Phys. Rev. E* 80 (2009): 021702-1–021702-8.

Taushanoff, S., K. V. Le, J. Williams, R. J. Twieg, B. K. Sadashiva, H. Takezoe, and A. Jákli. Stable amorphous blue phase of bent-core nematic liquid crystals doped with a chiral material. *J. Mater. Chem.* 20 (2010): 5893–5898.

Tschierske, C. and D. J. Photinos. Biaxial nematic phases. *J. Mater. Chem.* 20 (2010): 4263–4294.

Ungar, G., V. Percec, and M. Zuber. Liquid crystalline polyethers based on conformational isomerism. 20. Nematic-nematic transition in polyethers and copolyethers based on 1-(4-hydroxyphenyl)-2-(2-R-4-hydroxyphenyl)ethane with R = fluoro, chloro, and methyl and flexible spacers containing an odd number of methylene units. *Macromolecules* 25 (1992): 75–80.

Vaupotič, N., M. Čepič, M. A. Osipov, and E. Gorecka. Flexoelectricity in chiral nematic liquid crystals as a driving mechanism for the twist-bend and splay-bend modulated phases. *Phys. Rev. E* 89 (2014): 030501(R)-1–030501(R)-5.

Vaupotič, N., S. Curk, M. A. Osipov, M. Čepič, H. Takezoe, and E. Gorecka. Short-range smectic fluctuations and the flexoelectric model of modulated nematic liquid crystals. *Phys. Rev. E* 93 (2016): 022704-1–022704-5.

Vaupotič, N., S. Curk, M. A. Osipov, M. Čepič, H. Takezoe, and E. Gorecka. Short-range smectic fluctuations and the flexoelectric model of modulated nematic liquid crystals. *Phys. Rev. E* 93 (2016): 022704.

Vaupotič, N., J. Szydlowska, M. Salamonczyk, A. Kovarova, J. Svoboda, M. Osipov, D. Pociecha, and E. Gorecka. Structure studies of the nematic phase formed by bent-core molecules. *Phys. Rev. E* 80 (2009): 030701(R)-1–030701(R)-4.

Virga, E. G. Double-well elastic theory for twist-bend nematic phases. *Phys. Rev. E* 89 (2014): 052502-1–052502-10.

Wang, L., L. Yu, X. Xiao, Z. Wang, P. Yang, W. He, and H. Yang. Blue phase liquid crystals induced by bent-shaped molecules based on 1,3,4-oxadiazole derivatives. *Liq. Cryst.* 39 (2012a): 99–103.

Wang, L., L. Yu, X. Xiao, Z. Wang, P. Yang, W. He, and H. Yang. Effects of 1.3.4-oxadiazoles with different rigid cores on the thermal and electro-optical performances of liquid crystalline blue phases. *Liq. Cryst.* 39 (2012b): 629–638.

Warrier, S. R., D. Vijayaraghavan, and N. V. Madhusudana. Evidence for a nematic–nematic transition in thin cells of a highly polar compound. *Europhys. Lett.* 44 (1998): 296–301.

Wiant, D., J. T. Gleeson, N. Eber, K. Fodor-Csorba, A. Jákli, and T. Toth-Katona. Nonstandard electroconvection in a bent-core nematic liquid crystal. *Phys. Rev. E* 72 (2005): 041712-1–041712-12.

Wild, J. H., K. Bartle, N. T. Kirkman, S. M. Kelly, M. O'Neill, A. T. Stirner, and R. P. Tuffin, Synthesis and investigation of nematic liquid crystals with flexoelectric properties. *Chem. Mater.* 17 (2005): 6354–6360.

Williams, R. Domains in liquid crystals. *J. Chem. Phys.* 39 (1963): 384–388.

Wright, D. C. and N. D. Mermin. Crystalline liquids: The blue phases. *Rev. Mod. Phys.* 61 (1989): 385–432.

Xiang, Y., J. W. Goodby, V. Gortz, and H. F. Gleeson. Revealing the uniaxial to biaxial nematic liquid crystal phase transition via distinctive electroconvection. *Appl. Phys. Lett.* 94 (2009): 193507-1–193507-3.

Xiang, Y., Y.-K. Liu, A. Buka, N. Eber, Z.-Y. Zhang, M.-Y. Xu, and E. Wang. Electric-field-induced patterns and their temperature dependence in a bent-core liquid crystal. *Phys. Rev. E* 89 (2014): 012502-1–012502-9.

Xiang, Y., M.-J. Zhou, M.-Y. Xu, P. Salamon, N. Eber, and A. Buka. Unusual polarity-dependent patterns in a bent-core nematic liquid crystal under low-frequency ac field. *Phys. Rev. E* 91 (2015): 042501-1–042501-9.

Yelamaggad, C. V., I. S. Shashikala, G. Liao, D. S. S. Rao, S. K. Prasad, Q. Li, and A. Jákli. Blue phase, smectic fluids, and unprecedented sequences in liquid crystal dimers. *Chem. Mater.* 18 (2006): 6100–6102.

Yoon, H. G., S.-W. Kang, R. Y. Dong, A. Marini, K. A. Suresh, M. Srinivasarao, and S. Kumar. Nematic biaxiality in a bent-core material. *Phys. Rev. E* 81 (2010): 051706-1–051706-7.

Yoshizawa, A. Material design for blue phase liquid crystals and their electro-optical effects. *RSC Adv.* 3 (2013): 25475–25497.

Yoshizawa, A., Y. Kogawa, K. Kobayashi, Y. Takanishi, and J. Yamamoto. A binaphthyl derivative with a wide temperature range of a blue phase. *J. Mater. Chem.* 19 (2009): 5759–5764.

Yu, L. J. and A. Saupe. Observation of a biaxial nematic phase in potassium laurate-1-decanol–water mixtures. *Phys. Rev. Lett.* 45 (1980): 1000–1003.

Zakharov, A. V. and R. Y. Dong. The flexoelectric effect in nematic liquid crystals: A statistical–mechanical approach. *Eur. Phys. J. E* 6 (2001): 3–6.

Zep, A., S. Aya, K. Aihara, K. Ema, D. Pociecha, K. Madrak, P. Bernatowicz, H. Takezoe, and E. Gorecka. Multiple nematic phases observed in chiral mesogenic dimers. *J. Mater. Chem. C* 1 (2013): 46–49.

Zheng, Z., D. Shen, and P. Huang. Wide blue phase range of chiral nematic liquid crystal doped with bent-shaped molecules. *New J. Phys.* 12 (2010): 113018-1–113018-10.

Zheng, Z., D. Shen, and P. Huang. The liquid crystal blue phase induced by bent-shaped molecules with different terminal chain lengths. *New J. Phys.* 13 (2011): 063037-1–063037-8.

Zhu, C., M. R. Tuchband, A. Young, M. Shuai, A. Scarbrough, D. M. Walba, J. E. Maclennan, C. Wang, A. Hexemer, and N. A. Clark. Resonant carbon K-edge soft X-ray scattering from lattice-free heliconical molecular ordering: Soft dilative elasticity of the twist-bend liquid crystal phase. *Phys. Rev. Lett.*, 116 (2016): 147803–147821.

5 Chirality in Bent-Shaped Mesogens

For more than 160 years after the discovery of molecular handedness by Pasteur, the interest in chiral molecular and super-molecular systems has been growing continuously. Biological matter is essentially chiral. Chirality is strongly correlated with the properties and the functions of many biological systems. Biologically active chiral materials with opposite handednesses may even function as drugs and poisons. A good example is thalidomide. Chirality is also responsible for a rich pattern-forming behavior in other physicochemical systems.

Bent-shaped liquid crystals often exhibit a spontaneous breaking of reflection symmetry and formation of chirality-segregated domains, although molecules are achiral. They often occur in such phases as B_2, B_4, B_7, the dark conglomerate (DC), and the twist-bend N_{TB}. In some phases, mesogens even form micro- and nano-scaled self-assembled helical and filament-like structures.

In addition, achiral bent-shaped molecules have an ability to enhance their chiral properties, when they are added to conventional chiral phases of rod-shaped molecules, such as cholesteric (Ch or N*) and smectic C* (SmC*) phases. In this way, bent-shaped molecules sometimes behave as chiral molecules despite being assumed to be nonchiral or racemic systems. This fact is very special in the sense that such phenomena occur in fluidic liquid crystalline phases. Spontaneous deracemization or spontaneous chiral resolution of racemic molecules has been known to occur in crystals and at interfaces, but bent-shaped liquid crystals (LCs) are a first true fluid system to exhibit the spontaneous chiral symmetry breaking in bulk. This chapter is devoted to describe such intriguing chirality-related phenomena emerging in bent-shaped mesogens.

5.1 ORIGIN OF CHIRALITY

There are three different kinds of sources of molecular chirality: (a) central chirality, (b) axial chirality, and (c) plane chirality (Figure 5.1) (Takezoe 2012). Central chirality is mostly due to the existence of a chiral carbon with four different substituents. Any atom that has four different groups bonded to it, such as S, N, Si, and P, could be also a chiral center. Axial chirality originates from twisted isomers of molecules, between which a sufficiently high energy barrier exists, preventing the chiral conformational interconversion in ambient conditions. Plane chirality arises in molecules with a chiral plane (O–C–C–Br in Figure 5.1c), where the substituent group destroys a perpendicular plane of symmetry. Generally, bent-shaped molecules bear neither chiral carbon nor axial chirality, so they are racemic or achiral (nonchiral), even if they have two twisted metastable conformations. But chiral structures may also occur

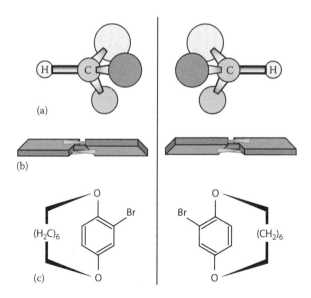

FIGURE 5.1   Images of (a) central chirality, (b) axial chirality, and (c) plane chirality.

when the constituent molecules are achiral. In this case, a spontaneous reflection symmetry breaking occurs resulting in a chiral organization of nonchiral molecules. This kind of chirality can be designated as *structural chirality*.

Link et al. (1997) clarified that tilted bent molecules make the constituent layer chiral (see Section 3.4). Depending on the stacking of chiral layers, the overall structure can be *homochiral* when all layers have the same handedness. The equal amounts of left- and right-handed layers give a *racemic* structure (Selinger 2003).

Both ferroelectric and antiferroelectric phases with homochiral and racemic layer structures can exist in the B_2 phase. An important question arises: whether or not the layer chirality and molecular axial chirality are correlated with each other. Before answering this question, some experiments to study the conformation of bent-shaped molecules are described in the following.

One of the important experiments relies on nuclear magnetic resonance (NMR). At least three groups performed measurements on the same compounds. Sekine et al. (1997) first measured ^{13}C NMR in P12-O-PIMB and found that the signal due to carbonyl carbons appears as a doublet both in B_2 and B_4 phases. They concluded that the two carbonyls are under different electronic circumstances because of conformational chirality of the molecule. Kurosu et al. (2004) made more detailed ^{13}C NMR studies in P14-O-PIMB, which shows the direct B_2–B_4 phase transition, and confirmed the previous result (Figure 5.2). They concluded that the bent-core molecules assume the twisted conformation; two carbonyl carbons of ester moieties are twisted away from each other by rotating out of the molecular core plane with opposite dihedral angles. This twisted molecular conformation is attributable to the origin of the chirality of the B_2 and B_4 phases in the achiral bent-core molecular system. However, Walba et al. (2005) examined the carbon doublets in the B_2 phase of P9-O-PIMB by NMR and denied the evidence. Later, Xu et al. (2006) performed ^{13}C, ^{2}H NMR including

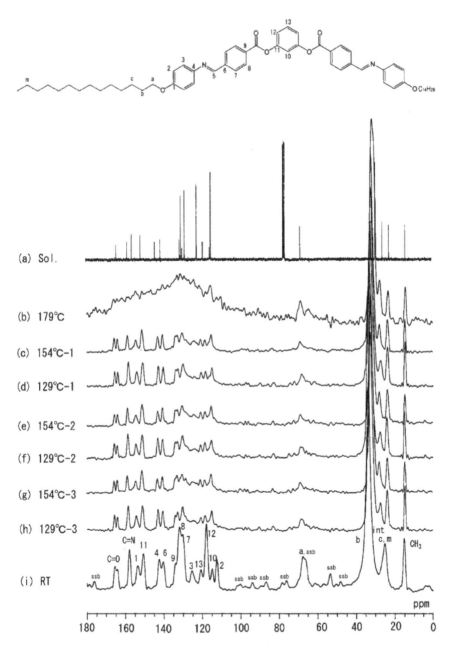

FIGURE 5.2   Temperature dependence of ^{13}C NMR spectra in P14-O-PIMB (shown at the top) at various temperatures (a)-(i). (Reprinted with permission from Kurosu, H., Kawasaki, M., Hirose, M., Yamada, M., Kang, S., Thisayukta, J., Sone, M., Takezoe, H., and Watanabe, J., Solid-state ^{13}C NMR study of chiral twisted conformation attributable to chirality in smectic phases of achiral banana-shaped molecules, *J. Phys. Chem. A*, 108, 4674. Copyright 2004 by the American Chemical Society.)

CP ^{13}C and ^{13}C NMR 2D and confirmed the doublet in B$_2$ and crystalline phase of a different compound (Pbis11BB), concluding that these doublets must come from the asymmetric lateral wings in the molecule and not from the dynamic effect, since there were no cross-peaks between them in the exchange spectrum. Thus, not only crystal B$_4$ but also B$_2$ phases show a static twisted molecular core with an asymmetry in the orientation of the two lateral wings. More recent advanced NMR measurements (Yamada et al. 2013) also support the conclusion.

Polarized Fourier transform infrared (FTIR) spectra were measured in the antiferroelectric B$_2$ phase of a bent-core mesogen P10-PIMB (Zennyoji et al. 2001). The measurements were made using a 50×200 μm rectangular area in a large uniform circular domain made by applying a dc electric field. Because the layers are curved, the dichroic ratio R $(=A_{max}/A_{min})$, where A_{max} and A_{min} are maximal and minimal absorption coefficients, respectively, is always smaller than what we expected. Figure 5.3 shows polar plots of the absorption peaks for the C=O stretching (1736 cm^{-1}), phenyl ring stretching (1600 cm^{-1}), and ester C–O–C asymmetric stretching (1258 cm^{-1}) both in homochiral and racemic domains. Without applying a field, the phenyl stretching and C–O–C asymmetric stretching peaks appear at 0°, which corresponds to the layer normal direction, as expected. By applying a dc field, the maximum absorption of the phenyl stretching and C–O–C asymmetric stretching peaks rotates by about 35°, which is the same as the apparent tilt angle obtained by texture observation. In contrast, the maximum for the C=O stretching appears at 80°, which is neither parallel nor perpendicular to that of C–O–C under a field. This situation is possible if and only if the two ester groups are twisted with respect to the central phenyl ring. The angle 80° is almost parallel to the layer, so the overall molecular orientation is concluded as shown in Figure 5.4. Gorecka et al. (2000) also observed similar polarized FTIR results in a rod-shaped SmC* material doped with a bend-core material.

Vibrational circular dichroism (VCD) spectroscopy, which is CD in an IR wavelength region, is also an effective method to examine the molecular chirality. Figure 5.5 shows FTIR and VCD spectra of a homochiral B$_4$ domain in a twin dimer (Choi et al. 2007). Strong VCD signals are observed in a C=O stretching of ester (1 in Figure 5.5a) and a C–O stretching of ester (7, 8, and 10 in Figure 5.5a) vibrations. This implies that the central ester group is strongly related to the chiral structure. The chiral conformation suggested by different experiments mentioned earlier was also confirmed by Monte Carlo simulation (Earl and Wilson 2003, Earl et al. 2005). It was found that bent-core molecules adopted a number of conformations with extremely high magnitudes of chirality associated with them.

Let us return to the previous question, whether or not the layer chirality and molecular axial chirality are related to each other. Niwano et al. (2004) gave an answer, confirming a strong correlation of chirality between B$_2$ and B$_4$ phases. This paper has a misconception in the structure of the B$_4$ phase; B$_4$ was considered as twist grain boundary (TGB)-like phase with a sub-μm pitch and antiferroelectric polar order. However, the experimental results are convincing except for CD measurements: CD measurements have to be carried out using ultraviolet (UV)-absorption-saturation-free thin cells such as a few-hundred nm (Otani et al. 2009). Otherwise, the information below 400 nm would be erroneous because of strong fundamental absorption.

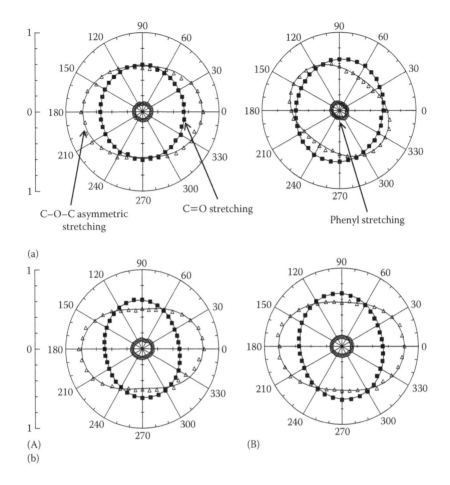

FIGURE 5.3 Polar plot of absorption peaks for C=O stretching, phenyl stretching, and C–O–C asymmetric stretching. Profiles from (a) homochiral and (b) racemic domains (A) in the absence and (B) under 11.9 V/μm are shown. (Reprinted from Zennyoji, M. et al., *Mol. Cryst. Liq. Cryst.*, 366, 693, 2001. With permission from Taylor & Francis.)

Some important experimental results are as follows: (1) once the chiral domains are formed in the B_4 phase, the domain shape and chirality are conserved after successive transformation between B_2 and B_4 in a pure achiral bent-core mesogen; (2) in a bent-core mesogen doped with a chiral analogue, any layer structure formed in the B_2 phase is converted to the homochiral SmC_AP_A after temperature cycling through the B_4 phase; (3) the resulting homochiral SmC_AP_A structure is stable on repeated temperature cycling between B_2 and B_4. The results (1) and (3) demonstrate a strong correlation of chirality between B_2 and B_4 phases. The result (2) indicates that the conversion between SmC_SP_F in B_4 and SmC_AP_A in B_2 always occurs at the B_2–B_4 phase transition. This must be an important issue, which should be considered theoretically.

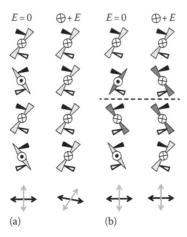

$$E = 0 \qquad \oplus + E \qquad E = 0 \qquad \oplus + E$$

(a)   (b)

FIGURE 5.4 Molecular orientation in (a) homochiral and (b) racemic domains. Gray and black arrows at the bottom indicate the transition moments of the C–O–C asymmetric stretching and the C=O stretching, respectively. (Redrawn from Zennyoji, M. et al., *Mol. Cryst. Liq. Cryst.*, 366, 693, 2001.)

5.2 ENHANCED CHIRALITY BY DOPING CHIRAL PHASES WITH NONCHIRAL BENT-CORE MOLECULE

When a nonchiral liquid crystal is doped with chiral molecules, chiral phase is induced and the chirality is enhanced with increasing doping rate. For instance, when the nematic LC is doped with a chiral dopant, the chiral nematic (N*) or cholesteric phase is induced and the helical pitch becomes shorter with higher amounts of the chiral dopant. The opposite is true: When chiral LCs such as N* are doped with a nonchiral dopant, the helical pitch becomes longer. This is quite reasonable. However, quite a striking doping effect was reported by Thisayukta et al. (2002): Doping the N* material with nonchiral bent-core molecules enhanced the chirality, that is, the pitch became shorter. They used three types of bent-core molecules, which were added to a cholesterol derivative. In all cases, the reciprocal optical pitch ($1/nP$), where n is an average refractive index and P is a pitch, increased with increasing content of bent-core molecules, as shown in Figure 5.6. In contrast, usual dopant effect was observed when a rod-shaped terephthalbis p-butylaniline (TBBA) molecule was used as a dopant—the pitch becomes longer. They suggested the asymmetric twist conformation in the ester linkage group connecting the central core with the side wings as the origin of the enhanced twisting power. Namely, one of the twist conformations is stabilized by the chiral host and produces a feedback enhancing the chirality of the host.

The interaction energy between the left-handed bent-core conformation and the chiral host molecule U_{LH} is different from that between the right-handed bent-core conformation and the same chiral host molecule U_{RH}. The nonzero difference $\Delta U = U_{LH} - U_{RH}$ induces finite enantiomeric excess (*ee*) in bent-core molecules, resulting in an increased amount of chiral molecules in the system. If this effect overcomes the dilution effect, unusual chiral enhancement occurs. Actually, according to Earl and

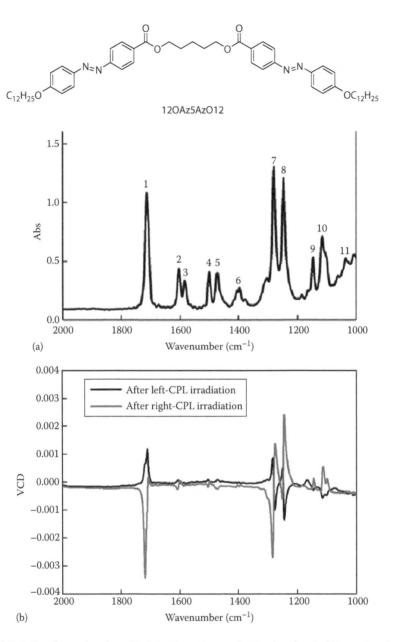

FIGURE 5.5 **(See color insert.)** (a) Absorption and (b) vibrational CD spectra from homochiral domains in the B_4 phase of a twin dimer (12OAz5AzO12) obtained by left- and right-circular-polarized-light (CPL) light irradiations. For numbers in (a), please refer to the text. (Reprinted from Choi, S.-W., Kawauchi, S., Tanaka, S., Watanabe, J., and Takezoe, H., Vibrational circular dichroism spectroscopic study on circularly polarized light-induced chiral domains in the B_4 phase of a bent mesogen, *Chem. Lett.*, 36, 1018–1019, 2007. With permission from the Chemical Society of Japan.)

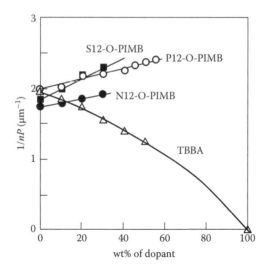

FIGURE 5.6 Reciprocal optical pitch ($1/nP$) of a cholesteric liquid crystal as a function of dopant ratio of three bent-core molecules and one rod-shaped molecule. (Redrawn from Thisayukta, J. et al., *J. Am. Chem. Soc.*, 124, 3354, 2002.)

Wilson (2003), even a small increase in the proportion of left-handed or right-handed chiral conformations sampled by bent-core molecules in the N* phase would have a relatively large effect on the twist of the system due to the extremely high helical twisting powers (HTPs) of the chiral conformation.

A similar effect was also observed in chiral smectic phases (SmC* and SmC$_A^*$ phases) (Gorecka et al. 2003). The twisting power of smectic phases made of rod-like molecules, SmC* and SmC$_A^*$ phases, is enhanced by adding a small amount of achiral bent-shaped molecules; the helical pitch becomes shorter with increasing dopant ratio. Figure 5.7 shows the temperature dependence of the selective reflection wavelength. Only by adding 1% of the bent-core molecule, the SmC$_A^*$ phase is stabilized (the temperature range of SmC$_A^*$ expands), and the pitch becomes short both in SmC* and SmC$_A^*$ at the same temperatures. The same scenario as in the N* phase is valid, that is, the bent-core molecule is not flat but twisted (conformational chirality), and one of the twisted forms is favored by the chiral host, enhancing the chirality of the host. Gorecka et al. (2003) suggested additional possible scenario: the chiral host induces a tilted and polar-ordered structure of bent-shaped molecules, and the dopant with induced chirality transfers its chirality back to the host, enhancing the chirality. By considering the remarkably larger chirality enhancement in smectics compared to N* (compare Figure 5.6 and inset of Figure 5.7), both scenarios may be effective in SmC* and SmC$_A^*$.

The same conclusion of the chirality enhancement was derived by a different analysis in the SmC* phase by Archer and Dierking (2005, 2010). They experimentally determined the full Landau potential of a SmC* material doped with bent-core molecules; temperature- and electric-field-dependent tilt angle and polarization measurements were analyzed according to the generalized Landau model of ferroelectric

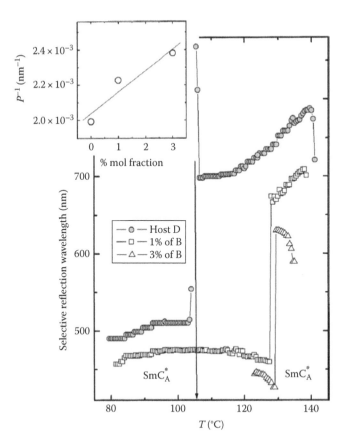

FIGURE 5.7 Selective reflection wavelength (optical pitch) of a tilted chiral smectic (host) showing the SmC* and SmC$_A^*$ phases for a pure host and mixtures containing 1% and 3% of a bent-core molecule. (Reprinted with permission from Gorecka, E., Čepič, M., Mieczkowski, J., Nakata, M., Takezoe, H., and Zeks, B., *Phys. Rev. E*, 67, 061704-1. Copyright 2003 by the American Physical Society.)

LCs. After minimizing the Landau expansion of the free energy density difference $g - g_0$ between SmA* and SmC* with respect to the total polarization, they obtained

$$g - g_0 = \frac{1}{2}\alpha\left(T - T_c^*\right)\Theta^2 + \frac{1}{4}b\Theta^4 + \frac{1}{6}c\Theta^6 - \frac{1}{2}\frac{(C\Theta + E)^2}{\left((1/\varepsilon_0\chi_0) - \Omega\Theta^2\right)}, \quad (5.1)$$

where

    Θ is the tilt angle

    α, b, c are constants

    ε_0 and χ_0 are dielectric permittivity of vacuum and the dielectric susceptibility in the medium, respectively

    Ω is the quadrupolar coupling coefficient

    E is an electric field amplitude applied

With all parameters known, this equation was used to visualize the full Landau potential under various temperatures, electric fields, and dopant concentrations. The Landau potential curves as a function of tilt angle under such parameters were shown. One of the chirality-related results is the electroclinic coefficient $e^* = \chi_0 \varepsilon_0 C /[\alpha(T - T_c^*)]$ as a function of dopant concentration. Here, C, the bilinear coupling coefficient, is pseudo-scalar in nature, so is e^*. As shown in Figure 5.8, e^* linearly increases as a function of dopant concentration, giving unambiguous evidence for the chiral induction capability of achiral bent-core molecules added to a chiral host material.

Choi et al. (2006a) reported that the amplification of the twisting power also occurs in the N* phase by introducing an achiral *rod-like* compound with an ester group. This is striking because now we know that the chirality enhancement occurs by adding achiral bent-core molecules (e.g., three bent-core molecules in Figure 5.6) but not by "rod-shaped" molecules (e.g., TBBA in Figure 5.6). So we need to seek another reason for the chirality enhancement. Kawauchi et al. (2007) intuitively explained this phenomenon based on the calculation of the potential energy of a phenyl benzoate molecule as a function of the dihedral angle: the phenyl benzoate molecule has two chiral conformations, as shown in Figure 5.9, but the potential barrier is very low. Hence, when the molecules are located in a chiral field, the potential minimum of one of the axially chiral conformers disappears and the molecule behaves as a chiral molecule (see cartoon shown at the bottom of Figure 5.9). In contrast, the potential surface of another molecule, benzylideneaniline, is less influenced by the chiral field. Actually, the addition of imine molecules to a cholesteric LC weakens the chirality and lengthens the pitch (Choi et al. 2006a).

Unusual chirality-induction effect of ester molecules was also observed as a spontaneous reflection symmetry breaking effect in the conventional SmB and SmE phases of achiral rod-shaped arylbenzoate esters (Jeong et al. 2009). Vibrational CD

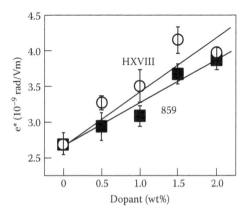

FIGURE 5.8   Electroclinic coefficient e^* as a function of dopant ratio of two different bent-core molecules. (Reprinted with permission from Archer, P. and Dierking, I., *Phys. Rev. E*, 72, 041713-1. Copyright 2005 by the American Physical Society.)

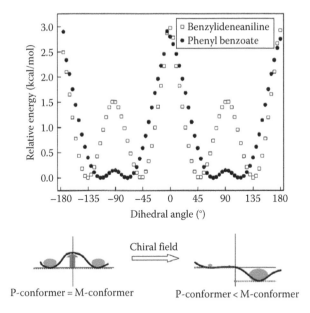

FIGURE 5.9 Relative potential energy of benzylideneaniline and phenyl benzoate as a function of dihedral angle. Schematic illustration of the potential change, when phenyl benzoate molecules are introduced in a chiral field, is shown at the bottom. (Reprinted from Kawauchi, S., Choi, S.-W., Fukuda, K., Kishikawa, K., Watanabe, J., and Takezoe, H., Why achiral rod-like compound with ester group amplifies chiral power in chiral mesophase, *Chem. Lett.*, 36, 750–751, 2007. With permission from the Chemical Society of Japan.)

measurements suggested the origin of chirality in the ester linkage. In this way, ester linkages have some particular features in chiral conformations, and many bent-core molecules possess ester linkages.

This phenomenon was studied theoretically by Earl et al. (2005), and the simplified explanation was given in the following (Takezoe 2012). The twist distortion free energy of the N* phase influenced by dopants is expressed as

$$F_d = \lambda\left(\mathbf{n}\mathrm{curl}\mathbf{n}\right) + \frac{1}{2}K_{22}\left(\mathbf{n}\mathrm{curl}\mathbf{n}\right)^2, \tag{5.2}$$

where the chirality strength λ is related to the twisting power and determined by the number of chiral molecules, and given by

$$\lambda = \lambda_0 - x_b\lambda_d + \Delta x\lambda_b, \tag{5.3}$$

where
λ_0 is the λ in the pure N* phase
$x_b = x_L + x_R$ and $\Delta x = x_L - x_R$ are the sum and the difference of the molar fraction of left-handed and right-handed bent-core conformers, x_L and x_R, in the N* solvent, respectively

The second term describes the dilution effect with a coefficient λ_d, and the third term describes the finite *ee* effect of bent-core molecules with a coefficient λ_b. If the effect of the third term is dominant compared to the second term, chirality enhancement occurs. Δx can be determined by minimizing the free energy ΔF

$$\Delta F\left(x_L, x_R\right) = kT\rho_0 x_L \ln x_L + kT\rho_0 x_R \ln x_R + \rho_0^2 x_L U_{LH} + \rho_0^2 x_R U_{RH} \tag{5.4}$$

Here, we assume x_L and x_R are small and the interaction between bent-core molecules is negligible. From that, we obtain Δx

$$\Delta x = x_b \tanh\left(\frac{\rho_0 \Delta U}{2kT}\right) \tag{5.5}$$

and a familiar expression

$$\frac{2\pi}{p} = \frac{\lambda}{K_{22}} \tag{5.6}$$

We finally obtain

$$\frac{2\pi}{p} = \frac{\lambda_0 + x_b\left[\lambda_b \tanh\left(\rho_0 \Delta U / 2kT\right) - \lambda_d\right]}{K_{22}} \tag{5.7}$$

This result means that a pitch shortening occurs with increasing the number of bent-core molecules, if the HTP caused by *ee* of bent-core molecules is much larger than the dilution effect.

Computer simulations (Wilson 2005) are another important tool to explore chiral properties such as HTP. An interesting result was reported by Earl et al. (2005). They studied the chiral order parameter χ (measured in Å^3), which predicts the sign and magnitude of the HTP of molecules, using Ferrarini et al.'s method (Ferrarini et al. 1996). For this purpose, a statistically significant and independent number of conformations, generated from an internal coordinate Monte Carlo simulation, are needed (Earl and Wilson 2003). The bent-shaped molecule N12-O-PIMB used in the doping experiment (Thisayukta et al. 2002) was simulated and χ values were obtained. As shown in the histogram (Figure 5.10), the extremely large magnitude of the χ associated with some of the conformations, which the bent-core molecule adopted, was found, although the overall average chirality order parameter was essentially zero within statistical error (Earl et al. 2005). HTP is approximately correspond to two times of χ (Earl and Wilson 2003). If one considers that the HTP of a commercially available chiral dopant CB15 (Merck) is of the order of 10 μm^{-1}, HTP of 10 μm^{-1} corresponds to χ of 5 Å^3. The result is surprising: the vast majority of conformations sampled by the bent-core molecule had $|\chi| > 100$ Å^3. Hence, only a small enantiomeric excess Δx is sufficient to induce a large chiral dopant effect of achiral bent-core molecules.

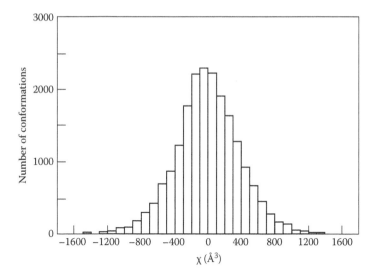

FIGURE 5.10 A histogram showing the number of conformations further showing certain twisting powers. The shape is symmetric, but the existence of molecular conformations having large twisting powers is shown. (Reprinted with permission from Earl, D.J., Osipov, M.A., Takezoe, H., Takanishi, Y., and Wilson, M.R., *Phys. Rev. E*, 71, 021706-1. Copyright 2005 by the American Physical Society.)

Some other anomalous doping effects are known too: the induction of antiferroelectric phases (Gorecka et al. 2000, Kishikawa et al. 2003), blue phases (Nakata et al. 2003), and frustrated twist grain boundary phases (Archer and Dierking 2006). These phenomena will be described in Section 8.1.

5.3 CHIRALITY CONTROL

Here we describe symmetry breaking (spontaneous chiral segregation, spontaneous deracemization), controlling chiral domains to break down the balance of the zero enantiomeric excess (*ee*), and chirality switching. Since the famous experiment by Pasteur (1848), chirality has been a major topic in chemistry: He showed that racemic molecules resolve spontaneously into chiral forms when they crystallize. We call them conglomerates, in which molecules form condensates comprised of only one enantiomer. The condensation into conglomerates occurs not only in crystals as Pasteur discovered, but also in a solution (Perez-Garcial and Amabilino 2002). There are several reports of chiral resolution in monolayers on different kinds of surfaces (Takezoe 2012), where molecules themselves are not of chiral conformation, but the 2D crystal packing motif is chiral. However, the 2D chiral form occurs essentially in a crystalline state. In the following Section 5.3.1, we will first describe reflection symmetry breaking in bent-shaped mesogenic phases, then introduce some methods to control chirality in Sections 5.3.2 and 5.3.3.

5.3.1 Spontaneous Symmetry Breaking

Among typical banana phases, B_1–B_7, B_2 (Link et al. 1997), B_4 (Hough et al. 2009b), and B_7 (Coleman et al. 2003) phases can be chiral. The Dark Conglomerate (DC) phase (Hough et al. 2009a) is another important phase showing symmetry breaking. As described in Section 3.3, Link et al. (1997) introduced a new concept of chirality in liquid crystals. When bent-shaped molecules tilt from the layer normal, the layer becomes chiral. The chirality is determined by the relationship among three vectors, layer normal $\mathbf{z}$, director $\mathbf{n}$, and bending tip (polar) direction of molecules $\mathbf{b}$. The idea is natural and clear; if you think $\mathbf{z}$, $\mathbf{n}$, and $\mathbf{b}$ to be corresponding to thumb, index, and middle fingers, you will easily recognize two, right- (positive) and left- (negative), handednesses. Simultaneous tilt and polar ordering is a common feature of phases formed from bent-shaped molecules. The chirality of all these phases originates from the layer structure with tilted molecules packed in a polar fashion, although molecular conformational chirality is likely related to the layer chirality, as described in Section 5.1.

The chirality of the B_2 (SmCP) phase is easily recognized by direct microscopy observation under an electric field, although curved layers, which are usually observed, make the determination a little complex. If the average director direction (largest refractive index axis or optical slow axis) is perpendicular to the layer under an electric field, the layer chirality is racemic, that is, every neighboring layer chirality is opposite, positive and negative. When it is tilted to the right (left) under an upward field, the system is homochiral and the chirality is negative (positive). One of the examples is shown in Figure 5.11 (Zennyoji et al. 2000): (a) homogeneously chiral domains obtained by applying a rectangular wave field, (b) racemic domains obtained by applying a triangular wave field, and (c) partially mixed domains, in which the layers with opposite chirality are partially mixed, by applying a dc field.

The recognition of symmetry breaking in the B_4 phase was more dramatically made by Berlin group (Druerke and Heppke 1997). Just by slightly rotating one of the crossed polarizers to decross them between which the sample is placed, they found that otherwise indistinguishable domains appear to be bright and dark, and the domains with different brightness were interchanged by decrossing the polarizers in opposite directions. This is just because of high optical rotatory power of each chiral domain, usually in the range of 1–3 deg/µm. One of the examples is shown in Figure 5.12 (Thisayukta et al. 2000). This method is very easy, so it became a standard method for confirming chiral segregated domains.

Spontaneous chiral symmetry breaking has also been observed in many systems other than bent-shaped molecules (Tschierske and Ungar 2016). Takanishi et al. showed two chiral domains in a fluid smectic phase of a racemic rod-shaped diastereomer with two stereogenic centers in its tail (Takanishi et al. 1999). Evidence for conglomerate formation was provided by striped domains with additional fine stripes. These stripes are tilted in two different directions with respect to the primary stripes. Observed electro-optic switching, which should not occur in nonchiral rod-like molecular systems, additionally supported the assumption about chiral microsegregation. Another example can be found in highly ordered smectic phases such as SmB and SmE of achiral rod-shaped 4-arylbenzoate esters, as mentioned

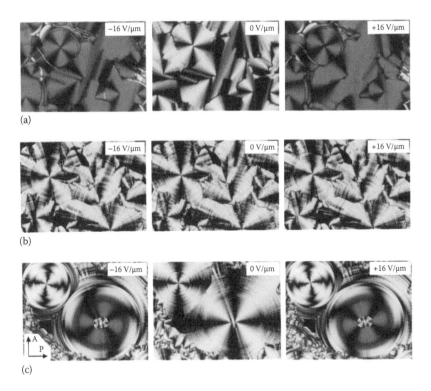

(a)

(b)

(c)

FIGURE 5.11 **(See color insert.)** Photomicrographs of (a) homochiral (b) racemic and (c) partially mixed domains in the absence (middle) and presence of positive (right) and negative (left) electric fields. (Reprinted from Zennyoji, M., Takanishi, Y., Ishikawa, K., Thisayukta, J., Watanabe, J., and Takezoe, H., Electrooptic and dielectric properties in bent-shaped liquid crystals, *Jpn. J. Appl. Phys.*, 39, 3536–3541, 2000. With permission from the Japan Society of Applied Physics.)

FIGURE 5.12 **(See color insert.)** Photomicrographs of the B_4 phase under oppositely decrossed polarizers. (Reprinted with permission from Thisayukta, J., Nakayama, Y., Kawauchi, S., Takezoe, H., and Watanabe, J., Distinct formation of a chiral smectic phase in achiral banana-shaped molecules with a central core based on a 2,7-dihydroxynaphthalene unit, *J. Am. Chem. Soc.*, 122, 7441. Copyright 2000 by the American Chemical Society.)

earlier (Jeong et al. 2009). Spontaneous deracemization was also observed in a disc-like molecule in the columnar phase (Nagayama et al. 2010), where the chirality originates from a twisted disc. Fast cooling from the Iso phase exhibits both chiral domains, but single chiral domains over the millimeter scale can be obtained by slow cooling. More recently, two isotropic phases were found to show phase-segregated domains: the cubic phase consisting of achiral double-swallow tailed tetracatenar compound (Dressel et al. 2014a, 2015) and the isotropic liquid of achiral and photoi-somerizable azobenzene-based rod-shaped molecules (Dressel et al. 2014b, Alaasar et al. 2016).

5.3.2 Controlling Chirality

It is natural for scientists to seek methods to unbalance the equal domain size of right- and left-handed chiral domains. The easiest, most frequently used and fool-proof method is doping with chiral molecules. Sekine et al. (1997) were the first to show the chiral doping effect. They observed the CD spectra of the scattered light in the B_4 phase of P8-O-PIMB with a small amount of the chiral dopant (2-methylbutyl p-(p-n-hexyloxybenzylidene amino)cinnamate [HOBAMBC] showing the SmC* phase). Finite dichroic ratio was observed in a mixture with 3 wt% HOBAMBC and increased in a mixture with a higher dopant ratio. Later, Thisayukta et al. (2000) used a chiral analog of a bent-core molecule with a 2,7-dihydroxynaphthalene cen-tral core as a chiral dopant and made CD measurements to show the chiral imbal-ance. They further used two different chiral bent-core molecules with chiral carbons in different positions of both wings (Thisayukta et al. 2001). As shown in Figure 5.13, CD intensity saturates in smaller content of the dopant molecule, when the chiral carbons are located near the core, than in the other chiral compound with the chiral carbons apart from the core. These results are in the B_4 phase. A similar effect was also confirmed in the N_{TB} phase (Emsley et al. 2013). Only the presence of ~2.4 wt% of a chiral solute (S)-1-phenylethanol in the mesogen CB7CB (see Figure 4.25) was found to give a homochiral domain using NMR.

Using chiral surface is another attempt. However, in case of nonfluidic phases such as B_4 and DC phases, chiral control was not possible once symmetry breaking occurred. In other words, surface must exert the chiral effect on bulk far from the sur-face. Then one can use only very thin cells with chiral surfaces. In this sense, the first attempt by Shiromo et al. (2005) did not provide a sufficient success, although 10% *ee* was obtained using rubbed chiral polyimide surfaces. Lee et al. (2013) succeeded in getting the B_4 phase with high *ee* by introducing temperature gradient. By form-ing a chiral surfactant monolayer at the water/liquid crystal interface in the isotropic phase and cooling it into the B_4 phase under the temperature gradient, monochiral domains were obtained. This is a technique of crystal growth, in the present case, homochiral domain growth. CD measurements clearly showed largely asymmetric chirality induction depending on L- or D-form of phospholipid.

A different type of chiral surface effect on chirality was reported by Jákli et al. (2001). They introduced mixtures of liquid crystals and reactive monomers into vacant cells and polymerized by UV light to produce polymer network. After removing LCs, bent-core LCs were introduced into the polymer network. If the

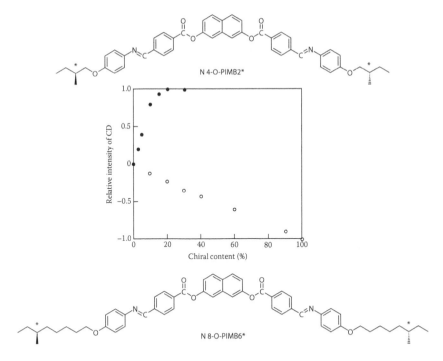

FIGURE 5.13 Dopant effect on chirality. Relative CD intensity as a function of concentrations of two different chiral dopants (shown above and below the graph) is shown. (From Thisayukta, J., Niwano, H., Takezoe, H., and Watanabe, J., Effect of chiral dopant on a helical Sm1 phase of banana-shaped N-*n*-O-PIMB molecules, *J. Mater. Chem.*, 11, 2717–2721, 2001. Reproduced by permission of The Royal Society of Chemistry.)

reactive monomers or LCs are chiral, chiral switching of the bent-core molecules was observed, although otherwise bent-core molecules showed racemic structures. A reference experiment was also made to confirm the effect. The polymer network was formed in the cholesteric phase or isotropic phase, resulting in uniform chirality in the former case. Thus, the long-range chirality transfer was confirmed to be due to the polymer network. Two effects can be considered, microscopic and macroscopic effects, corresponding to polymer chiral surface and helical polymer networks, respectively. According to the authors, the helix formation should play an important role. Hence, this is a chiral template effect rather than a chiral surface effect.

All the previously mentioned methods for biased reflection symmetry breaking use chiral species: such as chiral dopant, chiral molecule, and chiral surface. It is more interesting to produce chiral systems without using chiral species. Two different attempts have been successfully made. One is by using circularly polarized light, and the other is by using twisted nematic (TN) cells.

Irradiation with circularly polarized light has been used for deracemization in some liquid crystal systems (Huck et al. 1996, Burnham and Schuster 1999, Iftime et al. 2000, Wu et al. 2004). A nematic compound with racemic helical alkene was converted to chiral phase (cholesteric phase) by circular polarized UV light irradiation. However, the obtained *ee* was very small: only 0.07% *ee* (Huck et al. 1996) and 0.8% *ee* (Burnham and Schuster 1999). The experiment was also made using achiral SmA liquid crystalline azobenzene polymer (Iftime et al. 2000, Wu et al. 2004). The effect by circularly polarized light irradiation was clearly confirmed by CD and optical rotation. The obtained optical rotation was quite large, 2 deg/μm, although they did not report the *ee* value.

Compared with the previous works, the application of this technique to bent-shaped mesogens is very efficient and visually observable under an optical microscope. Choi et al. (2006b) prepared cells of a twin dimer (Figure 5.14, top) exhibiting the SmC_A–B_4 transition. Circularly polarized UV-light irradiation was

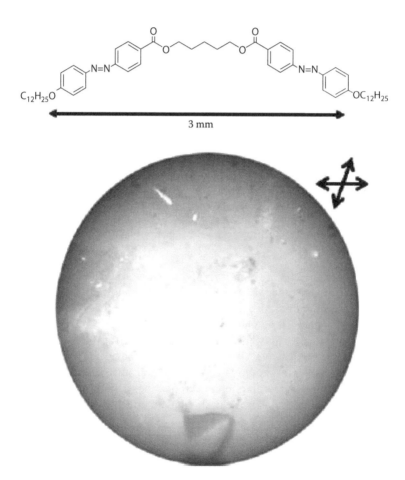

FIGURE 5.14 Photomicrograph showing a chirality-controlled (circularly polarized light irradiation) domain of bent-dimer shown at the top. (From Choi, S.-W., Izumi, T., Hoshino, Y., Takanishi, Y., Ishikawa, K., Watanabe, J., and Takezoe, H.: Circular-polarization-induced enantiomeric excess in liquid crystals of an achiral, bent-shaped mesogen. *Angew. Chem. Int. Ed.* 2006. 45. 1382–1385. Copyright Wiley-VCH Verlag GmbH & Co. KGaA. Reproduced with permission.)

made during the cooling process from SmC_A to B_4. CD spectra from a spot size of 10 mm show positive and negative signals (of the order of several hundreds of mdeg/μm) depending on the handedness of circularly polarized light. Moreover, the domain observation under decrossed polarizers clearly showed almost 100% *ee*, as shown in Figure 5.14. It was found in such domains that the CD intensity was more than 1 deg/μm. The origin of such large symmetry breaking effect was explained by the mechanism same as the reorientation of azo molecules by linearly polarized UV light. Namely, preferential conversion into a particular chiral conformation could occur in the switchable SmC_A phase under circularly polarized light because of a finite CD (absorption difference) between two chiral conformations. Since the accumulated conversion is fixed when the system is brought into the B_4 phase, the induced *ee* is preserved at room temperature. Hence, this homochiral sample can be used for device application.

Another efficient method without using chiral molecular species was also demonstrated by Choi et al. (2006c). They used a material (Figure 5.15 top), that shows the $N–B_4$ transition. The method is very simple. They introduced the material into TN cells with a twist angle less than 90° and then cooled it slowly to the B_4 phase. Chiral transfer occurs from a geometrical (extrinsic) one (twisted geometry) to a molecular and even macroscopic domain chirality (intrinsic). Figure 5.15a and b reveals microphotographs taken under decrossed polarizers,

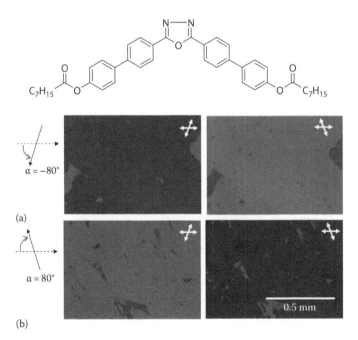

FIGURE 5.15 Photomicrographs showing chirality-controlled (using 80° twisted cells) domains of bent-dimer shown at the top. Opposite twists (a) and (b) give opposite chirality as indicated by bright and dark views under oppositely decrossed polarizers. (Images are similar to those in Choi, S.-W. et al., *Angew. Chem. Int. Ed.*, 45, 6503, 2006.)

obtained using 80° right- and left-handed TN cells. It is clear that whole areas are covered by almost homochiral domains, and the handedness can be controlled by the twist sense of the TN cells. They also showed that the CD intensity is saturated when the twist angle is more than 60°, certifying almost 100% *ee*. This method can be used only for materials, which show the N–B$_4$ transition. However, Ueda et al. (2013) showed that the choice of materials, which show the N–B$_4$ transition, can be widened by using mixtures of materials showing the

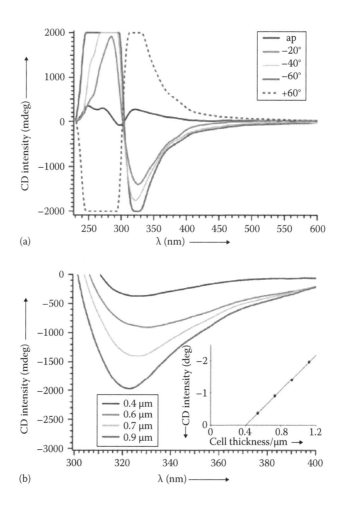

FIGURE 5.16   **(See color insert.)** CD spectra in the B$_4$ phase cooled from twisted nematic structures. (a) Cells with different twist angles including an opposite twist and (b) cells with different thicknesses. CD peak intensity against cell thickness is shown in an inset. (From Ueda, T., Masuko, S., Araoka, F., Ishikawa, K., and Takezoe, H.: A general method for the enantioselective formation of helical nanofilaments. *Angew. Chem. Int. Ed.* 2013. 125. 7001–7004. Copyright Wiley-VCH Verlag GmbH & Co. KGaA. Reproduced with permission.)

Iso–B_4 and Iso–N phase transitions. Figure 5.16 shows CD spectra measurements in the B_4 phase sandwiched in TN cells with (a) different twisting angles and (b) different thicknesses. These results tell more quantitatively that the CD signals are mirror images for right- and left-handed TN cells and the effect is from the bulk but not from the surface.

Reflection symmetry breaking and the control of the overall chirality are also dictated by the application of an electric field. The first attempt was made in the B_2 phase by Heppke et al. (1999). As mentioned in Section 3.3, there are four molecular orientation structures: two homochiral (SmC_SP_F and SmC_AP_A) and two racemic (SmC_SP_A and SmC_AP_F). The synclinic antiferroelectric SmC_SP_A state is generally stabilized when the B_2 phase appears on cooling. Normally, randomly oriented small domains are observed. However, by applying a dc electric field higher than the threshold field during slow cooling, large uniform circular domains of about 0.5 mm diameter were obtained (Zennyoji et al. 1999). After the application of a rectangular electric field ($300\ V_{pp}$, 10 Hz), an overall chiral state appears and stably exists after removing the field. Here both domains with opposite chirality exist and their total areas of opposite handednesses are approximately equal. The overall chiral texture can be converted into the racemic one by a triangular electric field ($300\ V_{pp}$, $f = 10$ Hz), and two oppositely tilted domains of the SmC_SP_A state are stabilized (Heppke et al. 1999). The same behavior was observed by other researchers (Jákli et al. 1998, Zennyoji et al. 2000). Such homochiral and racemic transformation can be made repeatedly by applying rectangular and triangular fields. Heppke et al. (1999) made dielectric and electro-optic measurements on each domain and found that the dielectric strength, relaxation time, and switching time are larger for the homochiral domains than for the racemic domains. Other groups (Ortega et al. 2003, Pelzl et al. 2006) made a different electric-field effect: If the electric-field intensity is above a certain threshold, then there is a transformation between the low and high birefringent phases with distinct chiral domains.

Two electric field effects on chirality control were reported recently. The first one is related to the DC phase (Nagaraj et al. 2014). As mentioned in Section 3.4, the DC phase appears below the isotropic phase with a dark texture under crossed polarizers. It is known that the DC phase can be converted to the B_2 phase by applying a sufficiently strong electric field. The DC phase observed in an oxadiazole-based bent-core molecule exhibits quite a different behavior particularly by applying an electric field. On increasing an alternating square electric field, the domains of opposite handedness become visible and they grow in size to be a homochiral state. Further increase of the field makes the state achiral under low birefringence conditions. The phenomenon seems to be interesting. Unfortunately, however, supporting data such as CD do not make sense. Further investigation is necessary for understanding the physics behind.

Another electric-field effect on chirality was reported by Deepa and Pratibha (2014). The experiments were made using 2,7-naphthylene bis[4-(3-methyl-4-n-tetradecyloxybenzoyloxy)]benzoate exhibiting two B_{1rev} phases. They found that if the compound is cooled under an ac or dc field from Iso, the B_{1rev} phases are eliminated and there is a direct transition into a DC phase with chiral domains. Since the morphology of the nucleating chiral domains is different under the ac and dc fields,

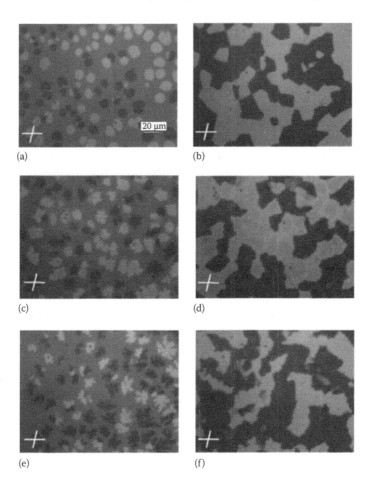

FIGURE 5.17   Textures of the DC_{ac} phase at (a), (c), and (e) 124°C and (b), (d), and (f) 116°C with increasing frequency. (Reprinted with permission from Deepa, G.B. and Pratibha, R., *Phys. Rev. E*, 89, 042504-1. Copyright 2014 by the American Physical Society.)

these DC phases are designated as DC_{ac} and DC_{dc} phases, respectively. The formation behavior also depends on the frequency applied during cooling from Iso for the formation of the DC_{ac} phase, as shown in Figure 5.17; (a), (c), and (e) are textures at 124°C with increasing frequencies and (b), (d), and (f) are those corresponding to the DC_{ac} phase at 116°C. As for the chiral domain formation, the dc field aids in a better chiral segregation leading to well-separated chiral domains (DC_{dc}). More interestingly, DC_{dc} domains thus formed show a finite *ee*, with a specific chirality irrespective of the field direction. However, these chiral domains can be reversed by reversing a field. The switching occurs with a single switching current peak under a triangular wave field, and the chiral domains retained after removing a field, that is, the chiral domains are bistable.

5.3.3 CHIRALITY SWITCHING

In general, chirality is an inherent molecular property and remains unchanged. In the ferroelectric SmC* and antiferroelectric SmC$_A^*$ phases, the constituent molecules at least in part are chiral and system chirality does not change. However, in the B$_2$ phase of the bent-core molecular system, chirality is not due the molecular chirality but is a structural one based on three specific axes, layer normal, molecular long axis, and bending directions. Hence, the interchange between right- and left-handed coordinate systems consisting of these three axes, that is, chirality switching, could occur. Actually, as mentioned earlier, the transformation between homochiral and racemic states is possible by applying proper fields. This means that layer chirality switching occurs, as clearly recognized by Schroder et al. (2004) for the first time. This can be made by the rotation of tilted molecules about their long axis. Schematic illustrations of the two switching processes, along molecular long axis and around the tilt cone, are shown in Figure 5.18a and b, respectively: (a) switching process associated with chirality switching and (b) switching process preserving (without changing) chirality.

The chirality switching (molecular rotation about the molecular long axis) was observed in a few compounds (Eremin et al. 2003, Schroder et al. 2004, Weissflog et al. 2005, Nakata et al. 2006, Pelzl et al. 2006, Geese et al. 2010). The detailed studies on the switching mechanism were made by Weissflog et al. (2005). It was clearly shown that the electro-optic response depends on temperature in the SmCP$_A$ phase; the chirality switching occurs at higher temperatures but switching preserving the chirality occurs at lower temperatures. They ascribed this temperature-dependent switching mechanism to the temperature dependence of the rotational viscosities related to two dynamics. Other researchers also pointed out some other factors for the chirality switching such as field strength and frequency.

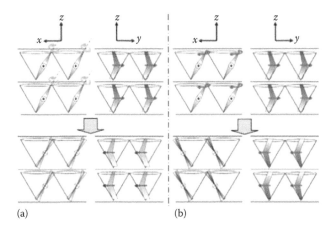

(a)   (b)

FIGURE 5.18 (**See color insert.**) Schematic illustration of two switching processes: (a) rotation about the molecular axis associated with chirality switching and (b) rotation around a cone, the same as the switching in the SmC* phase. The layer chirality is preserved. (Reprinted from Takezoe, H. and Takanishi, Y., Bent-core liquid crystals: Their mysterious and attractive world, *Jpn. J. Appl. Phys.*, 45, 597–625, 2006. With permission from the Japan Society of Applied Physics.)

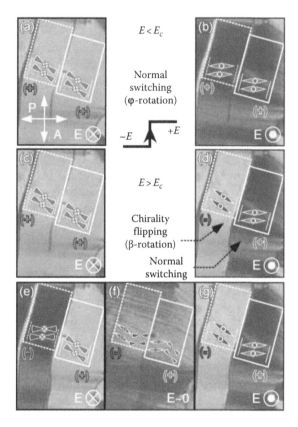

FIGURE 5.19 **(See color insert.)** Photomicrographs showing different switching modes. (a, b) Under reversed field E below E_c, normal switching around a cone is observed. (c, d) Under reversed field E slightly larger than E_c, chirality switching in part occurs. (e–g) Under a triangular field, chirality switching never occurs. (Reprinted with permission from Nakata, M., Shao, R.-F., Maclennan, J.E., Weissflog, W., and Clark, N.A., *Phys. Rev. Lett.*, 96, 067802-1. Copyright 2006 by the American Physical Society.)

Full description of theoretical interpretation was made by Nakata et al. (2006) by a nonequilibrium dissipative model of chiral smectic dynamics with anisotropic rotational viscosities. Their key experimental results are shown in Figures 5.19 and 5.20. Polarizing micrographs shown in Figure 5.19 clearly show two different switching modes; under a rectangular field $E < E_c$, normal switching occurs ((a) and (b)), and chirality switching in part occurs when E is slightly larger than E_c ((c) and (d)). Only normal switching was observed under a triangular field ((e)–(g)) and SmC_AP_A is stabilized in the absence of a field. Figure 5.20 shows an E–T diagram, which clearly shows the chirality switching zone at high temperatures and high electric fields. The rotation angles around a tilt cone and about a molecular long axis are defined as φ and β, respectively. The free energy U_b has equal minima at $\beta = 0$ and π for two chiral states separated by an energy barrier in between. Key factors to consider the dynamics of $\varphi(t)$ and $\beta(t)$

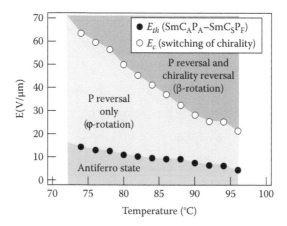

FIGURE 5.20 Experimentally obtained E–T diagram to show switching modes. At high E and T region, chirality switching occurs. β- and φ-rotations stand for rotations about the molecular long axis and around the cone, respectively. (Reprinted with permission from Nakata, M., Shao, R.-F., Maclennan, J.E., Weissflog, W., and Clark, N.A., *Phys. Rev. Lett.*, 96, 067802-1. Copyright 2006 by the American Physical Society.)

are the viscous anisotropy η_β/η_φ. With some appropriate assumptions, the following equations were obtained to simulate the dynamics:

$$\frac{d\varphi}{dt} = \frac{P_0 E \sin\varphi\cos\beta\sin^2\theta + (dU_b/d\beta)\cos\theta}{\eta_\varphi - \eta_\beta\cos^2\theta}, \tag{5.8}$$

$$\frac{d\beta}{dt} = \frac{P_0 E(\cos\varphi\sin\beta + \sin\varphi\cos\beta\cos\theta) - (dU_b/d\beta)}{\eta_\beta} - \frac{d\varphi}{dt}\cos\theta, \tag{5.9}$$

where

P_0 is the spontaneous polarization

E is an applied field

θ is the molecular tilt angle with respect to the layer normal

By using the Euler–Cauchy method, numerical solutions were obtained. The experimental observations shown in Figure 5.19 were well reproduced. In addition, the inverse E dependence of the switching time across the boundary between φ rotation and β rotation was obtained, being consistent with the experimental results. Figure 5.21 gives calculated dependence of E_c on barrier height U_{max} and on viscosity anisotropy. E_c depends linearly on U_{max} but diverges rapidly as viscous anisotropy is approached at $\eta_\beta/\eta_\varphi = 0.95$. Hence, if U_{max} is very high, and/or η_β is close to η_φ, chirality switching never occurs, which is normally the case in SmC or achiral SmCP.

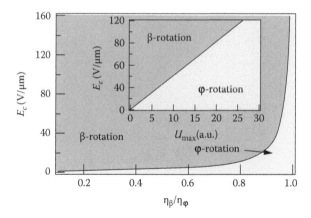

FIGURE 5.21 Simulation of a zone showing chirality switching (β-rotation) in E_c-viscous anisotropy space. The simulation result in E_c–U_{max} space is also shown. (Reprinted with permission from Nakata, M., Shao, R.-F., Maclennan, J.E., Weissflog, W., and Clark, N.A., *Phys. Rev. Lett.*, 96, 067802-1. Copyright 2006 by the American Physical Society.)

It is worth noting that chirality switching was also observed in bent-shaped dimers (unpublished data from Master thesis by N. Takada, Tokyo Institute of Technology), the B_{1rev} phase (Szydlowska et al. 2003), and the SmCP$_G$ bilayer phase (Bedel et al. 2004). The theory mentioned in the preceding text can also be applied to such examples.

REFERENCES

Alaasar, M., M. Prehm, Y. Cao, F. Liu, and C. Tschierske. Spontaneous mirror-symmetry breaking in isotropic liquid phases of photoisomerizable achiral molecules. *Angew. Chem. Int. Ed.* 55 (2016): 312–316.

Archer, P. and I. Dierking. Experimental determination of the full Landau potential of bent-core doped ferroelectric liquid crystals. *Phys. Rev. E* 72 (2005): 041713-1–041713-10.

Archer, P. and I. Dierking. Chirality enhancement through addition of achiral molecules. *Chem. Commun.* 46 (2010): 1467–1469.

Archer, P. and I. Dierking. A bent-core dopant-induced smectic A* twist state. *Liq. Cryst.* 33 (2006): 257–265.

Bedel, J. P., J. C. Rouillon, J. P. Marcerou, H. T. Nguyen, and M. F. Achard. Evidence for different polymorphisms with and without an external electric field in a series of bent-shaped molecules. *Phys. Rev. E* 69 (2004): 061702-1–061702-9.

Burnham, K. S. and G. B. Schuster. Transfer of chirality from circularly polarized light to a bulk material property: Propagation of photoresolution by a liquid crystal transition. *J. Am. Chem. Soc.* 121 (1999): 10245–10246.

Choi, S.-W., K. Fukuda, S. Nakahara, K. Kishikawa, Y. Takanishi, K. Ishikawa, J. Watanabe, and H. Takezoe. Amplification of twisting power in chiral mesophase by introducing achiral rod-like compound with ester group. *Chem. Lett.* 35 (2006a): 896–897.

Choi, S.-W., T. Izumi, Y. Hoshino, Y. Takanishi, K. Ishikawa, J. Watanabe, and H. Takezoe. Circular-polarization-induced enantiomeric excess in liquid crystals of an achiral, bent-shaped mesogen. *Angew. Chem. Int. Ed.* 45 (2006b): 1382–1385.

Choi, S.-W., S. Kang, Y. Takanishi, K. Ishikawa, J. Watanabe, and H. Takezoe. Intrinsic chirality in a bent-core mesogen induced by extrinsic chiral structures. *Angew. Chem. Int. Ed.* 45 (2006c): 6503–6506.

Choi, S.-W., S. Kawauchi, S. Tanaka, J. Watanabe, and H. Takezoe. Vibrational circular dichroism spectroscopic study on circularly polarized light-induced chiral domains in the B_4 phase of a bent mesogen. *Chem. Lett.* 36 (2007): 1018–1019.

Coleman, D. A., J. Fernsler, N. Chattham, M. Nakata, Y. Takanishi, E. Korblova, D. R. Link et al. Polarization-modulated smectic liquid crystal phases. *Science* 301 (2003): 1204–1211.

Deepa, G. B. and R. Pratibha. Chiral symmetry breaking dictated by electric-field-driven shape transitions of nucleating conglomerate domains in a bent-core liquid crystal. *Phys. Rev. E* 89 (2014): 042504-1–042504-9.

Dressel, C., F. Liu, M. Prehm, X.-B. Zeng, G. Ungar, and C. Tschierske. Dynamic mirror-symmetry breaking in bicontinuous cubic phases. *Angew. Chem. Int. Ed.* 53 (2014a): 13115–13120.

Dressel, C., T. Reppe, M. Prehm, M. Brautzsch, and C. Tschierske. Chiral self-sorting and amplicication in isotropic liquids of achiral molecules. *Nat. Chem.* 6 (2014b): 971–977.

Dressel, C., W. Weissflog, and C. Tschierske. Spontaneous mirror symmetry breaking in a re-entrant isotropic liquid. *Chem. Commun.* 51 (2015): 15850–15853.

Druerke, D. and G. Heppke. Presented at *Gordon Conference Liquid Crystals*, Tilton, New Hampshire, 1997, and *Banana-Shaped Liquid Crystal Workshop*, Berlin, Germany, 1997.

Earl, D. J., M. A. Osipov, H. Takezoe, Y. Takanishi, and M. R. Wilson. Induced and spontaneous deracemization in bent-core liquid crystal phases and in other phases doped with bent-core molecules. *Phys. Rev. E* 71 (2005): 021706-1–021706-11.

Earl, D. J. and M. R. Wilson. Predictions of molecular chirality and helical twisting powers: A theoretical study. *J. Chem. Phys.* 119 (2003): 10280–10288.

Emsley, J. W., P. Lesot, G. R. Luckhurst, A. Meddour, and D. Merlet. Chiral solutes can seed the formation of enantiomorphic domains in a twist-bend nematic liquid crystal. *Phys. Rev. E* 87 (2013): 040501(R)-1–040501(R)-4.

Eremin, A., S. Diele, G. Pelzl, and W. Weissflog. Field-induced switching between states of opposite chirality in a liquid-crystalline phase. *Phys. Rev. E* 67 (2003): 020702(R)-1–020702(R)-3.

Ferrarini, A., G. J. Moro, and P. L. Nordio. Simple molecular model for induced cholesteric phases. *Phys. Rev. E* 53 (1996): 681–688.

Geese, K., M. Prehm, and C. Tschierske. Bent-core mesogens with thiophene units. *J. Mater. Chem.* 20 (2010): 9658–9665.

Gorecka, E., M. Čepič, J. Mieczkowski, M. Nakata, H. Takezoe, and B. Zeks. Enhanced chirality by adding achiral molecules into the chiral system. *Phys. Rev. E* 67 (2003): 061704-1–061704-5.

Gorecka, E., M. Nakata, J. Mieczkowski, Y. Takanishi, K. Ishikawa, J. Watanabe, H. Takezoe, S. H. Eichhorn, and T. M. Swager. Induced antiferroelectric smectic-C_A phase by doping ferroelectric-C* phase with bent-shaped molecules. *Phys. Rev. Lett.* 85 (2000): 2526–2529.

Heppke, G., A. Jákli, S. Rauch, and H. Sawade. Eectric-field-induced chiral separation in liquid crystals. *Phys. Rev. E* 60 (1999): 5575–5579.

Hough, L. E., H. T. Jung, D. Kruerke, M. S. Heberling, M. Nakata, C. D. Jones, D. Chen et al. Helical nanofilament phases. *Science* 325 (2009b): 456–460.

Hough, L. E., M. Spannuth, M. Nakata, D. A. Coleman, C. D. Jones, G. Dantlgraber, C. Tschierscke et al. Chiral isotropic liquids from achiral molecules. *Science* 325 (2009a): 452–456.

Huck, N. P. M., W. F. Jager, B. de Lange, and B. L. Feringa. Dynamic control and amplification of molecular chirality by circular polarized light. *Science* 273 (1996): 1686–1688.

Iftime, G., F. L. Abarthet, A. Natansohn, and P. Rochon. Control of chirality of an azobenzene liquid crystalline polymer with circularly polarized light. *J. Am. Chem. Soc.* 122 (2000): 12646–12650.

Jákli, A., G. G. Nair, C. K. Lee, R. Sun, and L. C. Chien. Macroscopic chirality of a liquid crystal from nonchiral molecules. *Phys. Rev. E* 63 (2001): 061710-1–061710-5.

Jákli, A., S. Rauch, D. Lotzsch, and G. Heppke. Uniform textures of smectic liquid-crystal phase formed by bent-core molecules. *Phys. Rev. E* 57 (1998): 6737–6740.

Jeong, H. S., S. Tanaka, D. K. Yoon, S.-W. Choi, Y. H. Kim, S. Kawauchi, F. Araoka, H. Takezoe, and H.-T. Jung. Spontaneous chirality induction and enantiomer separation in liquid crystals composed of achiral rod-shaped α-arylbenzoate esters. *J. Am. Chem. Soc.* 131 (2009): 15055–15060.

Kawauchi, S., S.-W. Choi, K. Fukuda, K. Kishikawa, J. Watanabe, and H. Takezoe. Why achiral rod-like compound with ester group amplifies chiral power in chiral mesophase. *Chem. Lett.* 36 (2007): 750–751.

Kishikawa, K., N. Muramatsu, S. Kohmoto, K. Yamaguchi, and M. Yamamoto. Control of molecular aggregations by doping in mesophases: Transformation of smectic C phases to smectic C_A phases by addition of long bent-core molecules possessing a central strong dipole. *Chem. Mater.* 15 (2003): 3443–3449.

Kurosu, H., M. Kawasaki, M. Hirose, M. Yamada, S. Kang, J. Thisayukta, M. Sone, H. Takezoe, and J. Watanabe. Solid-state ^{13}C NMR study of chiral twisted conformation attributable to chirality in smectic phases of achiral banana-shaped molecules. *J. Phys. Chem. A* 108 (2004): 4674–4678.

Lee, G., R. J. Carlton, F. Araoka, N. L. Abbott, and H. Takezoe. Amplification of the stereo-chemistry of biomolecular adsorbates by deracemization of chiral domains in bent-core liquid crystals. *Adv. Mater.* 25 (2013): 245–249.

Link, D. R., G. Natale, R. Shao, J. E. Maclennan, N. A. Clark, K. Eva, and D. M. Walba. Spontaneous formation of macroscopic chiral domains in a fluid smectic phase of achiral molecules. *Science* 278 (1997): 1924–1927.

Nagaraj, M., K. Usami, Z. Zhang, V. Gortz, J. W. Goodby, and H. F. Gleeson. Unusual electric-field-induced transformations in the dark conglomerate phase of a bent-core liquid crystal. *Liq. Cryst.* 41 (2014): 800–811.

Nagayama, H., S. K. Varshney, M. Goto, F. Araoka, K. Ishikawa, V. Prasad, and H. Takezoe. Spontaneous deracemization of disk-like molecules in the columnar phase. *Angew. Chem. Int. Ed.* 49 (2010): 445–448.

Nakata, M., R.-F. Shao, J. E. Maclennan, W. Weissflog, and N. A. Clark. Electric-field-induced chirality flipping in smectic liquid crystals: The role of anisotropic viscosity. *Phys. Rev. Lett.* 96 (2006): 067802-1–067802-4.

Nakata, M., Y. Takanishi, J. Watanabe, and H. Takezoe. Blue phases induced by doping chiral nematic liquid crystals with nonchiral molecules. *Phys. Rev. E* 68 (2003): 041710-1–041710-6.

Niwano, H., M. Nakata, J. Thisayukta, D. R. Link, H. Takezoe, and J. Watanabe. Chiral memory on transition between the B_2 and B_4 phases in an achiral banana-shaped molecular system. *J. Phys. Chem. B* 108 (2004): 14889–14896.

Ortega, J., C. L. Folcia, J. Etxebarria, N. Gimeno, and M. B. Ros. Interpretation of unusual textures in the B_2 phase of a liquid crystal composed of bent-core molecules. *Phys. Rev. E* 68 (2003): 011707-1–011707-4.

Otani, T., F. Araoka, K. Ishikawa, and H. Takezoe. Enhanced optical activity by achiral rod-like molecules nanosegregated in the B_4 structure of achiral bent-core molecules. *J. Am. Chem. Soc.* 131 (2009): 12368–12372.

Pasteur, L. Relation qui peut exister entre la forme crystalline et la composition chimique, et sur la cause de la polarization rotatoire. *Ann. Chim. Phys.* 24 (1848): 442–459.

Pelzl, G., M. W. Schroder, A. Eremin, S. Diele, B. Das, S. Grande, H. Kresse, and W. Weissflog. Field-induced phase transitions and reversible field-induced inversion of chirality in tilted smectic phases of bent-core mesogens. *Eur. Phys. J. E* 21 (2006): 293–303.

Perez-Garcial, L. and D. B. Amabilino. Spontaneous resolution under supramolecular control. *Chem. Soc. Rev.* 31 (2002): 342–356.

Schroder, M. W., S. Diele, G. Pelzl, and W. Weissflog. Field-induced switching of the layer chirality in SmCP phases of novel achiral bent-core liquid crystals and their unusual large increase in clearing temperature under electric field application. *Chem. Phys. Chem.* 5 (2004): 99–103.

Sekine, T., T. Niori, M. Sone, J. Watanabe, S.-W. Choi, Y. Takanishi, and H. Takezoe. Origin of helix in achiral banana-shaped molecular systems. *Jpn. J. Appl. Phys.* 36 (1997): 6455–6463.

Selinger, J. V. Chiral and antichiral order in bent-core liquid crystals. *Phys. Rev. Lett.* 90 (2003): 165501-1–165501-4.

Shiromo, K., D. A. Sahade, T. Oda, K. Nihira, Y. Takanishi, K. Ishikawa, and H. Takezoe. Finite enantiomeric excess nucleated in an achiral banana mesogen by chiral alignment surfaces. *Angew. Chem. Int. Ed.* 44 (2005): 1948–1951.

Szydlowska, J., J. Mieczkowski, J. Matraszek, D. W. Bruce, E. Gorecka, D. Pociecha, and D. Guillon. Bent-core liquid crystals forming two- and three-dimensional modulated structures. *Phys. Rev. E* 67 (2003): 031702-1–031702-5.

Takanishi, Y., H. Takezoe, Y. Suzuki, I. Kobayashi, T. Yajima, M. Terada, and K. Mikami. Spontaneous enantiomeric resolution in a fluid smectic phase of a racemate. *Angew. Chem. Int. Ed.* 38 (1999): 2354–2356.

Takezoe, H. Spontaneous achiral symmetry breaking in liquid crystalline phases. *Top. Curr. Chem.* 318 (2012): 303–330.

Takezoe, H. and Y. Takanishi. Bent-core liquid crystals: Their mysterious and attractive world. *Jpn. J. Appl. Phys.* 45 (2006): 597–625.

Thisayukta, J., Y. Nakayama, S. Kawauchi, H. Takezoe, and J. Watanabe. Distinct formation of a chiral smectic phase in achiral banana-shaped molecules with a central core based on a 2,7-dihydroxynaphthalene unit. *J. Am. Chem. Soc.* 122 (2000): 7441–7448.

Thisayukta, J., H. Niwano, H. Takezoe, and J. Watanabe. Effect of chiral dopant on a helical Sm1 phase of banana-shaped N-*n*-O-PIMB molecules. *J. Mater. Chem.* 11 (2001): 2717–2721.

Thisayukta, J., H. Niwano, H. Takezoe, and J. Watanabe. Enhancement of twisting power in the chiral nematic phase by introducing achiral banana-shaped molecules. *J. Am. Chem. Soc.* 124 (2002): 3354–3358.

Tschierske, C. and G. Ungar. Mirror symmetry breaking by chirality synchronization in liquids and liquid crystals of achiral molecules. *ChemPhysChem* 17 (2016): 9–26.

Ueda, T., S. Masuko, F. Araoka, K. Ishikawa, and H. Takezoe. A general method for the enantioselective formation of helical nanofilaments. *Angew. Chem. Int. Ed.* 125 (2013): 7001–7004.

Walba, D. W., L. Eshdat, E. Korblova, and R. K. Shoemaker. On the nature of the B_4 banana phase: Crystal or not a crystal? *Cryst. Growth Des.* 5 (2005): 2091–2099.

Weissflog, W., U. Dunemann, M. W. Schroder, S. Diele, G. Pelzl, H. Kresse, and S. Grande. Field-induced inversion of chirality in $SmCP_A$ phases of new achiral bent-core mesogens. *J. Mater. Chem.* 15 (2005): 939–946.

Wilson, M. R. Progress in computer simulations of liquid crystals. *Int. Rev. Phys. Chem.* 24 (2005): 421.

Wu, Y., A. Natansohn, and P. Rochon. Photoinduced chirality in thin films of achiral polymer liquid crystals containing azobenzene chromophores. *Macromolecules* 37 (2004): 6801–6805.

Xu, J., R. Y. Dong, V. Domenici, K. Fodor-Csorba, and C. A. Veracini. ^{13}C and ^{2}H NMR study of structure and dynamics in banana B_2 phase of a bent-core mesogen. *J. Phys. Chem. B* 110 (2006): 9434–9441.

Yamada, K., S. Kang, K. Takimoto, M. Hattori, K. Shirata, S. Kawauchi, K. Deguchi, T. Shimizu, and J. Watanabe. Structural analysis of a banana-liquid crystal in the B_4 phase by solid-state NMR. *J. Phys. Chem. B* 117 (2013): 6830–6838.

Zennyoji, M., Y. Takanishi, K. Ishikawa, J. Thisayukta, J. Watanabe, and H. Takezoe. Partial mixing of opposite chirality in a bent-shaped liquid crystal molecular system. *J. Mater. Chem.* 9 (1999): 2775–2778.

Zennyoji, M., Y. Takanishi, K. Ishikawa, J. Thisayukta, J. Watanabe, and H. Takezoe. Electrooptic and dielectric properties in bent-shaped liquid crystals. *Jpn. J. Appl. Phys.* 39 (2000): 3536–3541.

Zennyoji, M., Y. Takanishi, K. Ishikawa, J. Thisayukta, J. Watanabe, and H. Takezoe. Molecular chirality due to twisted conformation in a bent-shaped liquid crystal studied by polarized FT-IR spectroscopy. *Mol. Cryst. Liq. Cryst. Sci. Technol.* 366 (2001): 693–701.

6 Nonlinear Optics in Bent-Shaped Mesogens

In dielectric media, polarization P is induced by an electric field E. When the field strength is low, the induced polarization is proportional to E. But in higher fields, P is a nonlinear function of E and it can be expanded in power series of E

$$P = P_0 + \chi^{(1)}E + \chi^{(2)}E^2 + \chi^{(3)}E^3 + \cdots, \qquad (6.1)$$

where

P_0 is the spontaneous polarization

$\chi^{(n)}$ is nth order electric susceptibility

Generally, $\chi^{(n)}$ is called nth order nonlinear electric susceptibility for $n = 2, 3, \ldots$, while $\chi^{(1)}$ is simply written by χ and is called linear electric susceptibility or simply electric susceptibility. The symmetry requires that $\chi^{(n)}$ with even n is nonzero only when the system is noncentrosymmetric within electric dipole approximation. Therefore, the second-order nonlinear optical (NLO) effect can be observed only in noncentrosymmetric systems, so that it provides a powerful tool to examine the existence of polar order. Second-harmonic generation (SHG) is the most simple second-order nonlinear optical effect and appears as light generation at a frequency 2ω by the incidence of intense light at a frequency ω. Here we restrict the description of the NLO phenomena only to the SHG. The second-order nonlinear polarization $P^{(2)}$ is given by

$$\mathbf{P}^{(2)} = \tilde{\chi}^{(2)}\mathbf{EE}$$
$$P_i^{(2)} = \sum \chi_{ijk}^{(2)}E_jE_k \qquad (6.2)$$

Here

$\mathbf{P}^{(2)}$ and $\mathbf{E}$ are vectors

$\tilde{\chi}^{(2)}$ is a third-rank tensor

The susceptibility tensor $\chi_{ijk}^{(2)}$ is symmetric in the second and third indices (intrinsic permutation's symmetry, $\chi_{ijk}^{(2)} = \chi_{ikj}^{(2)}$), which results in 18 independent components. This allows us to express the polarizability with the help of the d-matrix:

$$\begin{pmatrix} P_1^{(2)} \\ P_2^{(2)} \\ P_3^{(2)} \end{pmatrix} = \varepsilon_0 \begin{pmatrix} d_{11} & \cdots & d_{16} \\ \vdots & \ddots & \vdots \\ d_{31} & \cdots & d_{36} \end{pmatrix} \begin{pmatrix} E_1^2 \\ E_2^2 \\ E_3^2 \\ 2E_2E_3 \\ 2E_3E_1 \\ 2E_1E_2 \end{pmatrix}$$

6.1 MOLECULAR HYPERPOLARIZABILITY

Nonlinear polarizabilities on a molecular scale (for single molecules) can be described in a similar manner. The expansion of the molecular dipole moment **p** in powers of **E** yields

$$\mathbf{p} = \mathbf{m} + \alpha\mathbf{E} + \beta\mathbf{E}\mathbf{E} + \gamma\mathbf{E}\mathbf{E}\mathbf{E} + \cdots \qquad (6.3)$$

where

m is the permanent dipole
α, β, and γ are first-, second-, and third-order molecular polarizabilities

We sometimes call hyperpolarizability for nonlinear molecular polarizability. The macroscopic nonlinear susceptibility $\chi^{(2)}$ is given by

$$\chi^{(2)} = Nf\left(\omega_1\right)f\left(\omega_2\right)f\left(\omega_3\right)\langle\beta\rangle, \qquad (6.4)$$

where

N is the number density of molecules
$f(\omega_i)$ is the Lorentz local field factor at a frequency ω_i
The bracket $\langle\beta\rangle$ represents an average over molecular orientation in a unit volume

To have a large $\chi^{(2)}$, well-aligned higher components of the large first-order hyperpolarizability β is required. Two methods are available for the hyperpolarizability measurements: the electric-field-induced second harmonic (EFISH) method (Dworczak and Kieslinger 2000) and the hyper-Rayleigh scattering method (Clays and Persoons 1992).

As far as we know, there is only one report on the first-order hyperpolarizability tensor of bent molecules by means of the hyper-Rayleigh scattering method (Araoka et al. 1999). The β values of alkyl and alkoxy compounds are nearly the same and increase linearly with increasing chain length from ca. 10^{-29} esu to ca.

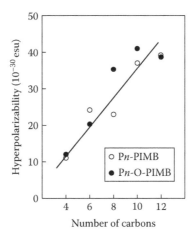

FIGURE 6.1 Hyperpolarizability as a function of the number of carbons in Pn-PIMB and Pn-O-PIMB.

4×10^{-29} esu, as shown in Figure 6.1. Using the β values and the dihedral angle of a bent molecule P12-O-PIMB, the tensor elements of β were also determined: $\beta_{zzz} = 15.3 \times 10^{-30}$ esu and $\beta_{zxx} = 61.5 \times 10^{-30}$ esu, where x is along the molecular long axis, z along the bending direction. This means that the nonlinear polarization along the z axis induced by an electric field applied along the x axis is much larger than that by an electric field along the z axis. Namely, a component along each arm $\beta_{\xi\xi\xi}$ gives a major contribution and is 86.1×10^{-30} esu. This value is comparable to that of intramolecular charge transfer molecules such as p-nitroaniline (about 30×10^{-30} esu).

Macroscopic nonlinear components d_{33} and d_{31} expressed by

$$d_{33} = Nf^3\beta_{zzz}$$
$$d_{31} = Nf^3\beta_{zxx}, \tag{6.5}$$

which were determined as 12.3 and 49.4 pm/V, respectively (Araoka et al. 1999). The d_{33} value agreed well with that determined using a liquid crystal (LC) cell by Macdonald et al. (1998) (10.1 pm/V). Further comparison will be made in Section 6.2.

More recently, designing molecules for NLO, special bent-core molecules have been synthesized (Pintre et al. 2010). Strong electron-withdrawing groups such as cyanocinnamate or nitro groups and an electron-donating group (piperazine moiety) were introduced. Unlike rod-shaped molecular systems, molecular hyperpolarizability along the bent (polar) axis does not average out to zero because of polar packing. Pintre et al. (2010) synthesized molecules shown in Figure 6.2 and obtained **m**β values of these molecules using EFISH. Although this method cannot determine β components, obtained **m**β values in the molecules A and B shown in Figure 6.2 are one order of magnitude larger than that in molecule C. Unfortunately, however, neither A nor B exhibits the appropriate mesophase but they can be used as a component of mixture systems.

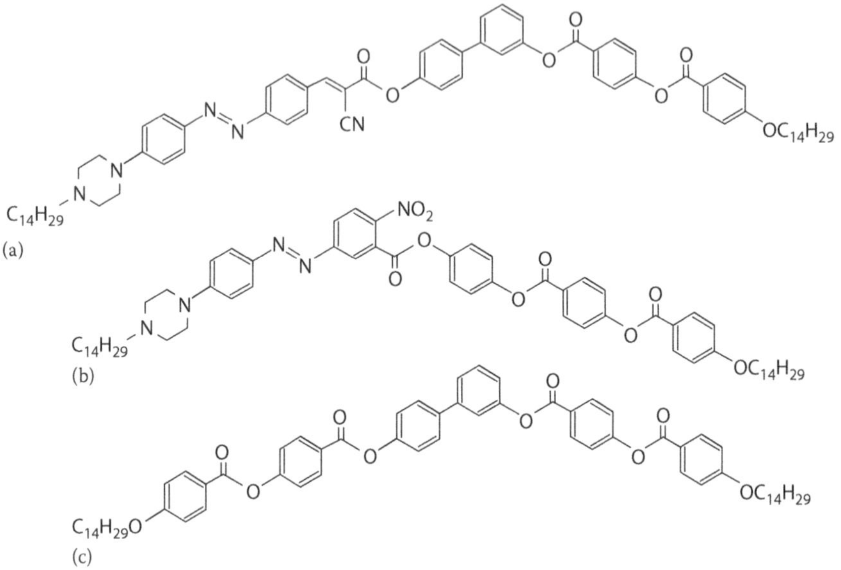

FIGURE 6.2 Chemical structures of bent-core molecules (a) A, (b) B, and (c) C for the EFISH measurements by Pintre et al. (2007).

6.2 TILTED B$_2$ AND B$_4$ PHASES

As mentioned earlier, SHG is a useful technique for detecting polar order. From a molecular point of view, bent-core molecules have high hyperpolarizability (see Section 6.1) and are promising materials for NLO applications. The first SHG measurement on bent-core molecules was made by two groups (Choi et al. 1998, Kentischer et al. 1998, Macdonald et al. 1998). All reports showed spontaneous SHG activity in the B$_4$ phase and the electric-field-induced SHG in the B$_2$ phase. Choi et al. (1998) used bent-shaped (P8-O-PIMB) and twin dimer (12OAM5AMO12) and measured SHG intensity as functions of an applied electric field and an incidence angle. Polarization switching by reversing the field was also confirmed by SHG interferometry. Macdonald et al. (1998) first determined nonlinear susceptibility components in P12-O-PIMB and P10-PIMB, obtaining a very large $d_{31} = 16.5$ pm/V in P12-O-PIMB.

Quantitative analyses of the second-order NLO susceptibility have been made in the B$_2$ phase by several groups. Ortega et al. (2000) measured SHG in the B$_2$ and B$_4$ phases of P12-O-PIMB. In order to determine the d components, the most important and difficult problem is to obtain a good alignment of the LC in cells. It is usually difficult to obtain well-aligned cells in the B$_2$ phase, unless bent-core mesogens have the N or SmA phase. Surface treatment such as rubbing is known to be inefficient, and one usually obtains circular domains in the B$_2$ phase. Ortega et al. (2000) took into account a random orientation of different domains and determined the nonlinear susceptibility components in P12-O-PIMB, in which d_{31} is about half the value by Macdonald et al. (1998). They also discussed the SHG in the B$_4$

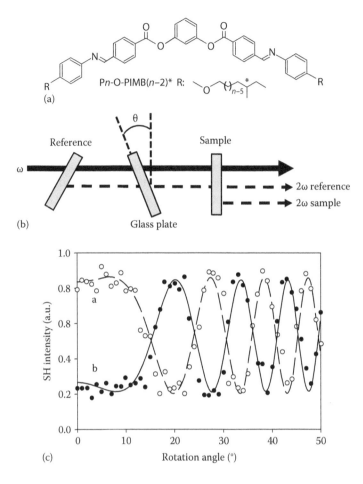

FIGURE 6.3 SHG interferometry measurement. (a) Chemical structure of Pn-O-PIMB (n – 2)*; (b) principle of SHG interferometry measurement; and (c) interferogram observed in P8-O-PIMB6* showing the SmC$_A$P$_F$ phase. (Reprinted from Nakata, M. et al., *Liq. Cryst.*, 28, 1301, 2001. With permission from Taylor & Francis.)

phase, but the quantitative analysis was first presented in the work by Araoka et al. (2005) where chirality was taken into account (SHG circular dichroism [SHG-CD], see Section 6.3).

The only treatment available to obtain uniformly aligned cells is field application parallel to the cell surfaces. Nakata et al. (2001) applied this alignment technique to the cells of P8-O-PIMB6* (SmC$_A$P$_F$) (Figure 6.3a) and performed SHG interferometry measurements (Figure 6.3b). By rotating a glass plate placed between a quartz plate and a sample cell, second-harmonic (SH) signals from the quartz and the sample interfere constructively and destructively because of the mutual phase shift between these SH lights caused by the refractive indices' difference at ω and 2ω. The interferograms shown in Figure 6.3c were taken in the absence of an external electric field after application of positive and negative in-plane electric fields. Two interferograms are mirror images or phase shifted by π, indicating bistable states with opposite polarizations.

Araoka et al. (2002) also used the same alignment technique and measured SHG in-plane anisotropy by the incidence of an Neodymium-Yttrium-Aluminum-Garnet (Nd-YAG) laser beam normal to the cell surface. The results are shown in Figure 6.4 as a function of the rotation angle of polarizers (a) parallel and (b) perpendicular to each other. According to a detailed analysis of the data, by taking account of optical anisotropy in the SHG active bulk, the best fit was obtained as shown by the solid curves in Figure 6.4. Nonlinear susceptibility tensor components, $d_{33} = 14.4$ pm/V and $d_{31} = 49.2$ pm/V, were obtained, where $X(1)$, $Y(2)$, and $Z(3)$ axes are along the smectic layer normal, cell surface normal, and applied electric field, respectively. These values are much larger than those of conventional SmC* ferroelectric liquid crystal materials.

Using Araoka's d components (Araoka et al. 2002), the hyperpolarizability tensor components of the molecule were also determined as $\beta_{zzz} = 14.5 \times 10^{-30}$ and $\beta_{zxx} = 67.6 \times 10^{-30}$ esu by assuming bent-core molecules densely packed into smectic layers. These values agree well with those of an achiral homologue P12-O-PIMB determined by hyper-Rayleigh scattering: $\beta_{zzz} = 15.3 \times 10^{-30}$ and $\beta_{zxx} = 61.5 \times 10^{-30}$ esu (1 esu $\approx 3.71 \times 10^{-21}$ $C^3 m^3/J$) (Araoka et al. 1999). The dihedral angle of the bent-core $\theta = 127.6°$ can be determined

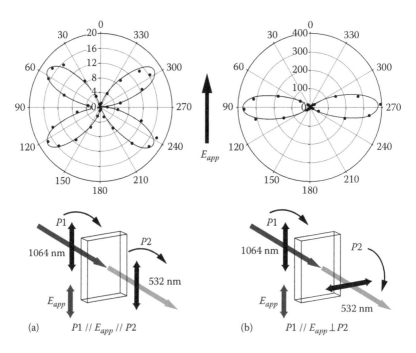

FIGURE 6.4 (See color insert.) SHG in-plane anisotropy measured using well-aligned sample P8-O-PIMB6* ($SmC_A P_F$) under an applied electric field (shown by arrow). (a) Under parallel polarizers and (b) perpendicular polarizers. The polar angles denote the angle between the polarization direction of the normally incident fundamental beam and the field direction. (Reprinted with permission from Araoka, F., Thisayukta, J., Ishikawa, K., Watanabe, J., and Takezoe, H., *Phys. Rev. E*, 66, 021705-1. Copyright 2002 by the American Physical Society.)

from a comparison between macroscopic and microscopic susceptibilities. This value is consistent with the values determined by other methods such as surface SHG of the Langmuir–Blodgett monolayer of P12-O-PIMB (KinoshitaLangmuir 1998), x-ray diffraction measurement (Pelzl et al. 1999, Ortega et al. 2000), and nuclear magnetic resonance (NMR) (Pelzl et al. 1999, Ortega et al. 2004) of P8-O-PIMB.

Ortega et al. (2004) used the same alignment technique and determined the nonlinear susceptibility components of a few bent-core LCs. The obtained nonlinear susceptibility components are similar to those obtained by the same group (Gallastegui et al. 2002) but are much smaller than the values by Araoka et al. (2002) and Macdonald et al. (1998). The only difference between P8-O-PIMB6* and Pn-O-PIMB is that the former is in the racemic SmC$_A$P$_F$ and the latter is in the homochiral SmC$_S$P$_F$. Data of nonlinear susceptibility components are available in other molecules such as bent-shaped molecules derived from biphenylene (Pintre et al. 2007) and polymers derived from bent-core mesogens (Gallastegui et al. 2005).

SHG has been applied to analyze more complicated molecular orientations in the B$_2$ phase. The measurements in Figure 6.4 were made under an electric field. However, the measurements using bistable polar structures after applying positive and negative electric fields (Figure 6.3) exhibited totally different SHG in-plane anisotropy, which was obtained by rotating parallel polarizers, as shown in Figure 6.5a (Araoka et al. 2004). This SHG pattern was analyzed by considering a splayed polarization orientation shown in Figure 6.5b including the effect of polarization charges. This structure could be realized by surface polar interaction and/or splayed polarization to avoid macroscopic polarization. Although the polarization along the surface normal is cancelled out, the inlayer polarization remains as one of the bistable states, resulting in SHG activity. The structure shown in Figure 6.5b is a

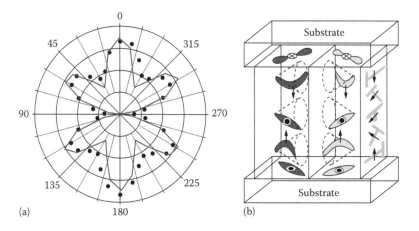

FIGURE 6.5 (a) SHG in-plane anisotropy measured by rotating parallel polarizers after terminating an electric field. Solid curve is a best fit using the structure shown in (b), where an in-plane polar order remains in contrast to no polarization along the surface normal direction. (Reprinted with permission from Araoka, F., Hoshi, H., and Takezoe, H., *Phys. Rev. E*, 69, 051704-1. Copyright 2004 by the American Physical Society.)

splayed state of the racemic SmC_AP_F, where the variations in the polarization in two adjacent layers are identical but the variations of the long molecular axis projected on the plane parallel to the substrate surface are opposite in the rotation sense from the top surface to the bottom in two adjacent layers. This structural feature explains the SHG activity and the absence of chirality. The solid curve is the best fit of the simulated results from the splayed state corresponding to the energy minimum. If a uniform splay is assumed, the characteristic six-peak pattern can still be obtained, but with a different shape. In this way, SHG is useful to analyze polarization (and in turn molecular) orientation.

Another tilted banana phase is the B_4 phase. At least four papers have been published on nonlinear optical properties of the B_4 phase (Araoka et al. 2003, 2005, 2013, Martinez-Perdiguero et al. 2009). The first two papers deal with chiral NLO effect, that is, electrogyration and SHG-CD, and will be discussed in Section 6.3. The B_4 phase was first assigned to a phase with a twist-grain-boundary (TGB) structure, in which the smectic layer discontinuously rotates about the axis parallel to the layer (Thisayukta et al. 2001). But later, a freeze fracture transmission electron microscopy (FFTEM) measurement unambiguously showed the helical structure (see Figure 3.30), which is different from the TGB structure; the rotation of the layer is not discontinuous as in TGB but continuous (Hough et al. 2009). SHG activity in the absence of an electric field was confirmed by several researchers (Choi et al. 1998, Macdonald et al. 1998, Ortega et al. 2000), but because of the lack of an alignment technique for helical nanofilaments, quantitative studies appeared much later. Martinez-Perdiguero et al. (2009) found that the SHG signal intensity in B_4 is two orders of magnitude weaker than that in B_2 under an electric field, refuting the SmC_AP_F local structure of B_4.

Araoka et al. (2013) succeeded in obtaining well-aligned helical nanofilaments (see Figure 8.10). They analyzed the SHG intensity profile as a function of the rotation angle of a pair of parallel polarizers. As shown in Figure 6.6b, the profile well

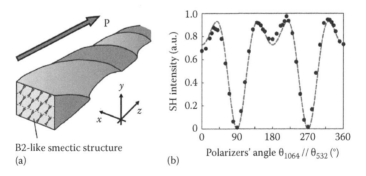

(a) B2-like smectic structure

(b)

FIGURE 6.6 (a) The structure of a helical nanofilament and (b) SHG intensity as a function of rotation angle of parallel polarizers using normally incident (y axis) light to a well-aligned cell in the B_4 phase without applying an electric field. Solid curve is a theoretical best fit. (From Araoka, F., Sugiyama, G., Ishikawa, K., and Takezoe, H.: Highly ordered helical nanofilament assembly aligned by a nematic director field. *Adv. Funct. Mater.* 2013. 23. 2701–2707. Copyright Wiley-VCH Verlag GmbH & Co. KGaA. Reproduced with permission.)

fitted to a theoretical expression based on the optical geometry (incidence and detection along the y axis) and the rotational average of the B_2 structure as the HNF structure (Figure 6.6a). The obtained ratio $d_{33}/d_{31} = 1.41$ is considerably larger than that in the B_2 phase ($d_{33}/d_{31} = 0.29$). This is because d_{31} in B_4 is much smaller than that in the B_2 phase because of the averaging due to the helical structure, while d_{33} is the same because we chose three axes parallel to both the helical axis and the bending direction (z axis in Figure 6.6a). Note that the normal beam incidence geometry used in the SHG experiment does not allow any chiral nonlinear optical effect. Full discussion of chiral nonlinear optical effect will be made in Section 6.3.

6.3 CHIRAL NONLINEAR OPTIC EFFECTS

Second-order NLO effect is possible only in systems with no inversion symmetry under the electric dipole (ED) approximation. That is why SHG has been successfully employed to examine polar structures in liquid crystals. More generally, however, NLO electric polarization can also be induced by the optical processes involving electric quadrapole (EQ) and magnetic dipole (MD) transitions. These processes in addition to the ED process are schematically shown in Figure 6.7 (Araoka and Takezoe 2004). EQ and MD transitions are allowed between states with the same parity, while ED transition occurs between states with opposite parity. Hence *eem*, *eeQ*, *mee*, and *Qee* processes are allowed even in centrosymmetric systems, where *e*, *m*, and *Q* stand for electric dipole, magnetic dipole, and electric quadrapole transitions, respectively. Although SHG due to such processes is negligibly weak compared with the *eee* process, these processes become dominant under special conditions. Besides the resonance condition, *eem* and *mee* processes involving the MD transition are particularly important in chiral systems, since MD is easily induced in chiral systems. If the SHG process, $\omega + \omega \to 2\omega$, involves *eem* or *mee* processes, SHG-CD is brought about. If the electro-optic effect, $\omega + 0 \to \omega$, involves *eem* or *mee* processes, the electrogyration (EG, chiral Pockels) effect arises, whereas the conventional Pockels effect arises for the *eee* process. Both chiral NLO effects have been clearly observed in the B_4 phase of bent-core molecules.

The EG effect is a first-order electro-optic effect and is a chiral analogue of the Pockels effect; the former is an electric-field-induced optical rotatory power, whereas the latter is an electric-field-induced birefringence. EG effect has been known to be present in certain crystal structures (Zheludev 1978), and one important example is

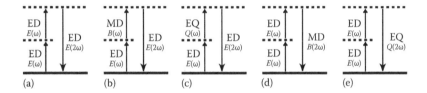

FIGURE 6.7 Five SHG processes: (a) *eee*; (b) *eem*; (c) *eeQ*; (d) *mee*; and (e) *Qee*, including ED (*e*: electric dipole), MD (*m*: magnetic dipole), and EQ (*Q*: electric quadrapole) transitions. Solid and dotted lines are real and virtual electronic levels.

the quartz crystal (Kobayashi et al. 1987). The electric-field-induced linear optical response is described through the constitutive equation of dielectric field strength **D** as

$$\mathbf{D} = \varepsilon\mathbf{E} + i\varepsilon_0\mathbf{G}\times\mathbf{E}. \tag{6.6}$$

Here **G** is the gyration vector parallel to the direction of light propagation. The overall retardation Δ induced by applying an electric field is given by

$$\Delta = \left[\delta^2 + (2\rho)^2\right]^{1/2} = \left[(\delta_0 + \delta_1)^2 + 4(\rho_0 + \rho_1)^2\right]^{1/2}, \tag{6.7}$$

where the subscripts 0 and 1 stand for non-perturbed and induced coefficients of birefringence δ and optical rotatory power ρ, respectively. Because of the linear response to an electric field, the field-induced refractive index change $\Delta n_{induced}$ is given by (Araoka et al. 2003)

$$\Delta n_{induced} = \left(\frac{\lambda}{2\pi}\right)\Delta_1 = \left(\frac{\lambda}{2\pi}\right)\left(\delta_1^2 + 4\rho_1^2\right)^{1/2} = n_0^3\gamma_{eff}E. \tag{6.8}$$

Measurements were conducted in two segregated chiral domains in the B_4 phase of achiral P8-O-PIMB (Araoka et al. 2003, He 2003). A linear relationship between $\Delta n_{induced}$ and E was obtained, leading to a large effective Pockels coefficient $\gamma_{eff} =$ 3.5 pm/V (He 2003). The dominant contribution to $\Delta n_{induced}$ is given by EG effect originating from the *mee* process ($\rho_1 = 1$ rad/cm), and the contribution of Pockels effect originating from the *eee* process ($\delta_1 = 0.07$ rad/cm) is relatively small (He 2003). Interestingly, the absolute values of ρ_1 are the same but the signs are opposite for the measurements in two chiral domains with the opposite handedness. It is also noted that ρ_1 in chiral P8-O-PIMB6* is five times larger than that in P8-O-PIMB.

Another important chiral NLO effect involving MD process is SHG-CD, which has been observed in some chiral systems (Kauranen et al. 1994, Verbiest et al. 1994). Bent-core molecules, which are nonchiral but form chiral domains, provide a very important example. SHG-CD, ΔI, is defined as

$$\Delta I = 2\frac{I_R - I_L}{I_R + I_L}, \tag{6.9}$$

where I_R and I_L are SHG intensities observed by the incidence of right and left circularly polarized light, respectively. A nonzero ΔI is obtained only when imaginary components of nonlinear susceptibility exist. Since the nonlinear susceptibility components for the ED process are real in the nonresonant region, nonzero ΔI in the nonresonant region certifies the existence of the MD process.

Experiments were conducted using a quarter-wave (QW) plate rotation method. The experimental setup is illustrated in Figure 6.8. With the rotation of the QW

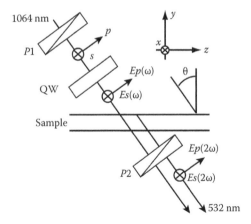

FIGURE 6.8 Schematic illustration for SHG-CD measurement. SHG signal by obliquely incident fundamental light through polarizers (*P1*) and a quarter wave plate (QW) is detected through a polarizer (*P2*). (Reprinted with permission from Araoka, F., Ha, N.Y., Kinoshita, Y., Park, B., Wu, J.W., and Takezoe, H., *Phys. Rev. Lett.*, 94, 137801-1. Copyright 2005 by the American Physical Society.)

plate between two polarizers (*P1* and *P2*), SHG intensity using four combinations of input and output polarizers was measured as a function of QW plate rotation angle. Figure 6.9 shows a set of data from different chiral domains (Araoka et al. 2005). The profiles of the signals are mirror images for (+) and (−) domains. Since the angles 45° and 225° correspond to I_R and the angles 135° and 315° to I_L, a large ΔI is detected: 0.35 and −0.39 for (+) and (−) domains, respectively, for s_{in}–p_{out} polarization combinations. Here, *s* and *p* polarizations are those perpendicular and parallel to the incidence plane, respectively, and in and out represent input and output, respectively.

The detailed analysis was made using two model structures; the bend (polar) direction is parallel or perpendicular to the helical axis, which has C_∞ and D_∞ point group symmetries, respectively. For D_∞ symmetry, at least the signal for 0° of s_{in}–s_{out} must be zero, clearly contradicting the experimental result. So the analysis was made using the model with C_∞ symmetry. The solid curves in Figure 6.9 are the best fit to all the experimental results. Here for the chiral components, the same absolute values with opposite signs were used for two different chiral domains. From the best fit, which shows satisfactory agreement, relative values of independent components of the nonlinear susceptibility were obtained. Surprisingly, many chiral and achiral components related to the MD process have almost half of the maximum component χ_{zzz}^{eee}.

Thus, the chiral nature of chiral domains spontaneously segregated from *achiral* bent-core molecules was clearly demonstrated using the chiral NLO effect. Bent-core mesogens provided us with an exceptional system showing large chiral NLO effects.

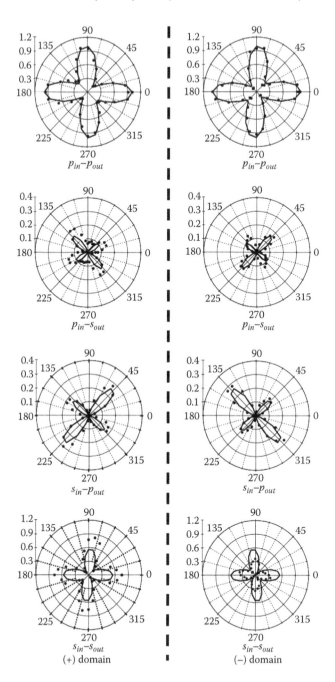

FIGURE 6.9 SHG-CD signal profiles (by rotating a QW plate) under four combinations of polarizer and analyzer for two chiral domains. Patterns in two chiral domains are symmetric with respect to the dotted lines. (Reprinted with permission from Araoka, F., Ha, N.Y., Kinoshita, Y., Park, B., Wu, J.W., and Takezoe, H., *Phys. Rev. Lett.*, 94, 137801-1. Copyright 2005 by the American Physical Society.)

REFERENCES

Araoka, F., N. Y. Ha, Y. Kinoshita, B. Park, J. W. Wu, and H. Takezoe. Twist-grain-boundary structure in the B4 phase of a bent-core molecular system identified by second harmonic generation circular dichroism measurement. *Phys. Rev. Lett.* 94 (2005): 137801-1–137801-4.

Araoka, F., H. Hoshi, and H. Takezoe. Splayed polarization in the ferroelectric phase of a bent-core liquid crystal as studied by optical second-harmonic generation. *Phys. Rev. E* 69 (2004): 051704-1–051704-4.

Araoka, F., B. Park, Y. Kinoshita, K. Ishikawa, H. Takezoe, J. Thisayukta, and J. Watanabe. Evaluation of the first-order hyperpolarizability tensor of bent molecules by means of hyper-Rayleigh scattering method. *Jpn. J. Appl. Phys.* 38 (1999): 3526–3529.

Araoka, F., G. Sugiyama, K. Ishikawa, and H. Takezoe. Highly ordered helical nanofilament assembly aligned by a nematic director field. *Adv. Funct. Mater.* 23 (2013): 2701–2707.

Araoka, F., Y. Takanishi, H. Takezoe, A. Kim, B. Park, and J. W. Wu. Electrogyration effect in a chiral bent-core molecular system. *J. Opt. Soc. Am. B* 20 (2003): 314–320.

Araoka, F. and H. Takezoe. Chiral nonlinear optic effect in a bent-core molecular system. *Ferroelectrics* 310 (2004): 3–9.

Araoka, F., J. Thisayukta, K. Ishikawa, J. Watanabe, and H. Takezoe. Polar structure in a ferroelectric bent-core mesogen as studied by second-harmonic generation. *Phys. Rev. E* 66 (2002): 021705-1–021705-5.

Choi, S.-W., Y. Kinoshita, B. Park, H. Takezoe, T. Niori, and J. Watanabe. Second-harmonic generation in achiral bent-shaped liquid crystals. *Jpn. J. Appl. Phys.* 37 (1998): 3408–3411.

Clays, K. and A. Persoons. Hyper-Rayleigh scattering in solution. *Rev. Sci. Instrum.* 63 (1992): 3285–3289.

Dworczak, R. and D. Kieslinger. Electric field induced second harmonic generation (EFISSH) experiments in the swivel cell: New aspects of an established method. *Phys. Chem. Chem. Phys.* 2 (2000): 5057–5064.

Gallastegui, J. A., C. L. Folcia, J. Etxebarria, J. Ortega, I. de Francisco, and M. B. Ros. Determination of the second order susceptibility tensor in banana-shaped liquid crystals. *Liq. Cryst.* 29 (2002): 1329–1333.

Gallastegui, J. A., C. L. Folcia, and J. Etxebarria, J. Ortega, N. Gimeno, and M. B. Ros. Fabrication and nonlinear optical properties of monodomain polymers derived from bent-core mesogens. *J. Appl. Phys.* 98 (2005): 083501-1–083501-5.

He, Y. Chiral Pockels effect in bent-shaped liquid crystal systems. Master thesis, Tokyo Institute of Technology, 2003 (in Japanese).

Hough, L. E., H. T. Jung, D. Kruerke, M. S. Heberling, M. Nakata, C. D. Jones, D. Chen et al. Helical nanofilament phases. *Science* 325 (2009): 456–460.

Kauranen, M., T. Verbiest, J. J. Maki, and A. Persoons. 2nd-harmonic generation from chiral surfaces. *J. Chem. Phys.* 101 (1994): 8193–8199.

Kentischer, F., R. Macdonald, P. Warnick, and G. Heppke. Second harmonic generation (SHG) investigations of different phases of banana shaped molecules. *Liq. Cryst.* 25 (1998): 341–347.

Kobayashi, J., T. Asahi, and S. Takahashi. Simultaneous measurements of electrogyration and electrooptic effects of alpha-quartz. *Ferroelectrics* 75 (1987): 139–152.

Macdonald, R., F. Kentischer, P. Warnick, and G. Heppke. Antiferroelectricity and chiral order in new liquid crystals of nonchiral molecules studied by optical second harmonic generation. *Phys. Rev. Lett.* 81 (1998): 4408–4411.

Martinez-Perdiguero, J., I. Alonso, C. L. Folcia, J. Etxebarria, and J. Ortega. Local structure of the B_4 phase studied by second harmonic generation and X-ray diffraction measurements. *J. Mater. Chem.* 19 (2009): 5161–5166.

Nakata, M., D. R. Link, F. Araoka, J. Thisayukta, Y. Takanishi, K. Ishikawa, J. Watanabe, and H. Takezoe. A racemic layer structure in a chiral bent-core ferroelectric liquid crystal. *Liq. Cryst.* 28 (2001): 1301–1308.

Ortega, J., J. A. Gallastegui, C. L. Folcia, J. Etxebarria, N. Gimeno, and M. B. Ros. Second harmonic generation measurements in aligned samples of liquid crystals composed of bent-core molecules. *Liq. Cryst.* 31 (2004): 579–584.

Ortega, J., N. Pereda, C. L. Folcia, J. Etxebarria, and M. B. Ros. Second-harmonic generation studies in the B2 and B4 phases of a banana-shaped liquid crystal. *Phys. Rev. E* 63 (2000): 011702-1–011702-7.

Pelzl, G., S. Diele, S. Grande, A. Jákli, Ch. Lischka, H. Kresse, H. Schmalfuss, I. Wirth, and W. Weissflog. Structural and electro-optical investigations of the smectic phase of chlorine-substituted banana-shaped compounds. *Liq. Cryst.* 26 (1999): 401–413.

Pintre, I. C., N. Gimeno, J. L. Serrano, M. B. Ros, I. Alonso, C. L. Folcia, J. Ortega, and J. Etxebarria. Liquid crystalline and nonlinear optical properties of bent-shaped compounds derived from 3,4′-biphenylene. *J. Mater. Chem.* 17 (2007): 2219–2227.

Pintre, I. C., J. L. Serrano, M. B. Ros, J. Martinez-Perdiguero, I. Alonso, J. Ortega, C. L. Folcia, J. Etxebarria, R. Alicante, and B. Villacampa. Bent-core liquid crystals in a route to efficient organic nonlinear optical materials. *J. Mater. Chem.* 20 (2010): 2965–2971.

Thisayukta, J., H. Takezoe, and J. Watanabe. Study on helical structure of the B₄ phase formed from achiral banana-shaped molecule. *Jpn. J. Appl. Phys.* 40 (2001): 3277–3287.

Verbiest, T., M. Kauranen, A. Persoons, M. Ikonen, J. Kurkelan, and H. Lemmetyinen. Nonlinear-optical activity and biomolecular chirality. *J. Am. Chem. Soc.* 116 (1994): 9203–9205.

Zheludev, I. S. Optical activity of crystals induced by an electric-field (electrogyration). *Ferroelectrics* 20 (1978): 51–59.

7 Banana Phases in Restricted Geometries

One of the key features of the confinement is that the confining surfaces break symmetry of the molecular order in their vicinity (Churaev et al. 1987, Oswald and Pieranski 2005, Doi 2013). The degree of translational and orientational order is strongly influenced by the presence of boundaries. Molecules in the surface layer of a liquid crystal may adopt a configuration that is different from the bulk (Israelachvili 1992). Because of their high surface-to-volume ratio, liquid crystals under confinement exhibit macroscopic properties different from those of macroscopic bulk. One example of the confinement effect in liquid crystals (LCs) is the surface-stabilized ferroelectric SmC* phase. Helical order in bulk of the chiral SmC* phase averages out the layer polarization established in each smectic layer (Figure 7.1a) (Lagerwall 1999). Confinement of the SmC* phase in a thin cell results in unwinding of the helix and stabilization of two states with a residual polarization and opposite tilts (Figure 7.1b) (Clark and Lagerwall 1980).

The research on liquid crystals in restricted geometries extends to various types of confinement, such as planar films, cylindrical filaments, and spherical droplets. Additionally, the effects of the substrate structure, curvature, and topography recently became an active topic of the current research. Liquid crystals in confined geometries are also interesting from the technological point of view. They play an important role in emerging electro-optical and photonic technologies. Confinement of liquid crystals not only enhances physical properties of LC devices but also opens new horizons for the development of novel multifunctional and photonic materials. Bent-shaped molecules are not exceptions in showing various confinement effects.

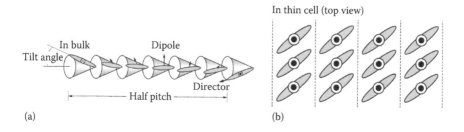

(a)   (b)

FIGURE 7.1 (a) The helielectric state in the bulk of the SmC* phase. The polarization precesses around the cone along the smectic layer normal. The net polarization is averaged out in this state and (b) a surface-stabilized state of the chiral SmC* phase in a thin cell. The layer polarizations (marked by the arrows in (a)) are aligned perpendicular to the substrate. The phase exhibits bistable switching.

7.1 FREELY SUSPENDED FILMS

Ability of smectic phases to form freely suspended films opened up a whole new area of interfacial research (Oswald and Pieranski 2006). Freely suspended films of thermotropic smectic liquid crystals are quite similar to soap films. Their first systematic studies date back to the pioneering works by Young et al. in the end of 1970 (Young et al. 1978). LC films must be necessarily stretched on a solid frame, which is usually confined in a heating chamber. Without the support, the films will inevitably collapse under the action of the surface tension. The elastic modulus is not high enough to oppose the surface tension and a supported film will collapse into a droplet. On the other hand, once formed, freely suspended films are very robust. They can be even pierced by a thin glass needle without breaking. A scheme of a freely suspended film is shown in Figure 7.2.

The film consists of a planar region with a nearly uniform thickness (number of layers), which is attached to the support through a meniscus (Picano et al. 2000, Zywocinski et al. 2000). Regions of different thicknesses, distinguished by the layer steps, are often found in smectic films. The thickness change across such steps is quantified by the thickness of smectic layers. Experimentally, the film thickness is determined by the analysis of a reflectivity spectrum. In case of an orthogonal incidence, the reflectivity $R(\lambda)$, as a function of wavelength λ, is determined by the refractive index n of the liquid crystal and the film thickness h:

$$R(\lambda) = \frac{p \sin^2(kh)}{1 + p \sin^2(kh)} \tag{7.1}$$

where $k = 2\pi n/\lambda$ and

$$p = \frac{(n^2 - 1)^2}{4n^2} \tag{7.2}$$

This dispersion of the reflectivity is responsible for a colorful appearance of films with the thicknesses in the range between 100 nm and about 1000 nm. When the

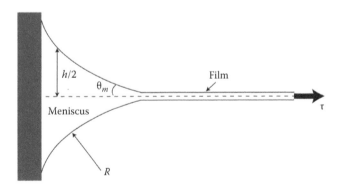

FIGURE 7.2 Scheme of a freely suspended film and its meniscus. The apparent contact angle θ_m is shown only schematically. On a microscopic scale the meniscus joins the film continuously.

films consist of just a few layers, the film thickness is determined by measuring reflectivity using a laser source.

What is responsible for such a remarkable stability of freely suspended films? To address this question, we shall consider the free energy of the film-meniscus system described in Picano et al. (2001). The free energy density depends on the film thickness profile $h(x)$ (Figure 7.2) and has four important contributions: (1) surface energy, which is proportional to the curvature of the film profile and given by the formula:

$$f_{surf} = 2\sigma\left(1+\frac{1}{2}\frac{dh}{dx}\right)^{1/2},$$ (7.3)

where σ is the surface tension; (2) the energy of interaction between the top and bottom surfaces of the film $f_d(h)$; (3) the energy of dislocations f_{dis}; and (4) the work of the hydrostatic pressure. The mechanical equilibrium condition can be obtained by minimizing the free energy with respect to h. This yields a pressure balance equation, which can be written as

$$\Delta P = \Pi_d - \frac{\sigma}{R}$$ (7.4)

where
   ΔP is the difference in hydrostatic pressure at the air/LC interface
   R is the local radius of curvature (Figure 7.2)
   $\Pi_d = -df_d/dh$ is the disjoining pressure

Disjoining pressure results from the interactions between the surfaces of the film (Derjaguin et al. 1978, Churaev et al. 1987). In smectic films, this interaction originates from the variation of the orientational order parameter near the film boundaries. In a film, far from the meniscus, the curvature term vanishes and it is the disjoining pressure that equilibrates the hydrostatic pressure difference (Picano et al. 2001). On the contrary, inside the meniscus, the disjoining pressure can be neglected and the film curvature equilibrates the pressure difference through the Laplace pressure. The meniscus is nearly tangential to the film when the thickness is larger than 100 smectic layers. In thin films, however, it makes a finite angle θ_m with the film plane. This so-called apparent contact angle θ_m (Figure 7.2) is directly related to the interaction energy f between the film surfaces and is given by the formula $\cos(\theta_m)=f_d(d)/2\sigma-1$, where d is the film thickness. A detailed theoretical study shows that the interaction energy between the surfaces of a smectic film originates mainly from the smectic order parameter difference between the surface layers and the bulk (Oswald and Pieranski 2003). The elastic layer compression is equilibrated by the disjoining pressure. Thus, the layers avoid collapse.

Smectic A phases appear black between crossed polarizers. The optical axis is perpendicular to the film. Uniaxial order appears evident if one observes such films under oblique incidence. In this case, films appear birefringent. In tilted phases, the optical axis tilts away from the layer normal (Figure 7.3a). Additionally, the phase becomes slightly biaxial. Typical *Schlieren* textures are observed in polarizing

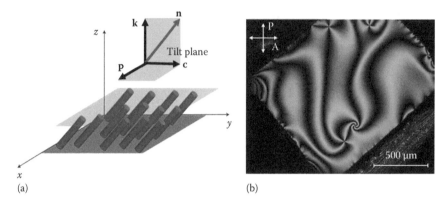

(a)   (b)

FIGURE 7.3 (a) A scheme of a smectic layer in a freely suspended film of the SmC* phase. The scheme shows the definition of the directors: **k** is the layer normal, **n** is the orientational director, **c** is the projection of **n** on the layer plane, and **p** is the polarization. The same configuration is valid for the SmCP phase in bent-core LCs and (b) a polarizing microscope image of a freely suspended SmCP film in a transmission mode. A pair of the topological defects with the strengths +1 and −1 can be recognized near the left bottom corner.

microscopy (Figure 7.3b). These textures manifest a long-range orientational order of the molecular tilt, which is given by the c-director $\mathbf{c}(x, y)$ (Pindak et al. 1980). Thus, freely suspended SmC films are examples of quasi-2D nematics. Yet the symmetry of the c-director is different from that of the nematic director **n**. There is no $\mathbf{c} \rightarrow -\mathbf{c}$ invariance for the c-director. As a result, mainly the defects with an integer topological strength can be found in this system (Figure 7.3b).

We now describe the behavior of freely suspended films of banana mesogens. The behavior of the c-director in an electric field also provides a new way to gain insight into the polar properties of the banana mesophases and also into the dynamics of the director relaxation. Hereafter, director stands for the c-director in this section, unless otherwise stated. Let us describe one example of freely suspended film of a bent-shaped molecule, which shows $SmCP_A$, $SmCP_F$, and SmA in the film form (Eremin et al. 2012a). Upon application of an electric field, director inversion walls appear. These are Neel-type walls, where the c-director continuously changes its orientation (Figure 7.4). Two different types of a response of the director field to an external electric field can be distinguished. In the pure dielectric response, only induced polarization is involved. Across the inversion wall, the director rotates by an angle of π (Figure 7.4a, bottom). When the polarity of the electric field is changed, the walls remain unchanged. Materials with a residual polarization exhibit the polar response. In a DC electric field, so-called 2π walls become stable, where the director rotates by an angle of 2π (Figure 7.4a, top). The π walls start forming pairs. An example showing both π and 2π walls in paraelectric and ferroelectric domains, respectively, is shown in Figure 7.4b.

In a general case, the dynamics of the director can be expressed (in the one-constant approximation) by the torque-balance equation:

$$\gamma_\varphi \dot{\varphi} = -PE \sin(\varphi) - \frac{1}{2} \varepsilon_0 \varepsilon_a E^2 \sin(2\varphi) + K \nabla^2 \varphi \qquad (7.5)$$

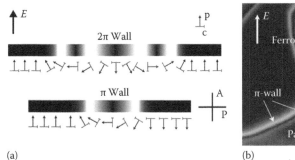
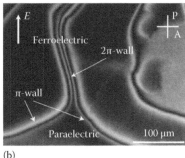

(a)   (b)

FIGURE 7.4 (a) A scheme of π- and 2π-director-inversion walls with the corresponding intensity profiles obtained between crossed polarizers and (b) photomicrograph showing both π and 2π walls. A 2π wall, that is stabilized by a dc electric field ($E = 15$ V/mm) in the ferroelectric SmCP$_F$ phase, "unzips" in the paraelectric SmC phase region, splitting into a pair of π walls on further cooling. (Reprinted with permission from Eremin, A., Floegel, M., Kornek, U., Stern, S., Stannarius, R., Nádasi, H., Weissflog, W. et al., *Phys. Rev. E*, 86, 051701-1. Copyright 2012 by the American Physical Society.)

where

γ_φ is the rotational viscosity

P is the polarization

$\varepsilon_a = \varepsilon_\| - \varepsilon_\perp$ is the in-plane dielectric anisotropy

K is the elastic constant (in the order of pN)

The equation expresses the balance between the viscous torque (left side) and the sum of the electric and elastic torques. The term $PE\sin(\varphi)$ stems from a ferroelectric interaction.

In ferroelectric phases, the first term dominates in a low-field. The width of the walls is proportional to the polar correlation length $\xi_p = (K/EP)^{1/2}$. In phases without the residual polarization, only dielectric reorientation occurs and π walls are stable in DC and AC fields. The characteristic length scale is $\xi_p = (K/\varepsilon_0\varepsilon_a)^{1/2}E$. Thus, the measurements of the width of the inversion walls provide means to determine the character of the polar order and estimate the polarization.

The synclinic and anticlinic characters of the tilt can be determined from the optical transmittance (or reflectance) under inclined incidence (Chattham et al. 2010, Eremin et al. 2012a). In the case of a synclinic structure, the transmittance is different when the film is tilted around the diagonal between the polarizers for the opposite angles $+\alpha$ and $-\alpha$. This occurs since the optical axis tilts toward the propagation direction of the light in one case, and opposite to it in the other. An anticlinic structure exhibits the same change of the transmittance upon the tilt in the opposite directions. This technique is especially useful in case of LCs made of bent-shaped mesogens, in which different types of polar smectics occur. An example of this analysis led to an observation of a polar leaning smectic phase in Chattham et al. (2015).

The antiferroelectric SmC$_S$P$_A$ and SmC$_A$P$_A$ phases show only a dielectric response in thin films since applied electric fields under normal conditions are too weak to

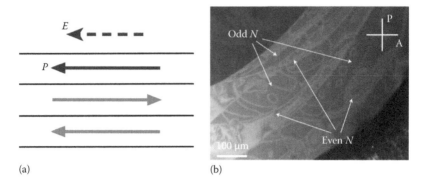

(a)   (b)

FIGURE 7.5   (a) Origin of the odd–even effect: in films with an odd number of layers, the polarization is not averaged out. The polarization of a non-compensated single layer couples to an external electric field and (b) director response to an applied electric field in a thin film with several layer steps. The image shows a clear odd–even effect depending on layer number N. The photograph was taken using depolarized reflection light microscopy (DRLM) technique. (Reprinted with permission from Eremin, A., Floegel, M., Kornek, U., Stern, S., Stannarius, R., Nádasi, H., Weissflog, W. et al., *Phys. Rev. E*, 86, 051701-1. Copyright 2012 by the American Physical Society.)

switch the film into the ferroelectric state. In very thin films, however, the antiferroelectric order gives rise to the so-called odd–even effect, where the character of the response depends on the number of layers in the film (Figure 7.5). The polar response is observed in thin films with an odd number of layers (Figure 7.5b). In this case, an external electric field couples to the surface polarization of an odd layer. In thick films, however, the polar response diminishes since the electric field coupling occurs only at a single uncompensated layer within a large number of layers. Surface polarization can also occur in thin films with an even number of layers. This occurs as a result of a symmetry breaking at the air/LC interface. Contrary to the bulk polarization, this surface polarization lies in the tilt plane. High residual polarization in freely suspended films of ferroelectric phases affects the director field around the topological defects. The polarization splay costs additional electrostatic energy and is avoided in high P_s materials, enhancing the effective bend elastic constant of the c-director. Avoidance of the polarization splay is responsible for the fluctuating thready texture of the $SmCP_F$ phase and narrowing of the areas of the director bend in the vicinity of topological defects.

Some higher ordered bent-core/bent-shaped liquid crystal (BLC) phases can form freely suspended films too. For example, B_5 phase exhibits a broken *Schlieren* texture as shown in Figure 7.6.

The SmCP phases typically exhibit *Schlieren* textures in freely suspended films, but the ferroelectric modifications also exhibit modulated structures of various morphologies such as stripes, pins, fingers, and labyrinths (Figures 7.7 and 7.8). These structures, visible by polarizing microscopy, can be formed by the c-director field $c(x, y)$. A periodic bend modulation of the c-director was reported in a $SmCP_F$ phase (Eremin et al. 2008b). The main motif of this modulation is an array of strings of the topological strength 0 separated by the stripes with a spontaneous bend deformation

FIGURE 7.6   Photomicrograph showing the transition from the SmCP into the B₅ phase in a freely suspended film: The *Schlieren* texture of the SmCP phase in the left bottom corner transforms into the broken *Schlieren* texture growing from the top right corner of the image. The image is taken by polarizing microscopy in transmission.

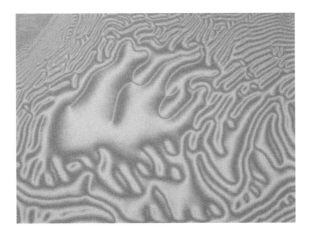

FIGURE 7.7   Ripple structure in the ferroelectric SmCP$_F$ phase. (Reprinted with permission from Eremin, A., Nemeş, A., Stannarius, R., and Weissflog, W., *Phys. Rev. E*, 78, 061705-1. Copyright 2008 by the American Physical Society.)

of the c-director. Since the bend deformation of the c-director is associated with chirality, the stripes of the alternating bend exhibit alternating handedness too. Hence, this kind of pattern was designated as ambidextrous.

Theoretically, ambidextrous bend deformation can be explained by the sign inversion of the bend elastic constant, which has also been proposed to take place in nematic phases of bent-core mesogens (Dozov 2001, Borshch et al. 2013, Chen et al. 2014).

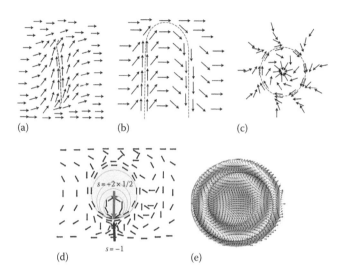

FIGURE 7.8 (**See color insert.**) Different morphologies observed in free-suspending films: (a) single string; (b) pin; (c) spiral with an $s = +1$ vortex in the center; (d) boojum; and (e) target pattern. The arrows show the **c**-director. (Reprinted with permission from Eremin, A., Nemeş, A., Stannarius, R., and Weissflog, W., *Phys. Rev. E*, 78, 061705-1. Copyright 2008 by the American Physical Society.)

In flat films, the free energy density can be expressed as

$$f = \frac{K_s}{2}\left(\nabla\cdot\mathbf{c}\right)^2 + \frac{K_b}{2}\left(\nabla\times\mathbf{c}\right)^2 + \alpha\left(\nabla\times\mathbf{c}\right) + \frac{K_{b4}}{4}\left(\nabla\times\mathbf{c}\right)^4 \qquad (7.6)$$

where
   K_s and K_b are splay and bend elastic constants, respectively
   α is the chiral order parameter and the forth-order term with the constant
      $K_{b4} > 0$ is required to maintain the stability of the modulated state

Instability occurs when K_b becomes negative. In 1D case, it leads to a formation of a striped pattern with a characteristic length-scale $R_0 = (K_{b4}/|K_b|)^{1/2}$ (Eremin et al. 2008a,b). Although this oversimplified model describes the pattern formation well, physical mechanisms of the instability remain still unclear.

   Another type of pattern observed in freely suspended films of bent-core mesogens was found in the polarization-modulated PM-SmCP phase (Eremin et al. 2013). These patterns are formed by layer dislocations, which organize into an intricate labyrinthine pattern (Figures 7.9 and 7.10). Labyrinth morphology can be found in a vast majority of physical, chemical, and biological systems such as superconductors, chemical reaction–diffusion systems, liquid crystals, and ferrofluids (Faber 1958, Möhwald 1990, Jackson et al. 1994, Goldstein et al. 1996, Rosensweig 1997). Striped patterns result from two or more competitive interactions favoring incompatible ground states (Seul and Andelman 1995). The evolution of these striped structures in 2D is accompanied by finger growth bifurcations and meandering. It is suggested

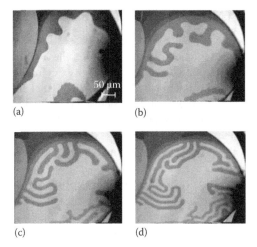

FIGURE 7.9 Onset of the instability and development of the labyrinthine pattern in a free-standing film during the SmCP-PM-SmCP transition in the absence of an external electric field at a constant heating rate of 1 K/min. The images (a-d) are taken with a time interval of 40 s. (Reprinted with permission from Eremin, A., Kornek, U., Stannarius, R., Weissflog, W., Nádasi, H., Araoka, F., and Takezoe, H., *Phys. Rev. E*, 88, 062512-1. Copyright 2013 by the American Physical Society.)

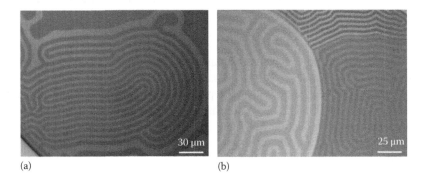

FIGURE 7.10 **(See color insert.)** Morphology of labyrinthine patterns: (a) and (b) Examples of labyrinthine patterns of edge dislocation in freely suspended films. (Reprinted with permission from Eremin, A., Kornek, U., Stannarius, R., Weissflog, W., Nadasi, H., Araoka, F., and Takezoe, H., *Phys. Rev. E*, 88, 062512-1. Copyright 2013 by the American Physical Society.)

that the main mechanism responsible for the formation of the labyrinthine structures in the PM-SmCP phase is the spontaneous polarization splay of the c-director. A linear term $\nabla \cdot \mathbf{P}$ in the free energy density contributes to the negative line tension of the dislocations and favors their elongation. The costs of the director distortions in the bulk compensate this tendency to result in an equilibrium pattern. Dynamic labyrinthine patterns of the director distortions were found in nonmodulated phases too. The origin of these patterns remains unclear.

7.2 FREELY SUSPENDED FIBERS AND FILAMENTS

Another type of structure encountered in several phases formed by bent-core meso-gens is a freely suspended filament. In this structure, the liquid crystal is suspended between two supports and the LC/air interface adopts a cylindrical shape. Along the cylindrical axis, the liquid crystal may still exhibit flow. Thicker filaments often consist of thinner fibers with the radius of the order of a micrometer.

Most fluids do not form free filaments due to the Rayleigh–Plateau instability. For a fluid cylinder of a radius R_0 and the length L, the free energy is given by the total surface multiplied by the surface tension σ. Small harmonic distortion of the cylinder shape given by a dependence $R(x) = R_0^* + \delta r \cos(qx)$ with a wavenumber q and the mean radius R_0^* (Figure 7.11) results in the change of the free energy at a constant volume given by:

$$\Delta F \approx \frac{\sigma \pi \delta r^2}{2R}\left(R^2 q^2 - 1\right). \tag{7.7}$$

The instability limit is given by the critical wavenumber $q_c = 1/R_0$. This corresponds to the maximal length of $L_{max} = 2\pi R_0$. Thus, a liquid cylinder with a slenderness ratio $L/R_0 > 2\pi$ is unstable. There are a few examples of non-Newtonian fluids like silk and polymer melts, where filaments are formed in a solidification process (Kerkam et al. 1991, Vollrath and Knight 2001). Fibers can also be prepared using electrospinning technique (Lagerwall et al. 2008, Lagerwall and Giesselmann 2010). Fiber-forming materials find a broad range of applications. Polymeric fibers are used in textile and optical industry.

A fiber structure can be envisioned in a smectic liquid crystal. A cylindrical structure of molecular layers wrapped around an axial core may stabilize a liquid crystal cylinder against the Rayleigh–Plateau instability (Jákli et al. 2003). However, stable cylindrical arrangement of layers is not typical for the most common smectic liquid crystal phases (SmA, SmC,…). These materials usually form flat freely suspended films, as mentioned in Section 7.1. There are a few exceptions when filaments are formed in isotropic solution like 8-cyanobiphenyl (8CB) filaments in alcohol (Naito et al. 1993, Todorokihara et al. 2004).

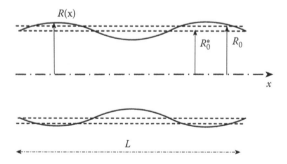

FIGURE 7.11 Schematics of a fluid cylinder with a fluctuating surface. Initial radius of the cylinder is R_0, the mean radius of the fluctuating cylinder R_0^* is slightly smaller than R_0.

First examples of freely suspended fibers were demonstrated in discotic liquid crystals triphenylene hexa-*n*-dodecanoate by van Winkle and Clark (1982). These fibers (strands) had a diameter smaller than 10 μm and the length of a few hundreds of microns, which significantly exceeds the critical slenderness ratio of Newtonian fluids.

The potential to stabilize a fluid cylinder by a 2D columnar structure has long been recognized. Kamenskiĭ and Kats (1983) investigated stability of discotic fibers by analyzing hydrodynamics equations. In their model, they used isotropic viscosity η and considered the fluid motion along and perpendicular to the fiber axis. The rigidity of the 2D columnar structure was characterized by a compression modulus *B*. The hydrodynamic equations include incompressibility condition for the fluid:

$$\frac{\partial v_z}{\partial z} + \frac{1}{r}\frac{\partial}{\partial r}(rv_r) = 0 \qquad (7.8)$$

Compression of the columnar lattice

$$\frac{\partial u_r}{\partial z} - v_r = \gamma B \frac{\partial}{\partial r}\left(\frac{1}{r}\frac{\partial}{\partial r}(ru_r)\right) \qquad (7.9)$$

Navier–Stokes equations for the longitudinal and transversal transport:

$$\rho\frac{\partial v_r}{\partial t} = -\frac{\partial P}{\partial r} + B\frac{\partial}{\partial r}\left(\frac{1}{r}\frac{\partial}{\partial r}(ru_r)\right) + \eta\left(\frac{\partial^2 v_r}{\partial r^2} + \frac{\partial}{\partial r}\left(\frac{1}{r}\frac{\partial}{\partial r}(rv_r)\right)\right) \qquad (7.10)$$

$$\rho\frac{\partial v_z}{\partial t} = -\frac{\partial P}{\partial z} + \eta\left(\frac{\partial^2 v_z}{\partial z^2} + \frac{\partial}{\partial r}\left(\frac{1}{r}\frac{\partial}{\partial r}(rv_z)\right)\right) \qquad (7.11)$$

For low viscosities, Kamenskiĭ and Kats (1983) obtained an estimation of the maximal length of the discotic cylinder

$$L \approx \frac{9.02R}{\sqrt{1-(2BR/\sigma)}} \qquad (7.12)$$

where
  R is the filament radius
  σ is the surface tension

The validity of this estimation was confirmed in experiments on hexa-*n*-dodecanoate, where the fiber diameter is ≈1 μm and the maximal length is 20 μm (with *B* ≈ 105 Pa) (Kamenskii et al. 1983 and van Winkle et al. 1982). According to this model, the primary factor responsible for the fiber stability is the compressibility of the columnar lattice in structured liquids. Although it seems that most smectic liquid crystals in the SmA and SmC phases should form freely suspended filaments or fibers, in fact they do not. Instead, they usually form only freely suspended films. The first example of freely suspended smectic filaments was found in bent-core liquid crystals in the B$_7$ phase

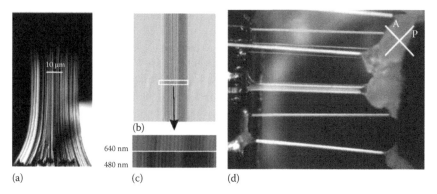

FIGURE 7.12 (**See color insert.**) Images of freely suspended filaments in the PM-SmCP phase. (a) and (d) Images of filaments under a polarizing microscope. The image in (a) clearly demonstrates the striped optical pattern along the filament axis. (b) Single filament observed in transmission without polarizers. (c) Details of (b) observed under monochromatic illumination. (Reprinted with permission from Eremin, A., Nemeş, A., Stannarius, R., Schulz, M., Nádasi, H., and Weissflog, W., *Phys. Rev. E*, 71, 031705-1. Copyright 2005 by the American Physical Society.)

(Pelzl et al. 1999, Jákli et al. 2000). Both columnar (B_1) and lamellar (PM-SmCP) types of the B_7 (polarization modulated) phases form stable filaments with an exceptionally large slenderness ratio exceeding 10,000 (Jákli et al. 2003, Stannarius et al. 2005). Liquid crystal filaments are birefringent and often exhibit a series of colored stripes appearing in polarizing microscopy in transmission (Figure 7.12a). The period of these stripes depends on the wavelength of the illumination (Figure 7.12a). This suggests that the optical stripes are a result of interference.

The structure of the filaments was studied by various techniques including x-ray diffraction, electron microscopy, and atomic force microscopy (Nemeş et al. 2006). Thick filaments usually consist of several fibrils or fibers with a diameter of approximately 3–4 μm (Figure 7.13a through d) (Nemeş et al. 2006, Bailey et al. 2010). The dispersion of the filament radii in the columnar B_7 phase is quite small. But in the lamellar-type PM-SmCP phase, fiber radii can vary from 3 to 10 μm. In the case of the PM-SmCP phase, the structure of single fibers is comprised by concentric rolls of smectic layers (Figure 7.13d) (Eremin et al. 2005). However, the direction of the layer undulation is not uniform. As was demonstrated by Coleman et al. using x-ray diffraction, the orientation of the layer undulation vector **m** with respect to the geometrical axis depends on the radius (Coleman et al. 2003). For thin fibers, it is nearly perpendicular to the fiber axis.

The polar structure of the fibers was investigated using second-harmonic generation (SHG) microscopy and interferometry techniques (Eremin et al. 2012b). A clear SHG signal was found from single fibers forming a thick filament in the PM-SmCP phase. This is an indication of a residual polarization in single fibers without an external electric field applied. Analysis of the optical polarization of the SHG signal showed that the electric polarization is along the fiber axis. In a thick filament, the polarization of the fibers has a random up/down distribution with a small excess in the direction of pulling.

Liquid crystalline fibers show very interesting mechanical properties, such as dependence of the filament tension on the radius and temperature. Most studied materials are with PM-SmCP and columnar-type B_7 phases. The filament tension

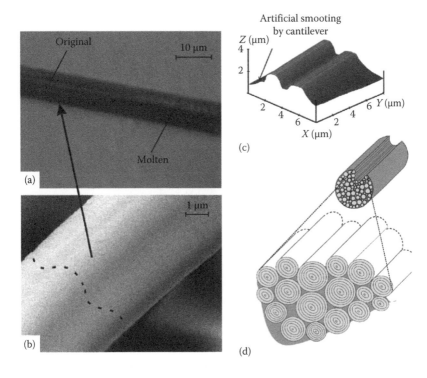

FIGURE 7.13 PM-SmCP filament quenched by injecting liquid nitrogen into the preparation box. (a) Optical microscope image after quenching; (b) backscattered SEM image of the region where the striped texture is still preserved. The dashed line emphasizes the suggested surface profile with an undulation; (c) height profile obtained by AFM in a tapping mode of a 8 μm × 8 μm area of a typical crystallized B_7 filament with 50 μm diameter. The filament has been placed on a glass substrate with its axis along y; and (d) a proposed model for the internal filament structure derived from the surface structure analysis. (From Nemeş, A., Eremin, A., Stannarius, R., Schulz, M., Nádasi, H., and Weissflog, W., Structure characterization of freestanding filaments drawn in the liquid crystal state, *Phys. Chem. Chem. Phys.*, 8, 469–476, 2006. Reproduced by permission of The Royal Society of Chemistry.)

was studied using direct measurements by a cantilever (Stannarius et al. 2005), vibration measurements using an electric excitation (Stannarius et al. 2005, Nemeş et al. 2006), and an acoustic excitation (Petzold et al. 2009) (see Figure 7.14).

Oscillation of a freely suspended filament can be described by a dynamic equation constructed from an inertial, elastic, surface tension and friction forces. A rough estimation of these terms given in Stannarius et al. (2005) shows that in the range of parameters typical for common LCs, like 8CB, the orientational elasticity brought about by distortions of the director field as well as the air friction can be neglected. The main contribution to the dynamic equation is expected to arise from the surface tension, which is proportional to the circumference of the filament cross section. In experiment, the vibration of the filaments in the PM-SmCP phase shows nearly linear dispersion where the ratio of the frequency ω to the wave number k (phase velocity) is constant. The phase velocity, however, is much larger than expected in

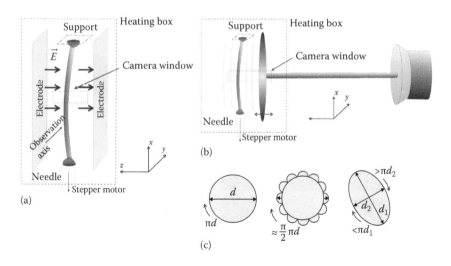

FIGURE 7.14 (a) Electric and (b) acoustic techniques for excitation of filament vibration. (c) Models of the filament cross section with corresponding circumferences: circular, corrugated circular, and elliptic. (From Petzold, J., Nemeş, A., Eremin, A., Bailey, C., Diorio, N., Jákli, A., and Stannarius, R., Acoustically driven oscillations of freely suspended liquid crystal filaments, *Soft Matter*, 5, 3120–3126, 2009. Reproduced by permission of The Royal Society of Chemistry.)

surface-tension-driven filaments with a circular cross section. Also the radius independence of the vibration frequency cannot be explained by a simple model where the dynamic is determined by the surface tension alone. These observations indicate (a) the filaments do not have a circular circumference and (b) there is an additional bulk contribution to the filament tension. The corrugated shape of the filament bundles consisting of several fibers results in a larger circumference, and hence a larger tension force than the force estimated from the apparent radius (Figure 7.13d). Detailed analysis of the filament shape revealed that in most cases thick filaments are not even nearly circular in cross section.

The most accurate measurement of the filament tension was made by using glass beads as gauge (Figure 7.15a) (Morys et al. 2012). Small micrometer-sized glass beads were attached to a horizontally suspended filament. The tension force was calculated from the measured deflection for different sizes and temperatures. It is assumed that the filament cross section has an elliptical shape with the semi axes a and b. The two contributions to the filament tension, that is, surface tension and a bulk tension, were accounted for in the expression:

$$\Sigma = \sigma P(a,b) + \Delta \pi a b \tag{7.13}$$

where
　$\sigma = 20\,\text{mN/m}$ is the LC/air interface tension assumed to be temperature independent
　P is the cross section circumference
　Δ is the bulk contributi.on.

The filament tension at an elevated temperature is shown in Figure 7.15b.

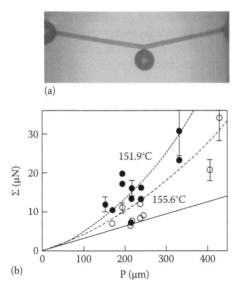

FIGURE 7.15 (a) A filament in the PM-SmCP phase with an attached bead used for the tension measurement and (b) Dependence of the filament tension S on the perimeter for all filaments. The solid line indicates the capillary force calculated with σ = 20 mN/m; the dotted and dashed curves are fits to Equation (7.13). Dependence of the net filament tension on the circumference of the filament cross section. (Reprinted with permission from Morys, M., Trittel, T., Eremin, A., Mrurphy, P., and Stannarius, R., *Phys. Rev. E*, 86, 040501(R)-1. Copyright 2012 by the American Physical Society.)

The tension in thin filaments is close to the values predicted from the surface tension. Thicker filaments, however, show systematically larger values that can be attributed to the bulk contribution (Figure 7.16b). Interestingly, the bulk contribution exhibits a strong temperature dependence: the bulk modulus η nearly triples over the temperature range of the PM-SmCP phase (Figure 7.16b).

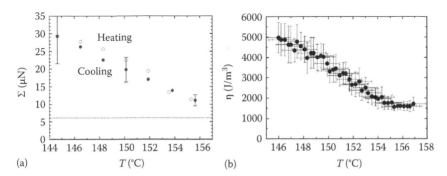

FIGURE 7.16 (a) Temperature dependent tensions Σ(T) of a thick filament during cooling (solid circles) and heating (open circles): length = 10.09 mm, a = 69 μm, b = 54 μm. (b) Temperature dependence of the coefficient η (Equation 7.2) of the volume contribution to the filament tension. (Reprinted with permission from Morys, M., Trittel, T., Eremin, A., Mrurphy, P., and Stannarius, R., *Phys. Rev. E*, 86, 040501(R)-1. Copyright 2012 by the American Physical Society.)

Several mechanisms were suggested to explain the exceptional stability of the filaments in the B_7 phase. These include layer elasticity, electrical self-interaction occurring in the ferroelectric state, and the out-of-plane polarization (Jákli et al. 2003). In fact, tilting of molecular plane away from the SmCP configuration leads to stabilization of the liquid crystal fibers. In this model, the symmetry of the B_7 phase is assumed to be C_1.

The main ingredient in these models is the presence of an effective bulk elastic modulus B, which, according to the experimental observations, lies in the kPa range, which is considerably smaller than in the compression modulus in conventional smectic phases (Jákli et al. 2003, Bailey et al. 2007, Eremin et al. 2012b).

It appears that the modulation is essential for stabilizing the filament structure. A transition into a nonmodulated phase results in melting and collapse of a filament

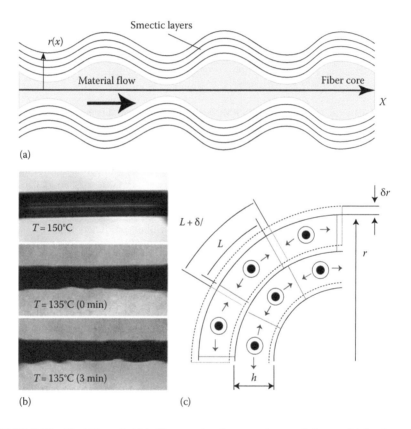

FIGURE 7.17 Unstable and stable fibers made of concentric smectic layers. (a) A scheme of an unstable fiber without layer decoration. Any change of the radius is allowed with the help of materials flow from the meniscus; (b) optical microscope images of unstable filaments (dark images) upon the transition into the nonmodulated SmCP phase (image width 600 μm); and (c) polarization-splay stripes in two subsequent smectic layers in a fiber. A slight increase of the curvature radius r results in a deviation of the stripe width δl from its equilibrium value. (Reprinted with permission from Eremin, A., Kornek, U., Stern, S., Stannarius, R., Araoka, F., Takezoe, H., Nádasi, H., Weissflog, W., and Jákli, A., *Phys. Rev. Lett.*, 109, 017801-1. Copyright 2012 by the American Physical Society.)

(Eremin et al. 2005, 2012b). This can be accounted for if one considers a periodic structure of the polarization-splay stripes on the surface of the smectic layer (Figure 7.17). They occur as a result of a topological constraint since the continuous splay cannot fill the whole smectic plane. The defect lines separate the stripes; and the stripe width is fixed by the temperature. Although strong distortions of the filament shape would lead to the nucleation of the new stripes, a small distortion with an amplitude δr (Figure 7.17c) of the cylindrical surface of the fiber results in the dilatation (or compression) of the stripes.

This deformation is coupled to the director distortion and results in the energy penalty. The free energy variation in response to a harmonic distortion becomes

$$\Delta F = \frac{\pi \sigma \delta r^2}{2 R_0} \left(R_0^2 q^2 + \frac{a R_0}{\pi \sigma} - 1 \right) \tag{7.14}$$

where the contribution of the stripe pattern with spontaneous polarization splay s_0 is given by the effective compression modulus $a = \pi K s_0^2 K \ln \left(R_0 / r_c \right)$ (r_c is the radius of the fiber core, K is the mean Frank elastic constant, and R_0 is the fiber radius). Assuming typical values for K and s_0, the instability occurs when the fiber radius reaches its critical value of approximately 4 μm. The effective compression modulus a lies in the range of 10–20 kPa. Thus, a periodic pattern of topological defects decorating smectic layers is capable of stabilizing filament structure.

7.3 STRUCTURE BY TOPOGRAPHIC CONFINEMENT

Topographic confinement of a complex liquid crystalline phase by a bounding surface gives rise to a great variety of new structures. Some of these structures, like labyrinthine patterns, have already been mentioned in Section 7.1 on freely suspended films. Here we discuss the structure formation at an LC/glass substrate.

An isotropic-B_7 phase transition in a bulk of a large droplet is accompanied by nucleation and growth of filaments with different morphologies such as screwlike and telephone-wire structures that can be distinguished in polarizing microscopy (Figure 7.18a). At lower temperature, they develop a disordered texture. In a thin layer on a glass substrate, however, screwlike and ribbonlike structures appear on cooling (Figure 7.18b). Such structures start to grow at a transition from the isotropic to the B_7 phase. The ribbons nucleate, elongate, and buckle, forming a maze texture. Some ribbons start coiling, forming telephone-wire structures. These structures use up the isotropic material from the thin layer. In thicker areas, the filament structures persist but they form a disordered spaghettilike pattern.

The effect of topological confinement reveals itself in the growth of the helical nanofilaments in the B_4 phase, as already described in Section 3.4.2. The structure of the helical filaments is not compatible with the flat interface since the filaments cannot either fill the space near surface nor form a full contact with the substrate. As a result, the structure of the phase is modified in the vicinity of the substrate as was demonstrated by Chen et al. using freeze fracture transmission electron microscopy (FFTEM) (Figure 7.19) (Chen et al. 2012). Close to the substrate, however, the

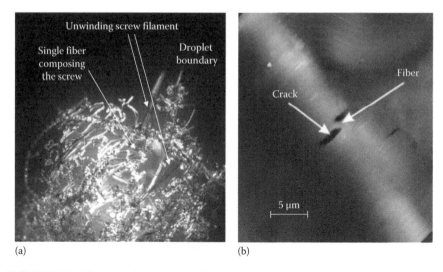

(a)   (b)

FIGURE 7.18 **(See color insert.)** (a) Optical image of a screwlike filament growing in a large isotropic droplet under crossed polarizers. As soon as it touches the boundary, a longitudinal stress along the screw leads to unwinding of the small fibers and (b) "defect" cracks occur in the screw filament. These cracks show that the filament is composed of a tightly coiled fiber. The scale bar is 2 μm. (Reprinted from Eremin, A. et al., *Liq. Cryst.*, 33, 789, 2006. With permission from Taylor & Francis.)

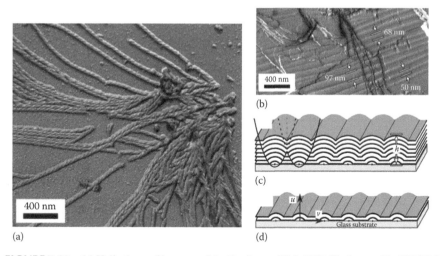

(a)   (d)

FIGURE 7.19 (a) Helical nanofilaments of the B$_4$ phase of P-9-OPIMB observed by FFTEM in a mixture of P-9-OPIMB with 75% 8CB quenched at 37°C; (b) Topography of the B$_4$ phase in the vicinity of the glass substrate. The FFTEM image shows fractures at different heights. (c) and (d) Model structures of parabolic focal conic arrays of smectic layers at different heights. The curves represent parabolas with their foci at the glass substrate. As the layers develop, they curve around the focus and flatten outside, giving alternating curved and flat regions. (From Chen, D., Heberling, M.-S., Nakata, M., Hough, L.E., Maclennan, J.E., Glaser, M.A., Korblova, E., Walba, D.M., Watanabe, J., and Clark, N.A.: Structure of the B$_4$ liquid crystal phase near a glass surface. *ChemPhysChem.* 2012. 13. 155–159. Copyright Wiley-VCH Verlag GmbH & Co. KGaA. Reproduced with permission.)

structure consists of alternating curved and flat regions. Farther from the substrate, the flat regions collapse and the structure transforms into a continuous modulation with a period of about 100 nm. This can be described as a 1D parabolic focal conic (PFC) array of smectic layers. The formation of PFCs at the surface is attributed to the flat topology of the substrate. Another type of structure found at the substrate in the B_4 phase is the bamboo-like structure (Chen et al. 2012). It is made of toric focal conic (TFC) domains with an axial defect line lying on the substrate and the circular defect line normal to the glass. The radial extension of bamboo-like structures is determined by the self-limiting growth.

Other interesting topographic confinements can be made using microdroplets of bent-core molecules and curved colloidal surface planes. Microdroplets of calamitic LCs have been studied since a long time ago (Trivedi 2011, Blanc 2013, Tkalec 2013, Lavrentovich 2014), but no such works have been reported in bent-core molecular systems. Well-known defect structures such as hedgehog, Saturn ring, and boojum in the vicinity of microparticles (Hirankittiwong 2014) may be different between calamitic and bent-core molecular systems. These are future problems.

REFERENCES

Bailey, C., E. C. Gartland, and A. Jákli. Structure and stability of bent core liquid crystal fibers. *Phys. Rev. E* 75 (2007): 031701-1–031701-9.

Bailey, C., M. Murphy, A. Eremin, W. Weissflog, and A. Jákli. Bundles of fluid fibers formed by bent-core molecules. *Phys. Rev. E* 81 (2010): 031708-1–031708-6.

Blanc, Ch., D. Coursault, and E. Lacaze. Ordering nano- and microparticles assemblies with liquid crystals, *Liq. Cryst. Rev.* 1 (2013): 1–27.

Borshch, V., Y. K. Kim, J. Xiang, M. Gao, A. Jákli, V. P. Panov, J. K. Vij et al. Nematic twist-bend phase with nanoscale modulation of molecular orientation. *Nat. Commun.* 4 (2013): 1–8.

Chattham, N., E. Korblova, R. Shao, D. M. Walba, J. E. Maclennan, and N. A. Clark. Triclinic fluid order. *Phys. Rev. Lett.* 104 (2010): 067801-1–067801-4.

Chattham, N., M. G. Tamba, R. Stannarius, E. Westphal, H. Gallardo, M. Prehm, C. Tschierske, H. Takezoe, and A. Eremin. Leaning-type polar smectic-C phase in a freely suspended bent-core liquid crystal film. *Phys. Rev. E* 91 (2015): 030502-1–030502-5.

Chen, D., M.-S. Heberling, M. Nakata, L. E. Hough, J. E. Maclennan, M. A. Glaser, E. Korblova, D. M. Walba, J. Watanabe, and N. A. Clark. Structure of the B_4 liquid crystal phase near a glass surface. *ChemPhysChem* 13 (2012): 155–159.

Chen, D., M. Nakata, R. Shao, M. R. Tuchband, and M. Shuai. Twist-bend heliconical chiral nematic liquid crystal phase of an achiral rigid bent-core mesogen. *Phys. Rev. E* 89 (2014): 022506-1–022506-5.

Churaev, N. V., B. V. Derjaguin, and V. M. Muller. *Surface Forces*. Springer, 1987.

Clark, N. A. and S. T. Lagerwall. Submicrosecond bistable electro-optic switching in liquid crystals. *Appl. Phys. Lett.* 36 (1980): 899–901.

Coleman, D. A., J. Fernsler, N. Chattham, M. Nakata, Y. Takanishi, E. Korblova, D. R. Link et al. Polarization-modulated smectic liquid crystal phases. *Science* 301 (2003): 1204–1211.

Derjaguin, B. V., Y. I. Rabinovich, and N. V. Churaev. Direct measurement of molecular forces. *Nature* 272 (1978): 313–318.

Doi, M. *Soft Matter Physics*. Oxford University Press, 2013.

Dozov, I. On the spontaneous symmetry breaking in the mesophases of achiral banana-shaped molecules. *Europhys. Lett.* 56 (2001): 247–253.

Eremin, A., M. Floegel, U. Kornek, S. Stern, R. Stannarius, H. Nádasi, W. Weissflog et al. Transitions between paraelectric and ferroelectric phases of bent-core smectic liquid crystals in the bulk and in thin freely suspended films. *Phys. Rev. E* 86 (2012a): 051701-1–051701-10.

Eremin, A., U. Kornek, R. Stannarius, W. Weissflog, H. Nádasi, F. Araoka, and H. Takezoe. Labyrinthine instability in freely suspended films of a polarization-modulated smectic phase. *Phys. Rev. E* 88 (2013): 062512-1–062512-8.

Eremin, A., U. Kornek, S. Stern, R. Stannarius, F. Araoka, H. Takezoe, H. Nádasi, W. Weissflog, and A. Jákli. Pattern-stabilized decorated polar liquid-crystal fibers. *Phys. Rev. Lett.* 109 (2012b): 017801-1–017801-5.

Eremin, A., L. Naji, A. Nemeş, R. Stannarius, M. Schulz, and K. Fodor-Csorba. Microscopic structures of the B7 phase: AFM and electron microscopy studies. *Liq. Cryst.* 33 (2006): 789–794.

Eremin, A., A. Nemeş, R. Stannarius, G. Pelzl, and W. Weissflog. Spontaneous bend patterns in homochiral ferroelectric SmCP films: Evidence for a negative effective bend constant. *Soft Matter* 4 (2008a): 2186–2191.

Eremin, A., A. Nemeş, R. Stannarius, M. Schulz, H. Nádasi, and W. Weissflog. Structure and mechanical properties of liquid crystalline filaments. *Phys. Rev. E* 71 (2005): 031705-1–031705-5.

Eremin, A., A. Nemeş, R. Stannarius, and W. Weissflog. Ambidextrous bend patterns in free-standing polar smectic-CP_F films. *Phys. Rev. E* 78 (2008b): 061705-1–061705-11.

Faber, T. E. The intermediate state in superconducting plates. *Proc. R. Soc. Lond. A* 248 (1958): 460–481.

Goldstein, R. E., D. J. Muraki, and D. M. Petrich. Interface proliferation and the growth of labyrinths in a reaction-diffusion system. *Phys. Rev. E* 53 (1996): 3933–3957.

Hirankittiwong, P., N. Chattham, J. Limtrakul, O. Haba, K. Yonetake, A. Eremin, R. Stannarius, and H. Takezoe. Optical manipulation of the nematic director field around microspheres covered with an azo-dendrimer monolayer. *Opt. Express* 22 (2014): 20087–20093.

Israelachvili, J. N. *Intermolecular and Surface Forces: With Applications to Colloidal and Biological Systems*. London, U.K.: Academic Press, 1992.

Jackson, D. P., R. E. Goldstein, and A. O. Cebers. Hydrodynamics of fingering instabilities in dipolar fluids, *Phys. Rev. E* 50 (1994): 298–307.

Jákli, A., D. Krüerke, and G. G. Nair. Liquid crystal fibers of bent-core molecules. *Phys. Rev. E* 67 (2003): 051702-1–051702-6.

Jákli, A., C. Lischka, W. Weissflog, G. Pelzl, and A. Saupe. Helical filamentary growth in liquid crystals consisting of banana-shaped molecules. *Liq. Cryst.* 27 (2000): 1405–1409.

Kamenskiĭ, V. G. and E. I. Kats. Stability of filaments of discotic liquid crystals. *JETP Lett.* 37 (1983): 261–264.

Kerkam, K., C. Viney, D. Kaplan, and S. Lombardi. Liquid crystallinity of natural silk secretions. *Nature* 349 (1991): 596–598.

Lagerwall, S. T. *Ferroelectric and Antiferroelectric Liquid Crystals*. New York: Wiley-VCH, 1999.

Lagerwall, J. P. F. and F. Giesselmann. Complex chirality at the nanoscale. *ChemPhysChem* 11 (2010): 975–977.

Lagerwall, J. P. F., J. T. McCann, E. Formo, G. Scalia, and Y. Xia. Coaxial electrospinning of microfibres with liquid crystal in the core. *Chem. Commun.* 42 (2008): 5420–5422.

Lavrentovich, O. D. Transport of particles in liquid crystals. *Soft Matter* 10 (2014): 9, 1264–1283.

Möhwald, H. Phospholipid and phospholipid-protein monolayers at the air/water interface. *Annu. Rev. Phys. Chem.* 41 (1990): 441–476.

Morys, M., T. Trittel, A. Eremin, P. Mrurphy, and R. Stannarius. Tension of freely suspended fluid filaments. *Phys. Rev. E* 86 (2012): 040501(R)-1–040501(R)-5.

Naito, H., M. Okuda, and Z.-C. Ou-Yang. Equilibrium shapes of smectic-A phase grown from isotropic phase. *Phys. Rev. Lett.* 70 (1993): 2912–2915.

Nemeş, A., A. Eremin, R. Stannarius, M. Schulz, H. Nadasi, and W. Weissflog. Structure characterization of free-standing filaments drawn in the liquid crystal state. *Phys. Chem. Chem. Phys.* 8 (2006): 469–476.

Oswald, P. and P. Pieranski. *Liquid Crystals: Concepts and Physical Properties*, CRC Press, Boca Raton, 2003.

Oswald, P. and P. Pieranski. *Nematic and Cholesteric Liquid Crystals: Concepts and Physical Properties Illustrated by Experiments*. Boca Raton, FL: Taylor & Francis/CRC Press, 2005.

Oswald, P. and P. Pieranski. *Smectic and Columnar Liquid Crystals Concepts and Physical Properties Illustrated by Experiments*. Boca Raton, FL: Taylor & Francis/CRC Press, 2006.

Pelzl, G., S. Diele, A. Jákli, C. Lischka, I. Wirth, and W. Weissflog. Helical superstructures in a novel smectic mesophase formed by achiral banana-shaped molecules. *Liq. Cryst.* 26 (1999): 135–139.

Petzold, J., A. Nemeş, A. Eremin, C. Bailey, N. Diorio, A. Jákli, and R. Stannarius. Acoustically driven oscillations of freely suspended liquid crystal filaments. *Soft Matter* 5 (2009): 3120–3126.

Picano, F., R. Holyst, and P. Oswald. Coupling between meniscus and smectic-A films: Circular and catenoid profiles, induced stress, and dislocation dynamics. *Phys. Rev. E* 62 (2000): 3747–3757.

Picano, F., P. Oswald, and E. Kats. Disjoining pressure and thinning transitions in smectic-A liquid crystal films. *Phys. Rev. E* 63 (2001): 021705-1–021705-9.

Pindak, R., C. Young, R. Meyer, and N. Clark. Macroscopic orientation patterns in smectic-C films. *Phys. Rev. Lett.* 45 (1980): 1193–1196.

Rosensweig, R. E.. *Ferrohydrodynamics*. Mineola, NY: Dover, 1997.

Seul, M. and D. Andelman. Domain shapes and patterns: The phenomenology of modulated phases. *Science* 267 (1995): 476–483.

Stannarius, R., A. Nemeş, and A. Eremin. Plucking a liquid chord: Mechanical response of a liquid crystal filament. *Phys. Rev. E* 72 (2005): 020702(R)-1–020702(R)-4.

Tkalec, U. and I. Musevic. Topology of nematic liquid crystal colloids confined to two dimensions. *Soft Matter* 9 (2013): 8140–8150.

Todorokihara, M., Y. Iwata, and H. Naito. Periodic buckling of smectic-A tubular filaments in an isotropic phase. *Phys. Rev. E* 70 (2004): 021701-1–021701-6.

Trivedi, R. and D. Engström. Optical manipulation of colloids and defect structures in anisotropic liquid crystal fluids. *J. Opt.* 13 (2011): 044001.

van Winkle, D. H. and N. A. Clark. Freely suspended strands of tilted columnar liquid-crystal phases: One-dimensional nematics with orientational jumps. *Phys. Rev. Lett.* 48 (1982): 1407–1410.

Vollrath, F. and D. Knight. Liquid crystalline spinning of spider silk. *Nature* 410 (2001): 541–548.

Young, C., R. Pindak, N. Clark, and R. Meyer. Light-scattering study of two-dimensional molecular-orientation fluctuations in a freely suspended ferroelectric liquid-crystal film. *Phys. Rev. Lett.* 40 (1978): 773–776.

Zywocinski, A., F. Picano, P. Oswald, and J. C. Geminard. Edge dislocation in a vertical smectic-A film: Line tension versus temperature and film thickness near the nematic phase. *Phys. Rev. E* 62 (2000): 8133–8140.

8 Mixing Effect in Bent-Shaped Mesogens

Two important papers (Gorecka et al. 2000, Pratibha et al. 2000), published in 2000, triggered a new trend of research in bent-shaped molecular systems. Mixing bent-shaped mesogens with other materials such as calamitic molecules, organic solvents, and metal nanoparticles reveals a variety of new phenomena. We will describe some of these phenomena in this chapter.

8.1 MIXTURES WITH CALAMITIC MOLECULES

Pratibha et al. (2000) showed a new type of phase transition associated with the orientational transition of bent-core molecules in an anisotropic SmA matrix of calamitic molecules and the induction of biaxial SmA_{2b} phases. The phase diagram of binary mixtures is shown in Figure 8.1 together with the compounds used at the bottom. The calamitic molecules exhibit the N and SmA_2 (bilayer SmA) phases, whereas the bent-core molecule the B_2 phase. With increasing molar fraction of calamitic molecules, B_2 changes to B_1 and further to B_6. The orientational transition of the bent-core molecules occurs in the mixtures between above and below 13 mol% of the bent-core molecule; the arrow axis is parallel to the layer normal in the SmA_{2b} phase, whereas it is parallel to the smectic layer in the B_6 phase (Madhusudana 2009). The behavior between 4 and 13 mol% of bent-core molecules is more interesting: the uniaxial (SmA_2) to biaxial (SmA_{2b}) phase transition, which is proved by the emergence of *Schlieren* texture, occurs at a temperature depending on the mixing ratio. Pratibha et al. (2000) suggested the location of bent-core molecules in a calamitic matrix based on the matched length of the core parts of bent-core and calamitic molecules. In the SmA_{2b} phase, one half of the bilayer favors the "up" while the other favors the "down" orientation of the bows, so that the medium is not longitudinally ferroelectric. The review (Madhusudana 2009) describes the mixing effect mentioned in the preceding text and the related topics.

The critical behavior at the SmA_2–SmA_{2b} phase transition was studied using two independent high-resolution calorimetric experiments using a slow temperature scan rate, 0.03 K/h (Sasaki et al. 2010). Some calorimetry results and the phase diagram obtained by the calorimetry are shown in Figure 8.2. A dramatic reduction in the heat anomaly at the SmA_2–SmA_{2b} transition as the mole fraction of the bent-shaped molecules X decreases (Figure 8.2b) can be explained assuming that the energy fluctuations around this transition are very sensitive to the ordering of the SmA_2 background. A tricritical point was found to be located between $X = 0.08$ and 0.09. Above this point, the SmA_2–SmA_{2b} transition is of the first order, whereas it is of the second order below this point. Full details of heat capacity data and the analysis are given in Sasaki et al. (2010).

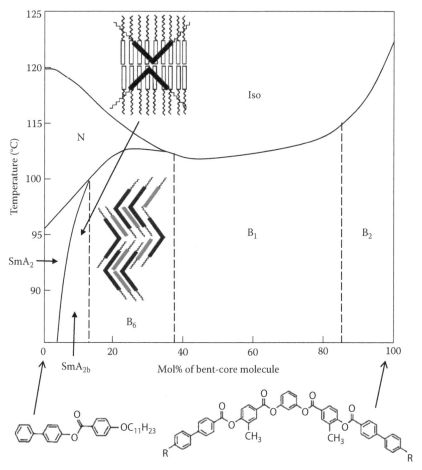

FIGURE 8.1 Phase diagram of binary mixtures of bent-core and calamitic liquid crystal molecules (shown in the bottom). The structures of two different biaxial smectic phases formed are illustrated in the figure. (Redrawn from Pratibha, R. et al., *Science*, 288, 2184, 2000.)

Pratibha et al. (2005) also used 8OCB as a rodlike molecule in the mixture. This mixture system is useful to conduct polarized FT-IR to explore the orientation relationship between a CN group of 8OCB and a C=O group of the bent core molecule shown in Figure 8.1. Based on the fourier-transform-infrared (FT-IR) spectroscopy in addition to x-ray diffraction (XRD), they found two phases with 2D (p2mg) structures instead of B_6 phase as shown in Figure 8.3: in partially bilayered 8OCB with 2D ordered bent-core molecules and B_1' and B_1, in which the arrow axes are perpendicular and parallel to the layer, respectively.

Another type of phase induction was shown by Gorecka et al. (2000). The idea is very simple. In the SmC(*) (SmC and/or SmC*) phase doped with bent-core mesogens, the molecular arms of bent-shaped liquid crystal (BLC) align the calamitic mesogens in locally anticlinic fashion as shown in Figure 8.4. This induces the $SmC_A(*)$ phase with alternating tilted layers and stabilizes $SmC_A(*)$. The doping-induced phase

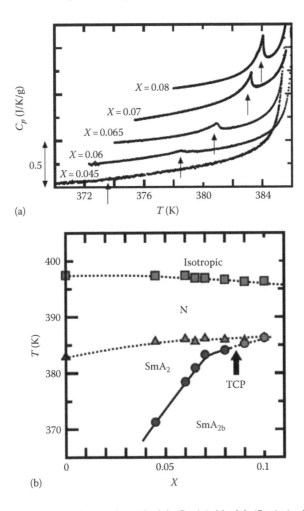

(a)

(b)

FIGURE 8.2 Critical behavior at the uniaxial (SmA$_2$)–biaxial (SmA$_{2b}$) phase transition observed in the region with small mole fractions of bent-shaped molecules. (a) Differential scanning calorimetry (DSC) charts observed by high-resolution calorimetric measurements; (b) a tricritical point exists at the phase diagram of a rod and bent binary mixture system. (Reprinted with permission from Sasaki, Y., Ema, K., Le, K.V., Takezoe, H., Dhara, S., and Sadashiva, B.K., *Phys. Rev. E*, 82, 011709-1. Copyright 2010 by the American Physical Society.)

change was examined by electro-optic and dielectric measurements using a calamitic molecule showing the Iso–SmC*–SmI* phase sequence and a variety of bent-core molecules. Dielectric measurements demonstrated the suppression of the Goldstone mode already at a weight fraction of bent-core dopant as low as 0.1–2 wt%. The temperature stability of SmC$_A^*$ increases quickly with increasing dopant concentration, exhibiting the direct Iso – SmC$_A^*$ phase transition. The stabilization of the anticlinic state by the bent-core mesogens was also manifested in the electro-optical effect where the electric field-induced SmC$_A^*$ – SmC* was shifted to higher fields. The same observations were also made in the nonchiral SmC phase (Kishikawa et al. 2003).

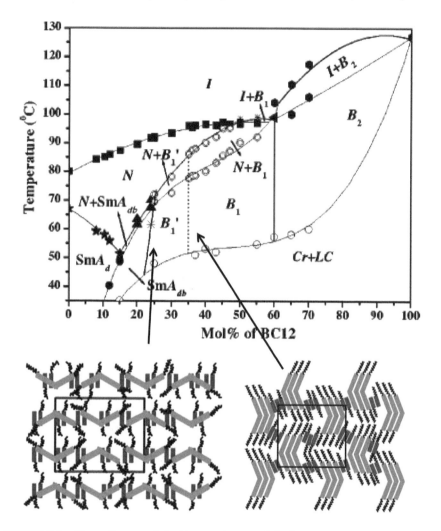

FIGURE 8.3 Phase diagram of the binary mixtures of the bent-core compound shown in Figure 8.1 and a rodlike compound 8OCB. Phase structures with 2D lattices are also shown. (Reprinted with permission from Pratibha, R., Madhusudana, N.V., and Sadashiva, B.K., *Phys. Rev. E*, 71, 011701-1. Copyright 2005 by the American Physical Society.)

Monte Carlo simulations were also made in the mixtures of bent-shaped spherocylinder dimer and rod-shaped spherocylinders (Maiti et al. 2002). The entropy-driven induction of SmC_A ordering in a SmA host was observed using bent-shaped dimers with appropriate bending angles. However, no evidence of the SmA_b phase was found. For smaller ratios of length/breadth of a bent-shaped dimer, entropy-driven transition from intralamellar to interlamellar nanophase segregation of bent-core molecules was also observed.

Doping a bent-core molecular system with rod-shaped molecules was made for a different purpose. To make the B_1 phase switchable, Kirchhoff and Hirst (2010)

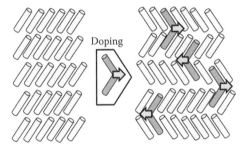

FIGURE 8.4 Concept of the SmC$_A$(*) phase induced by bent-core molecular doping.

introduced a ferroelectric chiral liquid crystal molecule. However, they observed an opposite effect: the ferroelectric dopant actually increased the threshold electric field and even resulted in stabilization of the non-switchable SmA phase at 5 wt% of the dopant.

More interesting and extensive studies in the bent- and rod-molecular mixtures were initiated by Takanishi et al. (2005). Bent-core molecules, which exhibit the B$_4$ (helical nanofilament) phase, and calamitic molecules, which exhibit the N phase, form phase-segregated structures in their mixtures. The original experiments were made using a prototype bent-core molecule, P8-O-PIMB, and a conventional calamitic molecule, 5CB (Takanishi et al. 2005). Later, another combination, P9-O-PIMB/8CB, was also used by the Boulder group in studies, for instance Chen et al. (2010).

The phase diagram of the P8-O-PIMB/5CB mixtures is shown in Figure 8.5. The B$_2$ and B$_3$ phases are destabilized easily by doping P8-O-PIMB with a small amount

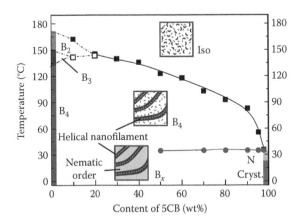

FIGURE 8.5 Phase diagram of the bent (P8-O-PIMB) and rod (5CB) mixtures. Cartoons showing three phases Iso, B$_4$, and B$_x$ are shown in the figure. Phase-segregated B$_4$ and B$_x$ are characterized by that 5CB is in the Iso and N phases, respectively. (Reprinted with permission from Otani, T., Araoka, F., Ishikawa, K., and Takezoe, H., Enhanced optical activity by achiral rod-like molecules nanosegregated in the B$_4$ structure of achiral bent-core molecules, *J. Am. Chem. Soc.*, 131, 12368. Copyright 2009 by the American Chemical Society.)

of 5CB and only the B_4 phase is stabilized. As illustrated in the diagram, the transition between B_4 and B_x exists nearly at the same temperature of the I–N transition of 5CB in the weight fraction of 5CB larger than 50%. Moreover, the XRD studies showed that the layer thickness does not change between B_4 and B_x and is independent of the 5CB fraction (Takanishi et al. 2005, Zhu et al. 2010). From these experimental facts, it is concluded that P8-O-PIMB and 5CB are phase segregated and the B_4–B_x transition is accompanied by the I–N transition of 5CB embedded in a framework of the B_4 structure of P8-O-PIMB molecules. The detailed thermal measurements indicate that the transition is of the first-order in pure 5CB ($X = 1.0$), it becomes critical in the $X = 0.8$ mixture, then finally shows a supercritical behavior for $X = 0.5$ (Takekoshi et al. 2006).

The phase separation was directly observed in the P9-O-PIMB/8CB mixtures by freeze fractured transmittance electron microscopy (FFTEM) as shown in Figure 8.6 (75% 8CB and pure P9-O-PIMB) (Chen et al. 2010). Helical nanofilaments can be seen; in the mixture no layer edges are visible, suggesting that the filaments are smoothly coated with 8CB, in contrast to pure P9-O-PIMB, where layer edges are easily identified. Chen et al. (2010) also found pretransitional ordering of 8CB at the nanofilament surface far above the Iso–N transition of 8CB. The phase separation was also observed using dynamic light scattering (Yamazaki et al. 2009). Figure 8.7 shows the log–log plot of scattering wave number (k_s) versus relaxation frequency ($1/\tau$), which was determined by exponential fitting of autocorrelation functions, in the B_x phase of binary mixtures (P8-O-PIMB/5CB) with various mixing ratios at 30°C. In pure 5CB, $1/\tau$ is quadratically proportional to k_s, as expected. However, the dependence is leveled off at a critical wave number (k_c), which depends on 5CB contents, and $1/\tau$ remains constant below k_c. This clearly indicates that the director fluctuation motion is localized in some coherent domains, that is, confinement of 5CB by helical nanofilament network. It is reasonable that k_c becomes larger with decreasing 5CB content, since k_c is inversely proportional to the coherent domain size, $\lambda = 2\pi/k_c$.

One of the most fascinating results in this binary mixture is the enhanced chirality discovered by Otani et al. (2009). By using a very thin cell such as 500 nm thick, one can observe circular dichroism (CD) spectra of the mixture without signal saturation. Figure 8.8 shows CD and absorption spectra in the mixture with 60 wt% 5CB in the B_4 phase. Note that the two CD spectra taken at two chiral domains are mirror images to each other. Comparing with the absorption spectra of P8-O-PIMB and 5CB, the

(a)   (b)

FIGURE 8.6 Freeze fractured transmittance electron microscopy images of a (a) P9-O-PIMB/8CB mixture and (b) pure P9-O-PIMB. (Reprinted with permission from Chen, D., Zhu, C., Shoemaker, R.K., Korblova, E., Walba, D.M., Glaser, M.A., Maclennan, J.E., and Clark, N.A., Pretransitional orientational ordering of a calamitic liquid crystal by helical nanofilaments of a bent-core mesogen, *Langmuir*, 26, 15541. Copyright 2010 by the American Chemical Society.)

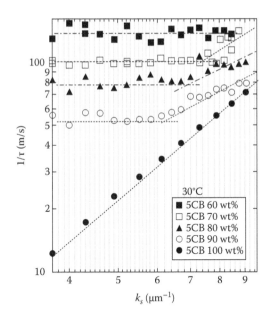

FIGURE 8.7 Log–log plot of scattering wave number k_s versus relaxation frequency $1/\tau$ obtained by dynamic light scattering in P8-OPIMB/5CB mixtures with 60, 70, 80, 90, and 100 wt% 5CB. (Reprinted from Yamazaki, Y., Takanishi, Y., and Yamamoto, J., Dynamic heterogeneity of a nanostructure in the hyper-swollen B_4 phase of achiral bent-core molecules diluted with rod-like liquid crystals, *Eur. Phys. Lett.*, 88, 56004-1–56004-4, 2009. With permission from IOC Publishing.)

CD originates from P8-O-PIMB. This is reasonable, since 5CB is still in the Iso phase. The temperature dependence of the CD intensity for four mixtures shows interesting behavior (Figure 8.9). In the mixtures with higher 5CB contents, CD intensity is remarkably enhanced when 5CB undergoes the Iso–N phase transition (B_4 to B_x). This suggests a superhelical structure formed by 5CB around helical nanofilaments of P8-O-PIMB (Otani et al. 2009). Actually, the CD peak was enhanced and blue-shifted toward the resonance wavelength of 5CB with decreasing temperature, which indicates the emerging contribution of the 5CB chiral structure (Araoka et al. 2011). It was also found by x-ray microbeam scattering that the nematic director of 5CB is almost parallel to the smectic layers and nanofilament axis, confirming the superhelical structure, that is, 5CB molecules orient along the groove of helical nanofilaments (Takanishi et al. 2014). It is also interesting that CD and optical rotation (OR) intensity can be controlled by applying an electric field; OR signal of about 0.7 deg/μm in the absence of a field can be almost diminished by applying about 15 V to a 700 nm thick cell of the mixture with 90 wt% 5CB (Araoka et al. 2011).

Some methods have been employed to align helical nanofilaments (Kim et al. 2014). A combinational use of topographic confinement, shear flow, and temperature gradients gave a successful alignment of helical nanofilaments (Yoon et al. 2011). However, the well-aligned area is still limited. Better alignment was achieved using a series of hierarchical self-assembly steps (Kim et al. 2014). P9-O-PIMB was

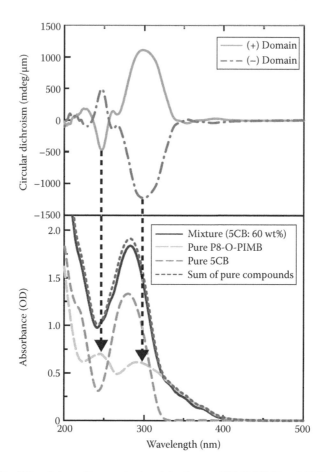

FIGURE 8.8 CD and absorption spectra in a 4:1 mixture of P8-O-PIMB and 5CB. Absorption spectra of pure P8-O-PIMB, 5CB, and their sum are also shown by dotted curves. (Reprinted with permission from Otani, T., Araoka, F., Ishikawa, K., and Takezoe, H., Enhanced optical activity by achiral rod-like molecules nanosegregated in the B_4 structure of achiral bent-core molecules, *J. Am. Chem. Soc.*, 131, 12368. Copyright 2009 by the American Chemical Society.)

introduced in the Iso phase into 5 µm thick nanoporous anodic aluminum oxide films, and then gradually cooled. Because of a heating stage attached to the film, temperature gradient parallel to the pore direction was established. Helical nanofilaments align along the pores with diameters between 20 and 70 nm. The filaments within a single core have the same chirality, but overall chirality control is not possible without introducing chiral dopants.

Another sophisticated method for alignment control of helical nanofilaments is the use of mixtures of bent-core and calamitic molecules. For this purpose, the Iso–B_4 phase transition must occur in the N phase of calamitic molecules. Araoka et al. (2013) used 4′-cyano-(1,1′-biphenyl)-4-yl-4-pentylbenzoate (5PCB), which exhibits N in a very high temperature range (T_{IN} = 240°C). Large domains with extinction crosses were observed (Figure 8.10a). Unlike a normal extinction cross, the

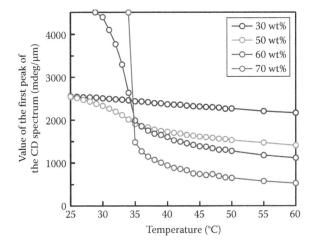

FIGURE 8.9 (**See color insert.**) CD intensity as a function of temperature in four mixtures of P8-O-PIMB/5CB. The temperature dependence behavior changes at the B_4–B_x transition particularly for two mixtures with higher 5CB contents. (Reprinted with permission from Otani, T., Araoka, F., Ishikawa, K., and Takezoe, H., Enhanced optical activity by achiral rod-like molecules nanosegregated in the B_4 structure of achiral bent-core molecules, *J. Am. Chem. Soc.*, 131, 12368. Copyright 2009 by the American Chemical Society.)

extinctions parallel and perpendicular directions are quite different. Precise interferometric second-harmonic generation (SHG) microscopy measurements provide a polar direction map, as shown by arrows in Figure 8.10b. Thus, not only the chirality but also the polarity in a single nanofilament domain is preserved in a domain formed from a single nucleus, being consistent with the growth model proposed by Chen et al. (2011). To enlarge the uniform nanofilament alignment, one should restrict the number of nucleation of domains. Temperature gradient along the rubbing direction can help this restriction. As a result, uniform chiral domains of several millimeters in length and 1 mm in width were obtained (Figure 8.10c) (Araoka et al. 2013). The domain becomes completely dark under crossed polarizers, suggesting complete alignment of the nematic director and helical nanofilaments. Under a decrossed condition, two striped domains with about 1 mm width can be recognized. These domains have opposite handedness that originated from different nuclei.

8.2 MIXTURES WITH ORGANIC SOLVENT: PHYSICAL GEL

As described in Section 8.1, nematic materials such as 5CB and 8CB can be included as a phase separated state in helical nanofilament networks in the B_4 phase. In other words, helical nanofilaments provide us with porous materials. Porous materials and nanoconfined systems using such materials are important both from scientific interest and applications to chemical and biological purposes. Inclusions are not necessarily liquid crystal materials, but can be a variety of organic solvents, and this complex system forms physical gel. Here we introduce two important works on physical gels made of liquid crystalline B_4 phase (Chen et al. 2013, Zep et al. 2013).

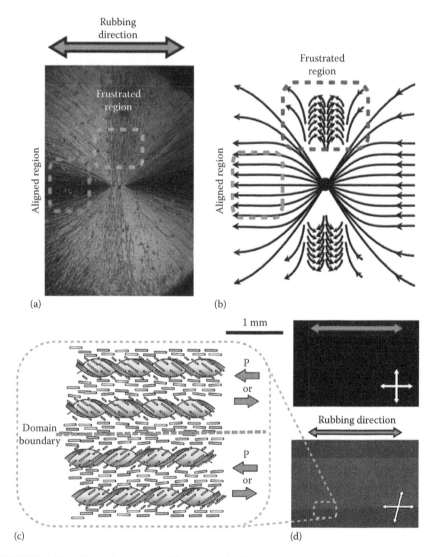

FIGURE 8.10 (**See color insert.**) Alignment of helical nanofilaments using a nematic field. (a) Texture of a defect of strength 1; (b) orientation map of the defect region; (c) cartoon showing the oriented two domains with different chiralities; and (d) microphotographs of the oriented domains under crossed and decrossed polarizers. (From Araoka, F., Sugiyama, G., Ishikawa, K., and Takezoe, H.: Highly ordered helical nanofilament assembly aligned by a nematic director field. *Adv. Funct. Mater.* 2013. 23. 2701–2707. Copyright Wiley-VCH Verlag GmbH & Co. KGaA. Reproduced with permission.)

Chen et al. (2013) used a variety of liquid crystal guests showing N, SmA, and Col$_h$, and a polymer and materials without LC phases. In all mixtures with P9-O-PIMB, the various effects of nanoconfinement were observed. For instance, in contrast to the other systems, where the guest x-ray scattering peak is usually broadened by nanoconfinement in the B$_4$ nanofilament network, the x-ray diffraction from P3HT (conducting polythiophene polymer) in the nanofilament network is slightly sharper than that in the neat sample, indicating a longer correlation length. In this case, upon cooling, P3HT phase separates first from the uniform isotropic melt, forming crystalline nanowhiskers. Different nanoconfinement effect was also observed in a simple alkane (hexadecane). Unlike the other mixtures, the layer spacing of the nanofilament swells from 47.96 to 48.89 A. This suggests the intercalation of hexadecane into the alkyl tail of P9-OPIMB, which is identical to the hexadecane.

Actually, these systems are the first organogels made of bent-core LC molecules. The helical nanofilament network is able to absorb a large amount of solvent. One example, a 98% hexadecane–P9-O-PIMB mixture, is shown in Figure 8.11 (Chen et al. 2013). The presence of interdigitated helical nanofilaments in the gel is directly observed by TEM after evaporation of the hexadecane, xerogel (Figure 8.11, right). The growth of helical nanofilament clusters can be observed as white clusters with the naked eye in a 99.994% hexadecane mixture when cooled from Iso to room temperature. Thus, the helical nanofilaments in organic solvents form a gel-like, 3D nanoporous network. The B_4 gel is distinct from conventional organogel, in which 3D networks are formed by the self-assembly of low-molecular-weight gelators through noncovalent interactions. In contrast, helical nanofilaments grow under a self-limiting condition caused by the elastic energy in the B_4 gel. Thus, the helical nanofilament network can be used as a template for making porous materials with chiral surfaces, which could be employed, for example, as enantioselective catalyst supports.

Another intriguing system was reported by Zep et al. (2013). They used twin dimers shown in Figure 8.12 (top) and several solvents, and studied gelation ability. It was found that all the studied bent dimers were excellent gelators for various organic fluids: toluene, nitrobenzene, 3-methylcyclohexanone, (−)- and (+)-menthone, chloroform, and dichloromethane. However, no gelation was observed for polar organic solvents (N,N-dimethylformamide (DMF), ethanol, and methanol). The thermo-reversible sol–gel

 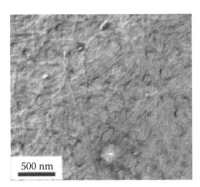

500 nm

FIGURE 8.11 Photo image of organogel of a 98% hexadecane–P9-O-PIMB mixture and its TEM image. (Reprinted from Chen, D., Zhu, C., Wang, H., Maclennan, J.E., Glaser, M.A., Korblova, E., Walba, D.M., Rego, J.A., Soto-Bustamante, E.A., and Clark, N.A., Nanoconfinement of guest materials by helical nanofilament networks of bent-core mesogens, *Soft Matter*, 9, 462–471, 2013. Reproduced by permission of The Royal Society of Chemistry.)

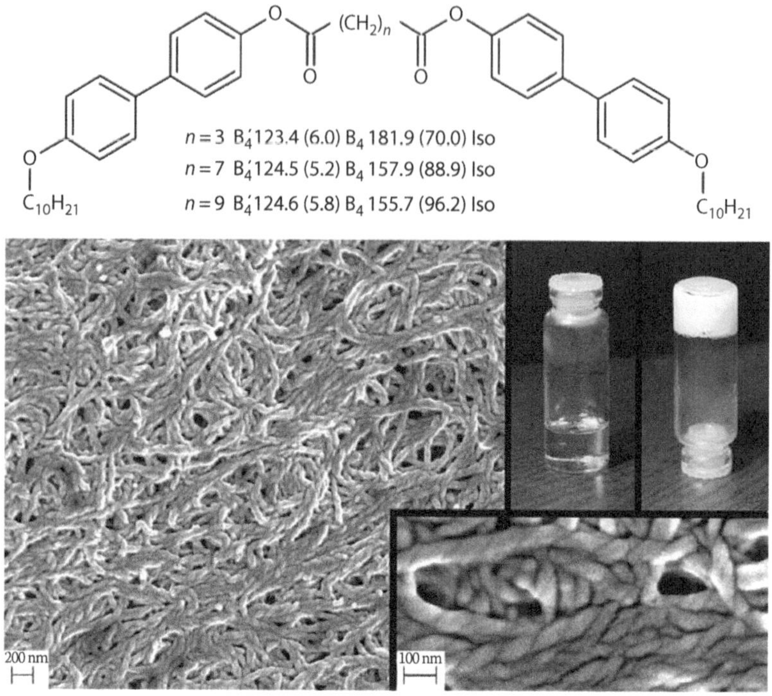

$n = 3$ B_4'123.4 (6.0) B_4 181.9 (70.0) Iso
$n = 7$ B_4'124.5 (5.2) B_4 157.9 (88.9) Iso
$n = 9$ B_4'124.6 (5.8) B_4 155.7 (96.2) Iso

FIGURE 8.12 SEM images with photo images of dimer molecules shown at the top. (From Zep, A., Salamonczyk, M., Vaupotič, N., Pociecha, D., and Gorecka, E., Physical gels made of liquid crystalline B_4 phase, *Chem. Commun.*, 49, 3119–3121, 2013. Reproduced by permission of The Royal Society of Chemistry.)

transition was detected by DSC and optically. The transition temperature depends on solvent and is much lower than the Iso–B_4 phase transition temperature of the pure materials. SEM observation clearly reveals helical tubular filaments (Figure 8.12).

8.3   MIXTURES WITH METAL NANOPARTICLES

Together with increasing attention on nanomaterials, many liquid crystal scientists also started to work on mixtures of LCs and nanoparticles and LC ligands grafted onto nanoparticle surfaces for future applications. Fabre et al. (1990) were the first to show that the presence of nanoparticles dispersed into low molecular mass LCs can result in striking differences in the optical and electro-optic behavior of such systems. The advantage of using LCs is self-assembly of desired structures and possible manipulation by external stimuli. The topics have been reviewed by many authors (Hegmann et al. 2007, Bisoyi and Kumar 2011, Nealon et al. 2012, Stamatoiu et al. 2012, Lewandowski et al. 2014). Here we only describe the efforts on metal nanoparticles in bent-shaped liquid crystal systems.

Structured nanomaterials are desired because of potential applications based on their unusual optical properties, such as a negative refractive index or plasmon waveguiding. The most important purpose of using LCs in nanoparticle systems is to fabricate an ordered structure of nanoparticles. In this sense, bent-shaped molecules are interesting as LC molecules for grafting to nanoparticles and nanoclusters, since bent-shaped molecules exhibit a variety of complex phases and may provide different types of interactions including steric, polar, and chiral ones. Unfortunately, however, there are only a few papers reporting the systems including bent-shaped molecules (Marx et al. 2008, Wojcik et al. 2009).

Marx et al. (2008) studied the structure of gold nanoclusters and the physical properties of the dispersion of bent-shaped-molecular-grafted nanoclusters into bent-core mesogens. The strategy of this work is illustrated in Figure 8.13. Previously reported symmetric and nonsymmetric bent-core molecules, in which the methyl- or alkene-terminated hydrocarbon chains were replaced by bromo-, xanthate-, thioacetate-, and thio-terminations, were used for the studies. They found that while none of the symmetric derivatives display any LC behavior, the majority of the nonsymmetric derivatives do (B_1 and B_2 phases). Using two mono(thiol)-substituted bent-core derivatives, one of which shows a monotropic mesophase, the assembly of gold nanoclusters was investigated. By drop casting, monolayer-protected gold nanoparticles formed large areas of somewhat regular arrays, as shown in Figure 8.14a. As recognized by superimposed molecular images in an enlarged view (Figure 8.14b), the average spacing between nanoparticles is nearly the same as the molecular length of the bent-core molecule.

The preceding system shows no LC phases. To study nanoparticles in LC phases, Marx et al. (2008) prepared mixtures of nanoparticle and bent-core molecules. Since the

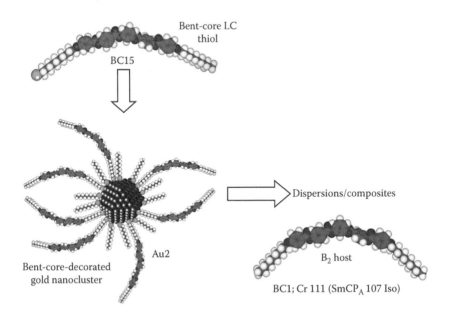

FIGURE 8.13 (**See color insert.**) Gold nanocluster decorated with bent-core molecules. (From Marx, V.M., Girgis, H., Heiney, P.A., and Hegmann, T., Bent-core liquid crystal (LC) decorated gold nanoclusters: Synthesis, self-assembly, and effects in mixtures with bent-core LC hosts, *J. Mater. Chem.*, 18, 2983–2994, 2008. Reproduced by permission of The Royal Society of Chemistry.)

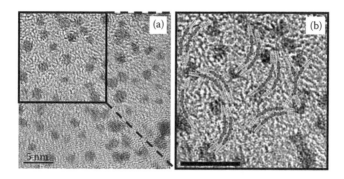

FIGURE 8.14 TEM images of gold nanocluster (Figure 8.13) obtained by drop casting solutions in toluene on carbon-coated copper grids. The squared area in (a) is enlarged in (b) with superimposed molecular images. (From Marx, V.M., Girgis, H., Heiney, P.A., and Hegmann, T., Bent-core liquid crystal (LC) decorated gold nanoclusters: Synthesis, self-assembly, and effects in mixtures with bent-core LC hosts, *J. Mater. Chem.*, 18, 2983–2994, 2008. Reproduced by permission of The Royal Society of Chemistry.)

nanoparticles are decorated by monolayered bent-core molecules, the nanoparticles can be reasonably well dispersed in bent-shaped LC hosts. It is clearly observed by the difference in the surface plasmon resonance wavelength that the different packing of the two bent-core hosts, which show $SmCP_A$ and Col_r, influences the aggregation of the bent-core-decorated nanoparticles. Electro-optic response was measured in the $SmCP_A$ phase of the nanoparticle-dispersed bent-core molecules with different concentrations of nanoparticles. It is interesting that the spontaneous polarization is a little larger and the response time is faster in the mixtures containing about 5 wt% nanoparticles, whereas the microscope textures are quite different.

Wojcik et al. (2009) made preliminary studies on the structure of bent-core-coated gold nanoclusters: short-range fcc or distorted icosahedral structures. The detailed studies are awaited. In this way, the studies on bent-core-molecule-coated nanoparticles are still quite primitive but contain a promising target for smart materials.

REFERENCES

Araoka, F., G. Sugiyama, K. Ishikawa, and H. Takezoe. Electric-field controllable optical activity in the nano-segregated system composed of rod- and bent-core liquid crystals. *Opt. Mater. Exp.* 1 (2011): 27–35.

Araoka, F., G. Sugiyama, K. Ishikawa, and H. Takezoe. Highly ordered helical nanofilament assembly aligned by a nematic director field. *Adv. Funct. Mater.* 23 (2013): 2701–2707.

Bisoyi, H. K. and S. Kumar. Liquid crystal nanoscience: An emerging avenue of soft-assembly. *Chem. Soc. Rev.* 40 (2011): 306–319.

Chen, D., J. E. Maclennan, R. Shao, D. K. Yoon, H. Wang, E. Korblova, D. M. Walba, M. A. Glaser, and N. A. Clark. Chirality-preserving growth of helical filaments in the B_4 phase of bent-core liquid crystals. *J. Am. Chem. Soc.* 133 (2011): 12656–12663.

Chen, D., C. Zhu, R. K. Shoemaker, E. Korblova, D. M. Walba, M. A. Glaser, J. E. Maclennan, and N. A. Clark. Pretransitional orientational ordering of a calamitic liquid crystal by helical nanofilaments of a bent-core mesogen. *Langmuir* 26 (2010): 15541–15545.

Chen, D., C. Zhu, H. Wang, J. E. Maclennan, M. A. Glaser, E. Korblova, D. M. Walba, J. A. Rego, E. A. Soto-Bustamante, and N. A. Clark. Nanoconfinement of guest materials by helical nanofilament networks of bent-core mesogens. *Soft Matter* 9 (2013): 462–471.

Fabre, P., C. Casagrande, M. Veyssie, V. Cabuil, and R. Massart. Ferrosmectics—A new magnetic and mesomorphic phase. *Phys. Rev. Lett.* 64 (1990): 539–542.

Gorecka, E., M. Nakata, J. Mieczkowski, Y. Takanishi, K. Ishikawa, J. Watanabe, H. Takezoe, S. H. Eichhorn, and T. M. Swager. Induced antiferroelectric smectic-C_A* phase by doping ferroelectric-C* phase with bent-shaped molecules. *Phys. Rev. Lett.* 85 (2000): 2526–2529.

Hegmann, T., H. Qi, and V. M. Marx. Nanoparticles in liquid crystals: Synthesis, self-assembly, defect formation and potential applications. *J. Inorg. Organomet. Polym. Mater.* 17 (2007): 483–508.

Kim, H., Y. H. Kim, S. Lee, D. M. Walba, N. A. Clark, S. B. Lee, and D. K. Yoon. Orientation control over bent-core smectic liquid crystal phases. *Liq. Cryst.* 41 (2014): 328–341.

Kirchhoff, J. and L. S. Hirst. Modification of the electro-optical properties of the B_1 liquid-crystal phase using a rodlike liquid-crystal dopant. *Phys. Rev. E* 82 (2010): 031701-1–031701-7.

Kishikawa, K., N. Muramatsu, S. Kohmoto, K. Yamaguchi, and M. Yamamoto. Control of molecular aggregations by doping in mesophases: Transformation of smectic C phases to smectic C_A phases by addition of long bent-core molecules possessing a central strong dipole. *Chem. Mater.* 15 (2003): 3443–3449.

Lewandowski, W., M. Wojcik, and E. Gorecka. Metal nanoparticles with liquid-crystalline ligands: Controlling nanoparticle superlattice structure and properties. *ChemPhysChem* 15 (2014): 1283–1295.

Madhusudana, N. V. On some liquid crystals made of banana-shaped molecules and their mixtures with rod-like molecules. *Liq. Cryst.* 36 (2009): 1173–1184.

Maiti, P. K., Y. Lansac, M. A. Glaser, and N. A. Clark. Induced anticlinic ordering and nanophase segregation of bow-shaped molecules in a smectic solvent. *Phys. Rev. Lett.* 88 (2002): 065504-1–065504-4.

Marx, V. M., H. Girgis, P. A. Heiney, and T. Hegmann. Bent-core liquid crystal (LC) decorated gold nanoclusters: Synthesis, self-assembly, and effects in mixtures with bent-core LC hosts. *J. Mater. Chem.* 18 (2008): 2983–2994.

Nealon, G. L., R. Greget, C. Dominguez, Z. T. Nagy, D. Guillon, J.-L. Gallani, and B. Bonnio. Liquid-crystalline nanoparticles: Hybrid design and mesophase structures. *Beilstein J. Org. Chem.* 8 (2012): 349–370.

Otani, T., F. Araoka, K. Ishikawa, and H. Takezoe. Enhanced optical activity by achiral rodlike molecules nanosegregated in the B_4 structure of achiral bent-core molecules. *J. Am. Chem. Soc.* 131 (2009): 12368–12372.

Pratibha, R., N. V. Madhusudana, and B. K. Sadashiva. An orientational transition of bent-core molecules in an anisotropic matrix. *Science* 288 (2000): 2184–2187.

Pratibha, R., N. V. Madhusudana, and B. K. Sadashiva. Two-dimensionally periodic phases in mixtures of compounds made of rodlike and bent-core molecules. *Phys. Rev. E* 71 (2005): 011701-1–011701-12.

Sasaki, Y., K. Ema, K. V. Le, H. Takezoe, S. Dhara, and B. K. Sadashiva. Critical behavior at transitions from uniaxial to biaxial phases in a smectic liquid-crystal mixture. *Phys. Rev. E* 82 (2010): 011709-1–011709-6.

Stamatoiu, O., J. Mirzaei, X. Feng, and T. Hegmann. Nanoparticles in liquid crystals and liquid crystalline nanoparticles. *Top. Curr. Chem.* 318 (2012): 331–393.

Takanishi, Y., G. J. Shin, J. C. Jung, S. W. Choi, K. Ishikawa, J. Watanabe, H. Takezoe, and P. Toledano. Observation of very large chiral domains in a liquid crystal phase formed by mixtures of achiral bent-core and rod molecules. *J. Mater. Chem.* 15 (2005): 4020–4024.

Takanishi, Y., H. Yao, T. Fukasawa, K. Ema, Y. Ohtsuka, Y. Takahashi, J. Yamamoto, H. Takezoe, and A. Iida. Local orientational analysis of helical filaments and nematic director in a nanoscale phase separation composed of rod-like and bent-core liquid crystals using small- and wide-angle x-ray microbeam scattering. *J. Phys. Chem. B* 118 (2014): 3998–4004.

Takekoshi, K., K. Ema, H. Yao, Y. Takanishi, J. Watanabe, and H. Takezoe. Appearance of a liquid crystalline nematic-isotropic critical point in a mixture system of rod- and bent-shaped molecules. *Phys. Rev. Lett.* 97 (2006): 197801-1–197801-4.

Wojcik, M., W. Lewandowski, J. Matraszek, J. Mieczkowski, J. Borysiuk, D. Pociecha, and E. Gorecka. Liquid-crystalline phases made of gold nanoparticles. *Angew. Chem. Int. Ed.* 48 (2009): 5167–5169.

Yamazaki, Y., Y. Takanishi, and J. Yamamoto. Dynamic heterogeneity of a nanostructure in the hyper-swollen B_4 phase of achiral bent-core molecules diluted with rod-like liquid crystals. *Eur. Phys. Lett.* 88 (2009): 56004-1–56004-4.

Yoon, D. K., Y. Yi, Y. Shen, E. D. Korblova, D. M. Walba, I. I. Smalyukh, and N. A. Clark. Orientation of a helical nanofilament (B_4) liquid-crystal phase: Topographic control of confinement, shear flow, and temperature gradients. *Adv. Mater.* 23 (2011): 1962–1967.

Zep, A., M. Salamonczyk, N. Vaupotič, D. Pociecha, and E. Gorecka. Physical gels made of liquid crystalline B_4 phase. *Chem. Commun.* 49 (2013): 3119–3121.

Zhu, C., D. Chen, Y. Shen, C. D. Jones, M. A. Glaser, J. E. Maclennan, and N. A. Clark. Nanophase segregation in binary mixtures of a bent-core and rodlike liquid-crystal molecule. *Phys. Rev. E* 81 (2010): 011704-1–011704-5.

9 Application of Bent-Shaped Mesogens

In spite of the richness of fascinating physical phenomena observed in bent-core liquid crystals (LC), no practical applications have yet been realized. Actually, only a few realistic application proposals have been made up to now. Direct applications are impeded by various problems. The most serious problem for the application is that a good alignment of smectic layers is very difficult to achieve, particularly in the switchable B_2 phase. Although reasonably good alignment was obtained by an in-plane field application (Nakata et al. 2001), it is still far from satisfaction for real practical applications. Another problem is relatively high LC temperature ranges. This problem, however, can be overcome by extensive efforts for synthesizing new materials and developing mixtures. In the following, we summarize studies aiming at practical applications.

9.1 TRIALS FOR DISPLAY APPLICATION

LC displays using nematic materials have been developed rapidly. We realize by looking at commercial LC displays that the technology is matured and no further problems seem to remain. Due to very strict engineers' viewpoints, however, the performance of the present LC displays is not satisfactory. One of the requirements is (1) a fast response. However, to achieve this, other factors, such as (2) high contrast ratio, (3) wide viewing angle, and (4) continuous gray level display, should not be deteriorated. One may easily realize that this is a significant and high hurdle to overcome. Actually, ferroelectric SmC* and antiferroelectric SmC_A^* displays were expected to be promising candidates for fast-responsive displays, and full screen displays were commercialized or manufactured by way of trial. Unfortunately, however, only a few ferroelectric and no antiferroelectric displays remain at present.

In such a situation, one of attractive ideas was proposed by Shimbo et al. (2006b). The display principle and electro-optic response by applying a square wave voltage are shown in Figure 9.1. The bent-core molecule shown in Figure 9.1a exhibits the $SmAP_R$ phase (see Section 3.2) (Pociecha et al. 2003). When cells with homeotropic-alignment surfaces are used, bent-core molecules orient perpendicular to the smectic layer with a random distribution of the polarization (bending) direction. As shown on the left side of Figure 9.1b and c (top view and side view), where arrows represent polarizations, the phase looks uniaxial, so that a perfect dark view is obtained between crossed polarizers in the absence of a field, resulting in a high contrast ratio (3000:1). By applying an in-plane field parallel to the layer, the polarization reorients toward the field direction, which is accompanied with the induction of birefringence. Under crossed polarizers, the transmittance increases and saturates to a certain level depending on the applied voltage (Figure 9.1d). The reorientation occurs due to the polar interaction between the field and polarization, so the switching speed is

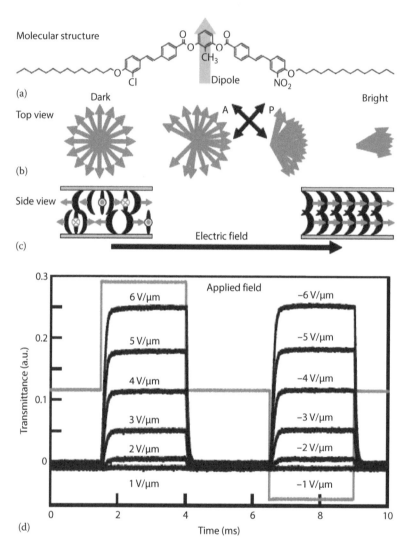

FIGURE 9.1   Display mode using the SmAP$_R$ phase. (a) Molecular structure for this particular mode. (b) and (c) Top and side views of polarization distribution; images from left to right correspond to the configurations with increasing applied electric field. (d) Electro-optic performance, in which fast switching and continuous grayscale display ability are shown. (Reprinted from Shimbo, Y., Takanishi, Y., Ishikawa, K., Gorecka, E., Pociecha, D., Mieczkowski, J., Gomola, K., and Takezoe, H., Ideal liquid crystal display mode using achiral banana-shaped liquid crystals, *Jpn. J. Appl. Phys.*, 45, L282–L284, 2006. With permission from the Japan Society of Applied Physics.)

remarkably fast; 40 µs for switching OFF and 240–130 µs for turning ON an electric field with increasing amplitudes. The reorientation is a continuous process, so that continuous gray levels can be displayed. Since the polarization director orients in a plane parallel to the layer, the viewing angle dependence of the transmittance is minimal. In this way, the present display mode using the $SmAP_R$ phase uses all the advantages of the existing LC display modes, such as vertical alignment mode (high contrast ratio), in-plane switching mode (wide viewing angle), ferroelectric LC mode (fast response), and V-shaped switching mode (continuous gray level) but with small threshold voltage. The ferroelectric coupling between the polarization and an electric field is practically operable because of the cooperative motion of bent-core molecules with quasi-long-range order of dipoles. By analyzing the optical second-harmonic generation (SHG) intensity as a function of field strength based on a two-dimensional Langevin process, it was found that a few hundreds of molecules form a polar domain (Shimbo et al. 2006a).

Despite so many attractive features, this mode has not been commercialized. This is because of the similar number of drawbacks: (1) narrow and high temperature range of $SmAP_R$, (2) layer deformation by high-field application, (3) small refractive index anisotropy, (4) high operating voltage required, and (5) considerable temperature dependence of physical parameters. To overcome (1), synthetic efforts have been made for broad temperature range (Gomola et al. 2009a,b, 2010). The temperature range was widened and lowered but is not enough to include room temperature. Problem (2) originates from a curved line of electric force caused by in-plane comb electrodes. Guo (2010) clearly demonstrated that the smectic layer is much stable in a bent-core mesogen with a single siloxane terminal chain compared with a conventional $SmAP_R$ compound. Figure 9.2 shows optical micrographs during (a and c) and after (b and d) applying 180 V to conventional (a and b) and siloxane-terminated (c and d) compounds. Small anisotropy (3) cannot be avoided because of the use of biaxiality. Compared with the birefringence in uniaxial molecules, 0.1–0.3, the maximum birefringence in $SmAP_R$ was simulated as 0.02. This means that we need thick cells for display application. Then, the problem (2) becomes more serious. The maximum electric field applied was 10 V/µm. This means 100 V/10 µm (electrode gap), which is too high for practical application (problem (4)). As mentioned in the preceding text, the ferroelectric coupling is relevant to the domain size. Although the applied voltage could be lowered by using larger sized domains, the response time becomes slower because of high viscosity. One of the examples of problem (5) is the domain size. With decreasing temperature, the domain size becomes large, influencing the switching speed.

Some other attempts aiming display applications have been made. However, none of them can compete with the current LC displays using the N phase. Alonso et al. (2007) proposed and demonstrated a display using a dark conglomerate phase. In the field OFF state, a dark conglomerate phase appears dark (like isotropic) under crossed polarizers. By applying an electric field, the smectic layers flatten (perpendicular to cell surfaces) to form a SmCP phase consisting of small domains with high birefringence ($\Delta n = 0.14$), giving a bright view. Fast switching between 100 and 300 µs was reported, but the contrast ratio was not very high, that is, 100–300:1 depending on SmCP states. Additional advantages are a wide viewing angle because of optical isotropy (OFF state) and random domains (ON state) and no necessity for alignment.

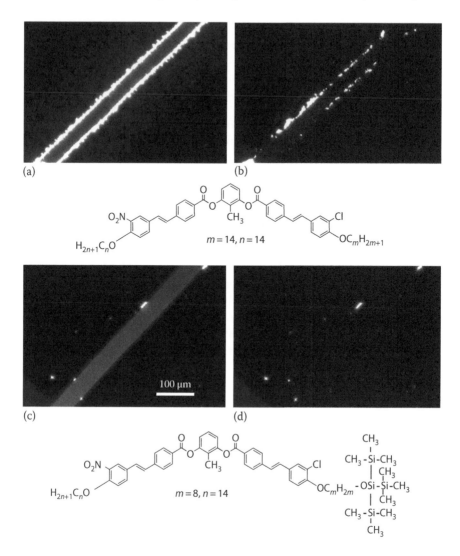

FIGURE 9.2 Performance under in-plane electric field application in two compounds. The images under (a) field ON and (b) OFF of a conventional compound show deterioration of alignment at the edges of the electrode. A siloxane-terminated compound under (c) ON and (d) OFF shows good performance without deterioration. (Reprinted from Guo, L., PhD thesis, Tokyo Institute of Technology, 2010.)

A scattering-type display using the SmCP phase was demonstrated by Jákli et al. (2002). The principle is based on the switching between synclinic states (SmC_SP_A and SmC_SP_F), which scatter light, and anticlinic states (SmC_AP_A and SmC_AP_F), which is transparent. Antiferroelectric and ferroelectric states appear in the absence and presence of an electric field, respectively. First in the absence of an electric field, the SmC_SP_A (racemic) state is stable and opaque. By applying a field, SmC_SP_A changes to SmC_AP_F (racemic) associated with transmittance increase, then changes

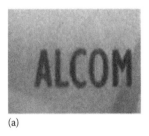 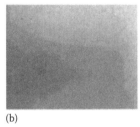

(a)   (b)

FIGURE 9.3 Light shutter showing (a) transparent (OFF) and (b) scattered (ON) states using an electric-field-induced change among the SmCP states. (Reprinted from Jákli, A. et al., *Liq. Cryst.*, 29, 377, 2002. With permission from Taylor & Francis.)

to SmC_SP_F (homochiral) by further field increase, which is very opaque. This process from racemic to homochiral states is associated with chirality switching (see Section 5.2). With decreasing field, SmC_SP_F relaxes to SmC_AP_A associated with increasing transmittance. Jákli et al. (2002) claim that the scattering is due to defects separating different chiral domains and/or oppositely tilted domains. The advantage of this device is low energy consumption due to bistable memory states with transparent and opaque views. Figure 9.3 demonstrates electrically switchable light shutter, although the "ON" state is not a memory state. In the "OFF" state, the chiral SmC_AP_A state is transparent so that a text behind the cell is readable, whereas in the "ON" state, the chiral SmC_SP_F state is opaque and the text cannot be seen. A fast switching speed like 100 μs has also been reported.

Electrotunable birefringence Δn (Nair et al. 2008) and refractive index n (Nagaraj et al. 2014) were also proposed for applications. Nair et al. (2008) used the SmC_A phase in mixtures of bent-core and rod-shaped molecules, which is induced and stabilized in a wide range of the mixing ratio. Since it appears below the N phase, good alignment is obtained, particularly if an ac electric field is applied during cooling. Electro-optic switching is based on the orientational change of the secondary director (bend and dipole direction), which induces Δn change of about 0.02. The switching speed is fast because of the rotational motion about the smectic layer normal. But still, this device is far from the practical application for displays as well. Nagaraj et al. (2014) reported a different electro-optic response. They used the dark conglomerate phase. By applying an electric field, a field-driven transformation takes place, which is associated with an average refractive index change but no change in the birefringence. But the change is only about 0.05 and is still a very primitive level for application.

A different type of device proposed by Mathews et al. (2011) is based on bent-core molecules bearing chiral carbons and an azo-linkage at each arm. The molecule shows the cholesteric phase exhibiting a selective reflection. The reflection band (reflection color) is tunable by external stimuli such as temperature, light, and an electric field. However, the same effect is well known in calamitic liquid crystals. It is hard to find a particular reason for using bent-shaped molecules instead of calamitic ones.

As mentioned in the preceding text, although some studies of bent-core molecules aiming display applications have been conducted, no commercial products were made, far from it, no demonstrations close to practical devices could be found. The situation is the same for other applications besides displays. However, for future studies, we will describe other application-oriented works in Section 9.2.

9.2   NON-DISPLAY APPLICATIONS

Because of large nonlinear susceptibility values (see Section 6.1) one can imagine the application to piezoelectric and pyroelectric devices, which can be used for transducers, sensors, and memory elements. However, we can find only a few reports on piezoelectric properties of polymers containing/consisting of bent-core mesogens (Sentman and Gin 2003, Charif et al. 2013).

Sentman and Gin (2003) first synthesized polymerizable and cross-linkable bent-core mesogens. They thermally cross-linked in the SmCP phase for 3 h under a static 10 V electric field and successfully obtained films with an ordered polar noncentrosymmetric network. They confirmed pyroelectric response and found a pyroelectric coefficient of 4.8 μC/(m^2 K). The first piezoelectric measurements were made by Jákli et al. (2009) using a bent-core small molecule containing an adamantine fragment at one of the terminal chains. The vibration-induced piezoelectric current shows complicated behavior against the frequency of vibration. But definitely the signal increases by two orders of magnitude by poling by high-enough electric fields, by which the electric-field-induced antiferroelectric (SmC$_A$P$_A$)–ferroelectric (SmC$_S$P$_F$) transition takes place. The authors claim that the piezoelectric constant is about 100 pC/N and is comparable to that of polyvinylidene fluoride (PVDF). Unfortunately, the piezoelectricity existed only at elevated temperatures and, due to the fluid nature of the layers and the macroscopically achiral structure of the film, the piezoelectricity decayed in a few minutes (Charif et al. 2013).

A large converse piezoelectric response was also observed in a thermoplastic elastomer containing bent-core liquid crystals (4-Cl substituted P10-O-PIMB) (Charif et al. 2013). Interestingly, the liquid crystal-polymer composite becomes aligned during compression molding leading to a vitrified macroscopic polarization without electric poling. The bent-core molecules were mixed with a linear poly(styrene-b-isobutylene-b-styrene) block copolymer. Films for the converse piezoelectric measurements were made by solvent casting followed by compression molding at 60°C, when bent-core molecules show the ferroelectric phase. The applied-voltage-induced film thickness change is shown in Figure 9.4. Because of the electrostriction effect in addition to the piezoelectric effect, the dependence is slightly deviated from the linear relationship. It is reasonable that the pure copolymer sample shows only the quadratic electrostriction effect. The piezoelectric charge constant d_{33} is largest in the composite containing 10 wt% bent-core molecules and is about 1 nm/V, which is larger than that of the commercially available piezoelectric ceramics such as lead zirconate titanate (PZT) and PVDF.

As mentioned in Section 7.2, one of the unique properties in bent-shaped molecular systems is the fiber-forming ability in some phases. Since the fiber is different from conventional ones in view of a cylindrical geometry and anisotropy, it is interesting to examine how light propagates along the fiber. For possible applications, it is extremely important not to suffer from light scattering, which often prevents LCs from the practical use. Fontana et al. (2009) used the B$_7$ phase of a nitro-substituted

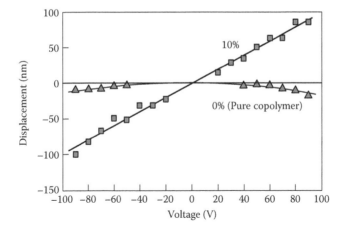

FIGURE 9.4 Piezoelectric response, that is, displacement against voltage in a block copolymer containing bent-core molecules. Note that the pure block copolymer shows the quadratic electrostriction effect (triangle). (Redrawn from Charif, A.C. et al., *RSC Adv.*, 3, 17446, 2013.)

P8-O-PIMB at the central core, the fiber of which is solid-like along the radial direction and liquid-like along the fiber. An interesting feature is a self-annealing effect, that is, shortly after pulling a new filament, defects appear as bright spots but move along the axis and eventually are expelled. In this way, defect-free fibers can be obtained even in the crystalline phase. The authors claim that the scattering of the light out of the filament surface due to director fluctuations, defects, or other mechanisms is insignificant. The transmitted power decays exponentially to about a half over 5 mm because of the absorption at 532 nm used, as shown in Figure 9.5.

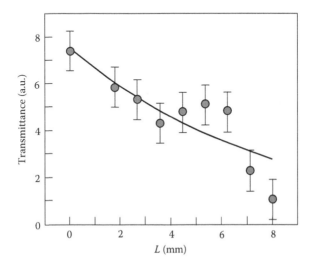

FIGURE 9.5 Transmission against the length of an LC fiber. (Redrawn from Fontana, J. et al., *Phys. Rev. E*, 80, 032701-1, 2009.)

The authors claim that this LC fiber is useful for optical fibers in the infrared range. However, it cannot be used as practical devices, unless some significant advantages over glass or even polymer fibers are added.

The B_4 or helical nanofilament (HNF) phase is a unique semicrystalline phase, as described in Section 3.4.2. Besides the unique structural features and physical properties such as polarity, chirality, and gel formation ability (see Section 8.2), promising applications for the design of superhydrophobic surfaces (Kim et al. 2013) and organic electronics (Callahan et al. 2014) have been demonstrated and shall be discussed in the following text.

One of the useful methods for the surface to stay dry is the use of surfaces with multiscale micro–nano hybrid structures like the one in lotus leaves: dual-scale structures consisting of coarse-scale (tens micrometer scale) rough surface coated by finer structures (hundreds nanometer scale). Such a dual surface, which enhances hydrophobicity, is provided by bent-core materials with the B_2–B_4 phase sequence. The method, which Kim et al. (2013) used, was very simple. Upon cooling P-9-OPIMB (NOBOW) (Figure 1.8) from Iso to B_2, micron-scale toric focal conic domains are formed at the surface. Upon further cooling into the B_4 phase, nano-scale HNFs cover the surface, resulting in a dual-scale structured surface, as confirmed by atomic force microscopy (AFM) and freeze fracture transmission electron microscopy (FFTEM) (Kim et al. 2013). Using such a surface, they demonstrated enhanced hydrophobicity (contact angle of about 102° for a water drop) compared with a simple lamellar surface (about 82°). Such surfaces with enhanced hydrophobicity may provide practical applications such as self-cleaning surfaces and anti-fogging coatings.

For their good processability, organic materials are promising for the applications to electronic devices such as organic photovoltaics, organic field-effect transistors, and organic light-emitting diodes. Liquid crystals provide additional advantages such as self-assembled structure and reasonably high electron mobility compared with amorphous materials such as polymers (Adam et al. 1994, Ohno et al. 2007). Here we introduce bent-core liquid crystal showing the HNF phase (P-9-OPIMB, Figure 1.8) as a good charge-generation interface (Callahan et al. 2014). The sample used was easily prepared just by spin-coating solutions of P-9-OPIMB and fullerenes: thin C_{60} film onto a thin film of HNFs. For noncontact photocurrent measurements, time-resolved microwave conductivity measurements (Ferguson et al. 2008) were performed. Since C_{60} is the sole absorbing species, all excitons are generated in the C_{60} layer, and charge separation occurs at the interface, leading to electron transport in the C_{60} layer and hole transport in the P-9-OPIMB layer. Figure 9.6 shows the photoconductivity transient of a HNF-C_{60} bilayer film and a pure C_{60} film excited at 470 nm. The conductivity of the bilayer film is 50 times larger than that of the pure C_{60} film. This means that the HNF surface acts as a good interface for photoinduced electron transfer because of high contact area of the HNF layer edge and the C_{60} acceptor. They also prepared a mixed layer of P-9-OPIMB and C_{60} and measured the photoconductivity using as-spun films and annealed film and discussed the relation to the heterostructure of these films (Callahan et al. 2014).

Last but not least, we describe another candidate phase for possible applications. Unlike many columnar B_1 group phases (see Section 3.5), which are characterized by a rectangular lattice, there exist other types of columnar phases consisting of

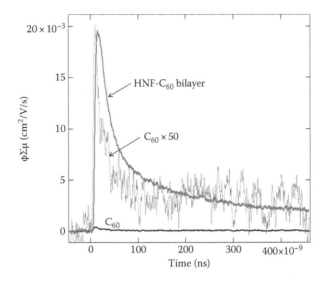

FIGURE 9.6 Photoconductivity (yield-mobility product) transients for films of pure C_{60} and bilayers of C_{60} on P-9-OPIMB HNF. (Reprinted with permission from Callahan, R.A., Coffey, D.C., Chen, D., Clark, N.A., Rumbles, G., and Walba, D.M., Charge generation measured for fullerene-helical nanofilament liquid crystal heterojunctions, *ACS Appl. Mater. Interfaces*, 6, 4823. Copyright 2014 by the American Chemical Society.)

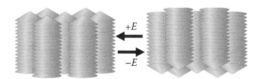

FIGURE 9.7 Polarly ordered columns of umbrella-shaped assemblies constructed by bent-core molecules.

bent-core molecules, which are characterized by a hexagonal lattice and an induced axial polarization (see Section 3.5). Gorecka et al. (2004) synthesized polycatenar bent-core molecules (see Figure 3.46) and confirmed an induced polarization along columns and its switchability. Three or four molecules construct an umbrella-type assembly (see Figure 3.46), forming columns (Figure 9.7). Although no macroscopic polarization exists in unperturbed state, polarization can be induced along the columns and reversed by a field application to the opposite direction. Figure 9.8 shows the temperature dependence of SHG in the compounds with $n = 12$ and 16 and the switching model at the bottom (Gorecka et al. 2006, Takezoe et al. 2006). For $n = 12$ compounds, SHG intensity increases with decreasing temperature and suddenly drops below the Col_h–$Col_h P_A$ phase transition, due to insufficient electric field for breaking the polarization-compensated structure. Another interesting phenomenon was seen in the compound with $n = 16$. Contrary to other homologues, the compound with $n = 16$ shows the continuous Col_h–$Col_h P_A$ phase transition, exhibiting a broad DSC

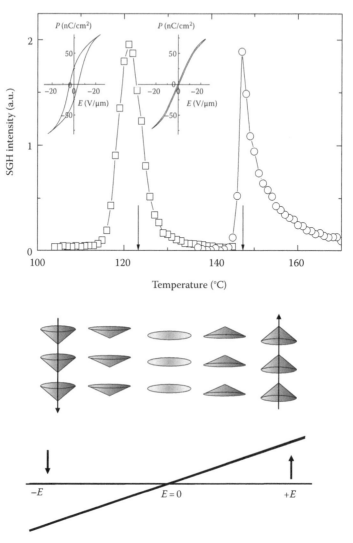

FIGURE 9.8 Temperature dependence of SHG intensity in polycatenar bent-core molecules with different end chains. *P–E* histograms for each compound are also shown in the insets. Cartoon of polarization switching is also shown at the bottom. (Reprinted with permission from Gorecka, E., Pociecha, D., Matraszek, J., Mieczkowski, J., Shimbo, Y., Takanishi, Y., and Takezoe, H., Polar order in columnar phase made of polycatenar bent-core molecules, *Phys. Rev. E*, 73, 031704-1. Copyright 2006 by the American Chemical Society.)

peak and continuous change of Δn (Gorecka et al. 2006). This is why the temperature dependence of SHG over the Col_h–$Col_h P_A$ phase transition is gradual (Figure 9.8). A similar behavior was also observed in the dielectric constant, which shows a maximum around the phase transition. The maximum increases its strength and is associated with a shift toward the lower temperature side with decreasing frequency. This behavior is reminiscent to that observed in relaxors, which have a structure consisting of randomly oriented polar domains. The compound with $n = 16$ seems to be the first liquid crystal relaxor.

Since the polarization is not retained in the absence of an electric field in the polycatenar compounds mentioned in the preceding text, they are not ferroelectric but paraelectric. The first columnar ferroelectric LC was discovered in a different system, which is also constructed by umbrella assemblies (Miyajima et al. 2012). The constituent molecule shown in Figure 9.9 is not categorized as a bent-core molecule, but actually this is a bent-shaped polycatenar molecule with an acute bending angle. Polarization originates from CN and hydrogen bonding. Sustainability and switchability of the polarization were confirmed mainly by SHG. The material can form a film by spin casting and a freely suspended film, and electric poling is possible even at room temperature (Miyajima et al. 2012, Araoka et al. 2013). Besides pyroelectric and piezoelectric device applications, one of the attractive applications is a high-density memory device. By taking account of intercolumnar distance (about 4.5 nm), the memory density of 40 Tb/in.^{2} is possible, if one column can be used as 1 bit (Figure 9.10).

In this way, although possible applications such as piezoelectric and memory devices utilizing polar phases have been proposed and demonstrated, all of them are still at an incubation stage.

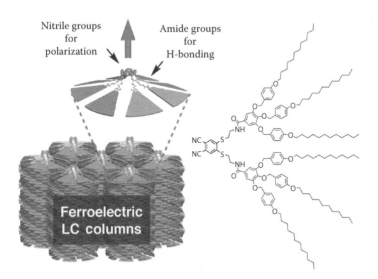

FIGURE 9.9 **(See color insert.)** Molecular structure (right) and columnar structure (left) in the ferroelectric columnar phase formed by bent-core molecules. (From Miyajima, D., Araoka, F., Takezoe, H., Kim, J., Kato, K., Takata, M., and Aida, T., Ferroelectric columnar liquid crystal featuring confined polar groups within core-shell architecture, *Science*, 336, 209–213, 2012. Reprinted with permission of AAAS.)

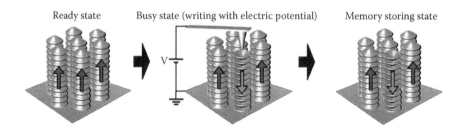

Ready state   Busy state (writing with electric potential)   Memory storing state

FIGURE 9.10   **(See color insert.)** Cartoon of a high-density memory device using the ferroelectric columnar LC phase. If the polarization switching and retaining in each column are possible, a high-density memory device is realized. (Reproduced from Takezoe, H., *Ferroelectrics*, 468, 1, 2014. With permission from Taylor & Francis.)

REFERENCES

Adam, D., P. Schuhmacher, J. Simmerer, L. Haussling, K. Siemensmeyer, K. H. Etzbach, H. Ringsdorf, and D. Haarer. Fast photoconduction in the highly ordered columnar phase of a discotic liquid-crystal. *Nature* 371 (1994): 141–143.

Alonso, I., J. Martinez-Perdiguero, J. Ortega, C. L. Folcia, and J. Etxebarria. Characteristics of a liquid crystal display using optically isotropic phases of bent-core mesogens. *Liq. Cryst.* 34 (2007): 655–658.

Araoka, F., S. Masuko, A. Kogure, D. Miyajima, T. Aida, and H. Takezoe. High-optical-quality ferroelectric film wet-processed from a ferroelectric columnar liquid crystal as observed by non-linear-optical microscopy. *Adv. Mater.* 30 (2013): 4014–4017.

Callahan, R. A., D. C. Coffey, D. Chen, N. A. Clark, G. Rumbles, and D. M. Walba. Charge generation measured for fullerene-helical nanofilament liquid crystal heterojunctions. *ACS Appl. Mater. Interfaces* 6 (2014): 4823–4830.

Charif, A. C., N. Diorio, K. Fordor-Csorba, J. E. Puskas, and A. Jákli. A piezoelectric thermoplastic elastomer containing a bent-core liquid crystal. *RSC Adv.* 3 (2013): 17446–17452.

Ferguson, A., N. Kopidakis, S. Shaheen, and G. Rumbles. Quenching of excitons by holes in poly(3-hexylthiophene) films. *J. Phys. Chem. C* 112 (2008): 9865–9871.

Fontana, J., C. Bailey, W. Weissflog, I. Janossy, and A. Jákli. Optical waveguiding in bent-core liquid-crystal filaments. *Phys. Rev. E* 80 (2009): 032701-1–032701-4.

Gomola, K., L. Guo, S. Dhara, Y. Shimbo, E. Gorecka, D. Pociecha, J. Mieczkowski, and H. Takezoe. Syntheses and characterization of novel asymmetric bent-core mesogens exhibiting polar smectic phases. *J. Mater. Chem.* 19 (2009a): 4240–4247.

Gomola, K., L. Guo, E. Gorecka, D. Pociecha, J. Mieczkowski, K. Ishikawa, and H. Takezoe. First symmetrical banana compounds exhibiting SmAP$_R$ mesophase and unique transition between two orthogonal polar phases. *Chem. Commun.* 2009 (2009b): 6592–6594.

Gomola, K., L. Guo, D. Pociecha, F. Araoka, K. Ishikawa, and H. Takezoe. An optically uniaxial antiferroelectric smectic phase in asymmetrical bent-core compounds containing a 3-aminophenol central unit. *J. Mater. Chem.* 20 (2010): 7944–7952.

Gorecka, E., D. Pociecha, J. Matraszek, J. Mieczkowski, Y. Shimbo, Y. Takanishi, and H. Takezoe. Polar order in columnar phase made of polycatenar bent-core molecules. *Phys. Rev. E* 73 (2006): 031704-1–031704-5.

Gorecka, E., D. Pociecha, J. Mieczkowski, J. Matraszek, D. Guillon, and B. Donnio. Axially polar columnar phase made of polycatenar bent-shaped molecules. *J. Am. Chem. Soc.* 126 (2004): 15946–15947.

Guo, L., Polar orthogonal smectic phases composed of bent-core liquid crystals. PhD thesis. Tokyo Institute of Technology, 2010.

Jákli, A., L.-C. Chen, D. Kruerke, H. Sawade, and G. Heppke. Light shutters from antiferroelectric liquid crystals of bent-shaped molecules. *Liq. Cryst.* 29 (2002): 377–381.

Jákli, A., I. C. Pintre, J. L. Serrano, M. B. Ros, and M. Rosario. Piezoelectric and electric-field-induced properties of a ferroelectric bent-core liquid crystal. *Adv. Mater.* 21 (2009): 3784–3788.

Kim, H., Y. Yi, D. Chen, E. Korblova, D. M. Walba, N. A. Clark, and D. K. Yoon. Self-assembled hydrophobic surface generated from a helical nanofilament (B_4) liquid crystal phase. *Soft Matter* 9 (2013): 2793–2797.

Mathews, M., R. S. Zola, D.-K. Yang, and Q. Li. Thermally, photochemically and electrically switchable reflection colors from soft-organized chiral bent-core liquid crystals. *J. Mater. Chem.* 21 (2011): 2098–2103.

Miyajima, D., F. Araoka, H. Takezoe, J. Kim, K. Kato, M. Takata, and T. Aida. Ferroelectric columnar liquid crystal featuring confined polar groups within core-shell architecture. *Science* 336 (2012): 209–213.

Nagaraj, M., V. Gortz, J. W. Goodby, and H. F. Gleeson. Electrically tunable refractive index in the dark conglomerate phase of a bent-core liquid crystal. *Appl. Phys. Lett.* 104 (2014): 021903-1–021903-5.

Nair, G. G., C. A. Bailey, S. Taushanoff, K. Fodor-Csorba, A. Vajda, Z. Varga, A. Bota, and A. Jákli. Electrically tunable color by using mixtures of bent-core and rod-shaped molecules. *Adv. Mater.* 20 (2008): 3138–3142.

Nakata, M., D. R. Link, F. Araoka, J. Thisayukta, Y. Takanishi, K. Ishikawa, J. Watanabe, and H. Takezoe. A racemic layer structure in a chiral bent-core ferroelectric liquid crystal. *Liq. Cryst.* 28 (2001): 1301–1308.

Ohno, A., A. Haruyama, K. Kurotaki, and J. Hanna. Charge-carrier transport in smectic mesophases of biphenyls. *J. Appl. Phys.* 102 (2007): 083711-1–083711-11.

Pociecha, D., M. Cepic, E. Gorecka, and J. Mieczkowski. Ferroelectric mesophase with randomized interlayer structure. *Phys. Rev. Lett.* 91 (2003): 185501-1–185501-4.

Sentman, A. C. and D. L. Gin. Polymerizable bent-core mesogens; switchable precursors to ordered, polar polymer materials. *Angew. Chem. Int. Ed.* 42 (2003): 1815–1819.

Shimbo, Y., E. Gorecka, D. Pociecha, F. Araoka, M. Goto, Y. Takanishi, K. Ishikawa, J. Mieczkowski, K. Gomola, and H. Takezoe. Electric-field-induced polar biaxial order in a nontilted smectic phase of asymmetric bent-core liquid crystal. *Phys. Rev. Lett.* 97 (2006a): 113901-1–113901-4.

Shimbo, Y., Y. Takanishi, K. Ishikawa, E. Gorecka, D. Pociecha, J. Mieczkowski, K. Gomola, and H. Takezoe. Ideal liquid crystal display mode using achiral banana-shaped liquid crystals. *Jpn. J. Appl. Phys.* 45 (2006b): L282–L284.

Takezoe, H. Historical overview of polar liquid crystals. *Ferroelectrics* 468 (2014): 1–17.

Takezoe, H., K. Kishikawa, and E. Gorecka. Switchable columnar phases. *J. Mater. Chem.* 16 (2006): 2412–2416.

Index